Applications of
MICROBIOLOGY

Applications of MICROBIOLOGY

Dr. Sarfaraz Ahmad

ANMOL PUBLICATIONS PVT. LTD.
NEW DELHI-110 002 (INDIA)

ANMOL PUBLICATIONS PVT. LTD.
Regd. Office: 4360/4, Ansari Road, Daryaganj,
New Delhi-110002 (India)
Tel.: 23278000, 23261597, 23286875, 23255577
Fax: 91-11-23280289
Email: anmolpub@gmail.com
Visit us at: www.anmolpublications.com

Branch Office: No. 1015, Ist Main Road, BSK IIIrd Stage
IIIrd Phase, IIIrd Block, Bengaluru-560 085 (India)
Tel.: 080-41723429 • Fax: 080-26723604
Email: anmolpublicationsbangalore@gmail.com

Applications of Microbiology

First Edition, 2011

ISBN 978-81-261-4729-8

PRINTED IN INDIA

Printed at Balaji Offset, Shahdara Delhi-32

Contents

Preface

Microbiology is a broad term which includes virology, mycology, parasitology, bacteriology and other branches. A microbiologist is a specialist in microbiology and these other topics. Microbiology is researched actively, and the field is advancing continually. It is estimated only about one percent of all of the microbe species on Earth have been studied. Although microbes were directly observed over three hundred years ago, the field of microbiology can be said to be in its infancy relative to older biological disciplines such as zoology and botany. Microorganisms may be tiny, but the relatively new science of microbiology is huge! Prokaryotic bacteria, eukaryotic fungi, and nonliving viruses are just some of the microbes that effect our health and environment. Some are beneficial, others pathogens or opportunistic. Microbiology is a field of biology that deals with minute, invisible living organisms. The principle objective of our site, Microbexpert.com is act as a virtual lab, and learning library for microbiology. Micro organisms are ubiquitous and play an important role in determining day to day activities of life. Microbiology is the study of *microorganisms*, which are microscopic, unicellular, and cell-cluster organisms. This includes eukaryotes such as fungi and protists, and prokaryotes. Viruses and prions, though not strictly classed as living organisms, are also studied. Microbiology typically includes the study of the immune system, or Immunology. Generally, immune systems interact with pathogenic microbes; these two disciplines often intersect which is why many colleges offer a paired degree such as "Microbiology and Immunology".

Whilst there are undoubtedly some who fear all microbes due to the association of some microbes with various human illnesses, many microbes are also responsible for numerous beneficial processes such as industrial fermentation (e.g. the production of alcohol, vinegar and dairy products), antibiotic production and as vehicles for cloning in higher organisms such as plants. Scientists have also exploited their knowledge of microbes to produce biotechnologically important enzymes such as Taq polymerase, reporter genes for use in other genetic systems and novel molecular biology techniques such as the yeast two-hybrid system. Bacteria can be used for the industrial production of amino acids. *Corynebacterium glutamicum* is one of the most important bacterial species with an annual production of more than two million tons of amino acids, mainly L-glutamate and L-lysine.

A variety of biopolymers, such as polysaccharides, polyesters, and polyamides, are produced by microorganisms. Microorganisms are used for the biotechnological production of biopolymers with tailored properties suitable for high-value medical application such as tissue engineering and drug delivery. Microorganisms are used for the biosynthesis of xanthan, alginate, cellulose, cyanophycin, poly(gamma-glutamic acid), levan, hyaluronic acid, organic acids, oligosaccharides and polysaccharide, and polyhydroxyalkanoates.

Microorganisms are beneficial for microbial biodegradation or bioremediation of domestic, agricultural and industrial wastes and subsurface pollution in soils, sediments and marine environments. The ability of each microorganism to degrade toxic waste depends on the nature of each contaminant. Since sites typically have multiple pollutant types, the most effective approach to microbial biodegradation is to use a mixture of bacterial species and strains, each specific to the biodegradation of one or more types of contaminants. There are also various claims concerning the contributions to human and animal health by consuming probiotics (bacteria potentially beneficial to the digestive system) and/or prebiotics (substances consumed to promote the growth of probiotic microorganisms). Recent research has suggested that microorganisms could be useful in the treatment of cancer. Various strains of non-pathogenic clostridia can infiltrate and replicate within solid tumours. Clostridial vectors can be safely administered and their potential to deliver therapeutic proteins has been demonstrated in a variety of preclinical models.

This book presents the very integrated approach with the possible latest information. Many of the topics have been dealt in greater details in the light of recent advances in the subject.

—*Author*

1

Introduction

Microorganisms may be tiny, but the relatively new science of microbiology is huge! Prokaryotic bacteria, eukaryotic fungi, and nonliving viruses are just some of the microbes that effect our health and environment. Some are beneficial, others pathogens or opportunistic.

Microbiology is a field of biology that deals with minute, invisible living organisms. The principle objective of our site, Microbexpert.com is act as a virtual lab, and learning library for microbiology. Micro organisms are ubiquitous and play an important role in determining day to day activities of life.

Microbiology is the study of *microorganisms*, which are microscopic, unicellular, and cell-cluster organisms. This includes eukaryotes such as fungi and protists, and prokaryotes. Viruses and prions, though not strictly classed as living organisms, are also studied. Microbiology typically includes the study of the immune system, or Immunology. Generally, immune systems interact with pathogenic microbes; these two disciplines often intersect which is why many colleges offer a paired degree such as "Microbiology and Immunology".

Microbiology is a broad term which includes virology, mycology, parasitology, bacteriology and other branches. A microbiologist is a specialist in microbiology and these other topics.

Microbiology is researched actively, and the field is advancing continually. It is estimated only about one percent of all of the microbe species on Earth have been studied. Although microbes were directly observed over three hundred years ago, the field of microbiology can be said to be in its infancy relative to older biological disciplines such as zoology and botany.

History

Ancient

The existence of microorganisms was hypothesized for many centuries before their actual discovery. The existence of unseen microbiological life was postulated by

Jainism which is based on Mahavira's teachings as early as 6th century BCE.. Paul Dundas notes that Mahavira asserted existence of unseen microbiological creatures living in earth, water, air and fire. Jain scriptures also describe nigodas which are sub-microscopic creatures living in large clusters and having a very short life and are said to pervade each and every part of the universe, even in tissues of plants and flesh of animals. The Roman Marcus Terentius Varro made references to microbes when he warned against locating a homestead in the vicinity of swamps "because there are bred certain minute creatures which cannot be seen by the eyes, which float in the air and enter the body through the mouth and nose and there cause serious diseases."

In 1546 Girolamo Fracastoro proposed that epidemic diseases were caused by transferable seedlike entities that could transmit infection by direct or indirect contact, or even without contact over long distances.

However, early claims about the existence of microorganisms were speculative, and not based on any data or observation. Actual observation and discovery of microbes had to await the invention of the microscope in the 17th century.

Modern

In 1676, Antonie van Leeuwenhoek observed bacteria and other microorganisms, using a single-lens microscope of his own design. While Van Leeuwenhoek is often cited as the first to observe microbes, Robert Hooke made the first recorded microscopic observation, of the fruiting bodies of molds, in 1665.. The first observation of microbes using a microscope is generally credited to the Dutch draper and haberdasher, Antonie van Leeuwenhoek, who lived for most of his life in Delft, Holland. It has, however, been suggested that a Jesuit priest called Athanasius Kircher was the first to observe microorganisms. He was among the first to design magic lanterns for projection purposes, so he must have been well acquainted with the properties of lenses. One of his book contains a chapter in Latin, which reads in translation – 'Concerning the wonderful structure of things in nature, investigated by Microscope. Here, he wrote 'who would believe that vinegar and milk abound with an innumerable multitude of worms.' He also noted that putrid material is full of innumerable creeping animalculae. These observations antedate Robert Hooke's Micrographia by nearly 20 years and were published some 29 years before van Leeuwenhoek saw protozoa and 37 years before he described having seen bacteria.

The field of bacteriology (later a subdiscipline of microbiology) was founded in the 19th century by Ferdinand Cohn, a botanist whose studies on algae and photosynthetic bacteria led him to describe several bacteria including *Bacillus* and *Beggiatoa*. Cohn was also the first to formulate a scheme for the taxonomic classification of bacteria and discover spores. Louis Pasteur and Robert Koch were contemporaries of Cohn's and are often considered to be the father of Microbiology and medical microbiology, respectively. Pasteur is most famous for his series of experiments designed to disprove the then widely held theory of spontaneous generation, thereby solidifying microbiology's identity as a biological science. Pasteur also designed methods for food preservation (pasteurization) and vaccines against several diseases such as anthrax, fowl cholera and rabies. Koch is best known for his contributions to the germ theory of disease, proving that specific diseases were caused by specific pathogenic microorganisms. He developed a series of criteria that have become known as the Koch's postulates. Koch was one of the first scientists to focus on the isolation of bacteria in pure culture resulting in his description of several novel bacteria including *Mycobacterium tuberculosis*, the causative agent of tuberculosis.

While Pasteur and Koch are often considered the founders of microbiology, their work did not accurately reflect the true diversity of the microbial world because of their exclusive focus on microorganisms having direct medical relevance. It was not until the late 19th century and the work of Martinus Beijerinck and Sergei Winogradsky, the founders of general microbiology (an older term encompassing aspects of microbial physiology, diversity and ecology), that the true breadth of microbiology was revealed.

Beijerinck made two major contributions to microbiology: the discovery of viruses and the development of enrichment culture techniques. While his work on the Tobacco Mosaic Virus established the basic principles of virology, it was his development of enrichment culturing that had the most immediate impact on microbiology by allowing for the cultivation of a wide range of microbes with wildly different physiologies. Winogradsky was the first to develop the concept of chemolithotrophy and to thereby reveal the essential role played by microorganisms in geochemical processes. He was responsible for the first isolation and description of both nitrifying and nitrogen-fixing bacteria.

Fields

The field of microbiology can be generally divided into several subdisciplines:

- Microbial physiology: The study of how the microbial cell functions biochemically. Includes the study of microbial growth, microbial metabolism and microbial cell structure.
- Microbial genetics: The study of how genes are organized and regulated in microbes in relation to their cellular functions. Closely related to the field of molecular biology.
- Cellular microbiology: A discipline bridging microbiology and cell biology.
- Medical microbiology: The study of the pathogenic microbes and the role of microbes in human illness. Includes the study of microbial pathogenesis and epidemiology and is related to the study of disease pathology and immunology.
- Veterinary microbiology: The study of the role in microbes in veterinary medicine or animal taxonomy.
- Environmental microbiology: The study of the function and diversity of microbes in their natural environments. Includes the study of microbial ecology, microbially-mediated nutrient cycling, geomicrobiology, microbial diversity and bioremediation. Characterization of key bacterial habitats such as the

rhizosphere and phyllosphere, soil and groundwater ecosystems, open oceans or extreme environments (extremophiles).

- Evolutionary microbiology: The study of the evolution of microbes. Includes the study of bacterial systematics and taxonomy.
- Industrial microbiology: The exploitation of microbes for use in industrial processes. Examples include industrial fermentation and wastewater treatment. Closely linked to the biotechnology industry. This field also includes brewing, an important application of microbiology.
- Aeromicrobiology: The study of airborne microorganisms.
- Food microbiology: The study of microorganisms causing food spoilage and foodborne illness. Using microorganisms to produce foods, for example by fermentation.
- Pharmaceutical microbiology: the study of microorganisms causing pharmaceutical contamination and spoil
- Agricultural microbiology: The study of agriculturaly important microorganisms.
- Soil Microbiology: The study of those microorganisms that are found in soil.
- Water microbiology: The study of those microorganisms that are found in water.
- Generation microbiology: The study of those microorganisms that have the same characters as their parents.
- Nano microbiology: The study of those microorganisms at nano level.

Benefits

Whilst there are undoubtedly some who fear all microbes due to the association of some microbes with various human illnesses, many microbes are also responsible for numerous beneficial processes such as industrial fermentation (e.g. the production of alcohol, vinegar and dairy products), antibiotic production and as vehicles for cloning in higher organisms such as plants. Scientists have also exploited their knowledge of microbes to produce biotechnologically important enzymes such as Taq polymerase, reporter genes for use in other genetic systems and novel molecular biology techniques such as the yeast two-hybrid system. Bacteria can be used for the industrial production of amino acids. *Corynebacterium glutamicum* is one of the most important bacterial species with an annual production of more than two million tons of amino acids, mainly L-glutamate and L-lysine.

A variety of biopolymers, such as polysaccharides, polyesters, and polyamides, are produced by microorganisms. Microorganisms are used for the biotechnological production of biopolymers with tailored properties suitable for high-value medical application such as tissue engineering and drug delivery. Microorganisms are used for the biosynthesis of xanthan, alginate, cellulose, cyanophycin, poly(gamma-glutamic acid), levan, hyaluronic acid, organic acids, oligosaccharides and polysaccharide, and polyhydroxyalkanoates.

Microorganisms are beneficial for microbial biodegradation or bioremediation of domestic, agricultural and industrial wastes and subsurface pollution in soils, sediments and marine environments. The ability of each microorganism to degrade toxic waste depends on the nature of each contaminant. Since sites typically have multiple pollutant types, the most effective approach to microbial biodegradation is to use a mixture of bacterial species and strains, each specific to the biodegradation of one or more types of contaminants.

There are also various claims concerning the contributions to human and animal health by consuming probiotics (bacteria potentially beneficial to the digestive system) and/or prebiotics (substances consumed to promote the growth of probiotic microorganisms).

Recent research has suggested that microorganisms could be useful in the treatment of cancer. Various strains of non-pathogenic clostridia can infiltrate and replicate within solid tumours. Clostridial vectors can be safely administered and their potential to deliver therapeutic proteins has been demonstrated in a variety of preclinical models.

In broadest sense, industrial microbiology, is concerned with all aspects of business that relate to microbiology. In a more restricted sense, industrial microbiology is concerned with:

(i) employing microorganisms to produce a desired product, and with

(ii) preventing microbes from diminishing the economic value of various products. This dual purpose is clearly seen in the food industry, a major area of industrial microbiology discussed in separate chapter on "Microbiology of foods, Milk and Dairy Products".

Various commercial products of economic value made by microbes are :

(i) medicines i.e. pharmaceuticals, including antibiotics, steroids, human protein, vaccines, and vitamins,

(ii) organic acids,

(iii) amino acids,

(vi) enzymes,

(v) alcohols,

(vi) organic solvents and

(vii) synthetic fuels. In addition to these, quite recently potential of microbes could also be realised in

(viii) recovery of metals from ores through bioleaching,

(ix) recovery of petrol, and

(x) single all protein production.

The term fermentation in industrial microbiology is used in a wider sense to include any chemical transformation of organic compounds carried out by using microbes and their enzymes.

Production methods in industrial microbiology bring together the raw materials (substrates), microorganisms (specific strains or microbial enzymes) and a controlled favourable environment (created in a ferment or) to produce the desired substance.

The essence of an industrial process is to combine the right organism, an in expensive substrate, and the proper environment to produce high yields of a desired product Microbes possess a wealth of metabolic equipment that brings about diverse chemical transformations.

This characteristic of microbes could be exploited in obtaining some valuable products of daily use. The cheap raw materials available in nature as a waste, may be converted into useful commercial products by the activity of microbes. Microbes thus serve a dual purpose. Firstly, they are good agents of disposal of these wastes, and secondly the resultant end products of their breakdown are seful commercial products.

Industrial microbiology is an important area of applied microbiology. It refers to the use of microorganisms in commercial enterprise and. Cheap raw materials are converted to valuable products through the metabolism of microbes. Microbes for this purpose could be exploited in different ways. For instance this includes:

(i) synthesis of fermentation products as acids, alcohols or other organic compounds,

(ii) transformation of one compound into another desired type,

(iii) the production of enzymes, antibiotics, or insecticides, or

(iv) the use of microbes themselves as food.

5 applications of microbiology

1) Medicine-fungi like Penicillium notatum are cultured in large scale to produce the antibiotic penicilin. non-dangerous parts of certain viruses like antigens on the capsid (in small amount) are used as vaccine.
2) food-Bacteria like Streptococcus thermophilus and Lactobacillus bulgaricus are used in the production of yoghurt. Yeast strains are used in bread making and beer production.
3) some bacteria are able to break down oil into water soluble products; used in clearing up oil spills in the sea.
4) agriculture-cultures containing nitrogen fixing bacteria like clostridium can be sprayed onto soil so that leguminous (or non-leguminous crops) can utilize nitrogen in air to make nitrates.
5) bacteria like E.coli are used in gene technology for their plasmids, on which desired genes can be added to form recombinant DNA.

Probiotics

Probiotics are live microorganisms thought to be healthy for the host organism. According to the currently adopted definition by FAO/WHO, probiotics are: "Live microorganisms which when administered in adequate amounts confer a health benefit on the host". Lactic acid bacteria (LAB) and bifidobacteria are the most common types of microbes used as probiotics; but certain yeasts and bacilli may also be helpful. Probiotics are commonly consumed as part of fermented foods with specially added active live cultures; such as in yogurt, soy yogurt, or as dietary supplements.

Etymologically, the term appears to be a composite of the Latin preposition *pro* ("for") and the Greek adjective (biotic), the latter deriving from the noun bios, "life".

At the start of the 20th century, probiotics were thought to beneficially affect the host by improving its intestinal microbial balance, thus inhibiting pathogens and toxin producing bacteria. Today, specific health effects are being investigated and documented including alleviation of chronic intestinal inflammatory diseases, prevention and treatment of pathogen-induced diarrhea, urogenital infections, and atopic diseases.

History of Probiotics

The original observation of the positive role played by certain bacteria was first introduced by Russian scientist and Nobel laureate Eli Metchnikoff, who in the beginning of the 20th century suggested that it would be possible to modify the gut flora and to replace harmful microbes with useful microbes. Metchnikoff, at that time a professor at the Pasteur Institute in Paris, produced the notion that the aging process results from the activity of putrefactive (proteolytic) microbes producing toxic substances in the large bowel. Proteolytic bacteria such as clostridia, which are part of the normal gut flora, produce toxic substances including phenols, indols and ammonia from the digestion of proteins. According to Metchnikoff these compounds were responsible for what he called "intestinal auto-intoxication", which caused the physical changes associated with old age. It was at that time known that milk fermented with lactic-acid bacteria inhibits the growth of proteolytic bacteria because of the low pH produced by the fermentation of lactose. Metchnikoff had also observed that certain rural populations in Europe, for example in Bulgaria and the Russian steppes who lived largely on milk fermented by lactic-acid bacteria were exceptionally long lived. Based on these facts, Metchnikoff proposed that consumption of fermented milk would "seed" the intestine with harmless lactic-acid bacteria and decrease the intestinal pH and that this would suppress the growth of proteolytic bacteria. Metchnikoff himself introduced in his diet sour milk fermented with the bacteria he called "Bulgarian Bacillus" and found his health benefited. Friends in Paris soon followed his example and physicians began prescribing the sour milk diet for their patients.

Bifidobacteria were first isolated from a breast-fed infant by Henry Tissier who also worked at the Pasteur Institute. The isolated bacterium named Bacillus bifidus communis was later renamed to the genus *Bifidobacterium*. Tissier found that bifidobacteria are dominant in the gut flora of breast-fed babies and he observed clinical benefits from treating diarrhea in infants with bifidobacteria. The claimed effect was bifidobacterial displacement of proteolytic bacteria causing the disease.

During an outbreak of shigellosis in 1917, German professor Alfred Nissle isolated a strain of *Escherichia coli* from the feces of a soldier who was not affected by the disease. Methods of treating infectious diseases were needed at that time when antibiotics were not yet available, and Nissle used the *Escherichia coli* Nissle 1917 strain in acute gastrointestinal infectious salmonellosis and shigellosis.

In 1920, Rettger demonstrated that Metchnikoff's "Bulgarian Bacillus", later called *Lactobacillus delbrueckii subsp. bulgaricus*, could not live in the human intestine, and the fermented food phenomena petered out. Metchnikoff's theory was disputable (at this stage), and people doubted his theory of longevity.

After Metchnikoff's death in 1916, the centre of activity moved to the United States. It was reasoned that bacteria originating from the gut were more likely to produce the desired effect in the gut, and in 1935 certain strains of *Lactobacillus acidophilus* were found to be very active when implanted in the human digestive tract. Trials were carried out using this organism, and encouraging results were obtained especially in the relief of chronic constipation.

The term "probiotics" was first introduced in 1953 by Kollath. Contrasting antibiotics, probiotics were defined as microbially derived factors that stimulate the growth of other microorganisms. In 1989 Roy Fuller suggested a definition of probiotics which has been widely used: "*A live microbial feed supplement which beneficially affects the host animal by improving its intestinal microbial balance*". Fuller's definition emphasizes the requirement of viability for probiotics and introduces the aspect of a beneficial effect on the host.

In the following decades intestinal lactic acid bacterial species with alleged health beneficial properties have been introduced as probiotics, including *Lactobacillus rhamnosus*, *Lactobacillus casei*, and *Lactobacillus johnsonii*.

Potential Benefits

Experiments into the benefits of probiotic therapies suggest a range of potentially beneficial medicinal uses for probiotics. For many of the potential benefits, research is limited and only preliminary results are available. It should be noted that the effects described are *not* general effects of probiotics. Recent research on the molecular biology and genomics of *Lactobacillus* has focused on the interaction with the immune system, anti-cancer potential, and potential as a biotherapeutic agent in cases of antibiotic-associated diarrhoea, travellers' diarrhoea, pediatric diarrhoea, inflammatory bowel disease and irritable bowel syndrome.

All effects can only be attributed to the individual strain(s) tested. Testing of a supplement does not indicate benefit from any other strain of the same species, and testing does not indicate benefit from the whole group of LAB (or other probiotics).

Managing Lactose Intolerance

As lactic acid bacteria actively convert lactose into lactic acid, ingestion of certain active strains may help lactose intolerant individuals tolerate more lactose than what they would have otherwise.

Prevention of Colon Cancer

In laboratory investigations, some strains of LAB (*Lactobacillus bulgaricus*) have demonstrated anti-mutagenic effects thought to be due to their ability to bind with heterocyclic amines, which are carcinogenic substances formed in cooked meat. Animal studies have demonstrated that some LAB can protect against colon cancer in rodents, though human data is limited and conflicting. Most human trials have found that the strains tested may exert anti-carcinogenic effects by decreasing the activity of an enzyme called β-glucuronidase (which can generate carcinogens in the digestive system). Lower rates of colon cancer among higher consumers of fermented dairy products have been observed in one population study.

Lowering Cholesterol

Animal studies have demonstrated the efficacy of a range of LAB to be able to lower serum cholesterol levels, presumably by breaking down bile in the gut, thus inhibiting its reabsorption (which enters the blood as cholesterol). Some, but not all human trials have shown that dairy foods fermented with specific LAB can produce modest reductions in total and LDL cholesterol levels in those with normal levels to begin with, however trials in hyperlipidemic subjects are needed.

Lowering Blood Pressure

Several small clinical trials have indicated that consumption of milk fermented with various strains of LAB may result in modest reductions in blood pressure. It is thought that this is due to the ACE inhibitor-like peptides produced during fermentation.

Improving Immune Function and Preventing Infections

LAB are thought to have several presumably beneficial effects on immune function. They may protect against pathogens by means of competitive inhibition (i.e., by competing for growth) and there is evidence to suggest that they may improve immune function by increasing the number of IgA-producing plasma cells, increasing or improving phagocytosis as well as increasing the proportion of T lymphocytes and Natural Killer cells.

Clinical trials have demonstrated that probiotics may decrease the incidence of respiratory tract infections and dental caries in children. LAB foods and supplements have been shown to aid in the treatment and prevention of acute diarrhea, and in decreasing the severity and duration of rotavirus infections in children and travelers' diarrhea in adults.

A 2010 study suggested that the anecdotal benefits of probiotic therapies as beneficial for preventing secondary infections, a common complication of antibiotic therapy, may be because keeping the immune system primed by eating foods enhanced with "good" bacteria may help counteract the negative effects of sickness and antibiotics. It was thought that antibiotics may turn the immune system "off" while probiotics turns it back on "idle", and more able to quickly react to new infections.

Helicobacter Pylori

LAB are also thought to aid in the treatment of *Helicobacter pylori* infections (which cause peptic ulcers) in adults when used in combination with standard medical treatments. However more studies are required into this area.

Antibiotic-associated Diarrhea

Antibiotic-associated diarrhea (AAD) results from an imbalance in the colonic microbiota caused by antibiotic therapy. Microbiota alteration changes carbohydrate metabolism with decreased short-chain fatty acid absorption and an osmotic diarrhea as a result. Another consequence of antibiotic therapy leading to diarrhea is overgrowth of potentially pathogenic organisms such as *Clostridium difficile.*

Probiotic treatment can reduce the incidence and severity of AAD as indicated in several meta-analyses. However, further documentation of these findings through randomized, double blind, placebo-controlled trials are warranted.

Efficacy of probiotic AAD prevention is dependent on the probiotic strain(s) used and on the dosage. Up to a 50% reduction of AAD occurrence has been found. No side-effects have been reported in any of these studies. Caution should, however, be exercised when administering probiotic supplements to immunocompromised individuals or patients who have a compromised intestinal barrier.

Reducing Inflammation

LAB and supplements have been found to modulate inflammatory and hypersensitivity responses, an observation thought to be at least in part due to the regulation of cytokine function. Clinical studies suggest that they can prevent reoccurrences of inflammatory bowel disease in adults, as well as improve milk allergies. They are not effective for treating eczema, a persistent skin inflammation. How probiotics counteract immune system overactivity remains unclear, but a potential mechanism is desensitization of T lymphocytes, an important component of the immune system, towards pro-inflammatory stimuli.

Improving Mineral Absorption

It is hypothesized that probiotic lactobacilli may help correct malabsorption of trace minerals, found particularly in those with diets high in phytate content from whole grains, nuts, and legumes.

Preventing Harmful Bacterial growth under Stress

In a study done to see the effects of stress on intestinal flora, rats that were fed probiotics had little occurrence of harmful bacteria latched onto their intestines compared to rats that were fed sterile water.

Irritable Bowel Syndrome and Colitis

B. infantis 35624, sold as Align, was found to improve some symptoms of irritable bowel syndrome in women in a recent study. Another probiotic bacterium, *Lactobacillus plantarum* 299v, was also found to be effective in reducing IBS symptoms. Additionally, a probiotic formulation, VSL#3, was found to be safe in treating ulcerative colitis, though efficacy in the study was uncertain. *Bifidobacterium animalis* DN-173 010 may help. For maintenance of remission of ulcerative colitis, Mutaflor (*E.coli* Nissle 1917) there are 3 controlled, randomized, double blind clinical studies which have proven equivalence of Mutaflor and mesalazine (5-ASAs).

Managing Urogenital Health

Several in vitro studies have revealed probiotics' potential in relieving urinary tract infections and bacterial vaginosis. Results have been varied on these studies, and in vivo studies are still required in this area to determine efficacy.

Disadvantages of Probiotics

In some specific situations (such as critically ill patients) they could be harmful. In a therapeutic clinical trial conducted by the Dutch Pancreatitis Study Group, the consumption of a mixture of six probiotic bacteria, increased the death rate of patients with predicted severe acute pancreatitis.

In a clinical trial conducted at the University of Western Australia, aimed at showing the effectiveness of probiotics in reducing childhood allergies, Dr Susan Prescott and her colleagues gave 178 children either a probiotic or a placebo for the first six months of their life. Those given the good bacteria were more likely to develop a sensitivity to allergens.

Some hospitals have reported treating lactobacillus septicaemia, which is a potentially fatal disease caused by the consumption of probiotics by people with lowered immune systems or who are already very ill.

There is no published evidence that probiotic supplements are able to replace the body's natural flora when these have been killed off; indeed bacterial levels in feces disappear within days when supplementation ceases.

Recent studies indicate that probiotic products such as yogurts could be a cause for obesity trends. However, this is contested.

Strains

Live probiotic cultures are available in fermented dairy products and probiotic fortified foods. However, tablets, capsules, powders and sachets containing the bacteria in freeze dried form are also available.

Some other bacteria mentioned in probiotic products: (yoghurt bacteria):

- *Lactobacillus bulgaricus*
- *Streptococcus thermophilus*
- *Lactobacillus bifidus*-became new genus *Bifidobacterium.*

Some fermented products containing similar lactic acid bacteria include:

- Pickled vegetables
- Fermented bean paste such as tempeh, miso and doenjang

- Kefir
- Buttermilk or Karnemelk
- Kimchi
- Pao cai
- Sauerkraut
- Soy sauce
- Zha cai.

Multi-probiotic

Research is emerging on the potential health benefits of multiple probiotic strains as a health supplement as opposed to a single strain. The human gut is home to some 400-500 types of microbes. It is thought that this diverse environment may benefit from multiple probiotic strains; different strains populate different areas of the digestive tract, and studies are beginning to link different probiotic strains to specific health benefits.

Synbiotics and Prebiotics

While related, prebiotics and synbiotics are not the same as probiotics.

Other Uses

Probiotics are used in industry to improve yields of pork and chicken production.

Research

Research about probiotics shows benefits.

A study in 2004 testing the immune system of students given either milk or Actimel over a 6 week exam period (3 weeks of studying, 3 weeks of exams) tested 19 different biomarkers. Of these 19 biomarkers only 2 were shown to be different between the two groups, increased production of lymphocytes and increased production of CD56 cells. The tests were not blind and show that certain probiotic strains may have no overall effect on the immune system or on its ability.

A 2007 study at University College Cork in Ireland showed that a diet including milk fermented with *Lactobacillus* bacteria prevented *Salmonella* infection in pigs.

A 2007 clinical study at Imperial College London showed that preventive consumption of a commercially available probiotic drink containing *L casei DN-114001*, *L bulgaricus*, and *S thermophilus* can reduce the incidence of antibiotic-associated diarrhea and *C difficile*-associated diarrhea. The efficacy and safety of a daily dose of *Lactobacillus acidophilus* CL1285 in the prevention of AAD was demonstrated by Montreal's Maisonneuve-Rosemont Hospital, in a clinical study of hospitalized patients. Current research is focusing on the molecular biology and genomics of *Lactobacillus* and bifidobacteria. The application of modern whole genome approaches is providing insights into bifidobacterial evolution, while also revealing genetic functions that explain their presence in the particular ecological environment of the gastrointestinal tract.

Prebiotic (nutrition)

Prebiotics are non-digestible food ingredients that stimulate the growth and/or activity of bacteria in the digestive system which are beneficial to the health of the body. They were first identified and named by Marcel Roberfroid in 1995. They are considered a functional food. Typically, prebiotics are carbohydrates (such as oligosaccharides), but the definition may include non-carbohydrates. The most prevalent forms of prebiotics are nutritionally classed as soluble fibre. To some extent, many forms of dietary fibre exhibit some level of prebiotic effect.

Roberfroid offered a refined definition in the 2007 Journal of Nutrition stating:

> *"A prebiotic is a selectively fermented ingredient that allows specific changes, both in the composition and/or activity in the gastrointestinal microflora that confers benefits upon host well-being and health."*

Additionally, in his 2007 revisit of Prebiotics, Roberfroid clarified that only two particular fructooligosaccharides fully meet this definition: oligofructose and inulin. Caution may be in order when referring to other substances as "prebiotic," and it is increasingly common to distinguish other similar substances by using terms such as "possible" or "likely" prebiotics. Many authorities also classify galactooligosaccharides (GOS) as "true" prebiotics. GOS is an animal-derived substance manufactured by extracting milk sugars from dairy products and using enzymes to break them down.

Researchers now also focus on the distinction between short-chain, long-chain, and full-spectrum prebiotics. "short-chain" prebiotics, e.g. oligofructose, contain 2-8 links per saccharide molecule, are typically fermented more quickly in the right-side of the colon providing nourishment to the bacteria in that area. Longer-chain prebiotics, e.g. Inulin, contain 9-64 links per saccharide molecule, and tend to be fermented more slowly, nourishing bacteria predominantly in the left-side colon. Full-spectrum prebiotics provide the full range of molecular link-lengths from 2-64 links per molecule, and nourish bacteria throughout the colon, e.g. Oligofructose-Enriched Inulin (OEI). The majority of research done on prebiotics is based on full-spectrum prebiotics, typically using OEI as the research substance.

Function

The prebiotic definition does not emphasize a specific bacterial group. Generally, however, it is assumed that a prebiotic should increase the number and/or activity of bifidobacteria and lactic acid bacteria. The importance of the bifidobacteria and the lactic acid bacteria (AKA lactobacillus or LABs) is that these groups of bacteria have several beneficial effects on the host, especially in terms of improving digestion (including enhancing mineral absorption) and the effectiveness and intrinsic strength of the immune system. A product that stimulates bifidobacteria is considered a bifidogenic factor. Some prebiotics may thus also act as a bifidogenic factor and vice versa, but the two concepts are not identical.

Sources

Traditional dietary sources of prebiotics include soybeans, inulin sources (such as Jerusalem artichoke, jicama, and chicory root), raw oats, unrefined wheat, unrefined barley and yacon. Some of the oligosaccharides that naturally occur in breast milk are believed to play an important role in the development of a healthy immune system in infants. It is becoming more common to properly distinguish between prebiotic substances and the food that contains them. References to almonds, honey and other foods (most commonly in promotional materials from growers of those foods) as "a prebiotic" are not accurate. No plant or food *is* a prebiotic: Wheat, honey and many other foods *contain* prebiotics to a greater or lesser extent, ranging from fairly large portions (chicory root, Jerusalem artichoke) to only trace quantities (thousands of other plant-based foods). Referring to a food as "a prebiotic" is no more accurate than calling a food "a vitamin."

Top 10 Foods Containing Prebiotics

Food	*Prebiotic Fibre Content by Weight*
Raw Chicory Root	64.6%
Raw Jerusalem Artichoke	31.5%
Raw Dandelion Greens	24.3%
Raw Garlic	17.5%
Raw Leek	11.7%
Raw Onion	8.6%
Cooked Onion	5%
Raw Asparagus	5%
Raw Wheat bran	5%
Whole Wheat flour, Cooked	4.8%
Raw Banana	1%

While there is no broad consensus on an ideal daily serving of prebiotics, recommendations typically range from 4-8g for general digestive health support, to 15g or more for those with active digestive disorders. Given an average 6g serving, below are the amounts of prebiotic foods required to achieve a daily serving of prebiotic fibre:

Food	*Amount of food to achieve 6g serving of prebiotics*
Raw Chicory Root	9.3g (about 1/3 oz)
Raw Jerusalem Artichoke	19g (about 3/4 oz)
Raw Dandelion Greens	24.7g (just under 1 oz)
Raw Garlic	34.3 g (about 1.2 oz)
Raw Leek	51.3g (about 1.8 oz)
Raw Onion	69.8g (about 2.5 oz)
Cooked Onion	120g (about 1/4 lb)
Raw Asparagus	120g (about 1/4 lb)
Raw Wheat Bran	120g (about 1/4 lb)
Whole Wheat Flour, Cooked	125g (about 1/4 lb)
Raw Banana	600g (about 1.3 lb)

Those wishing to ensure sufficient prebiotic intake should carefully consider the prebiotic content of their diet, as well as what caloric/nutritive load comes along with it: e.g. eating 1.3 pounds of bananas daily is likely to provide an excess of calories & sugars/carbohydrates to the diet. Prebiotic fibre supplements with minimal caloric/fat/sugar load are also available.

Prebiotic oligosaccharides are increasingly added to foods for their health benefits. Some oligosaccharides that are used in this manner are fructooligosaccharides (FOS), xylooligosaccharides (XOS), polydextrose and galactooligosaccharides (GOS). Some monosaccharides such as tagatose are also used sometimes as prebiotics.

In petfood also mannooligosaccharides are being used for prebiotic purposes.

Genetically engineering plants for the production of inulins has also become more prevalent., despite the still limited insight into the immunological mechanisms activated by such food supplementation.

Effects

Studies have demonstrated positive effects on calcium and other mineral absorption, immune system effectiveness, bowel pH, reduction of colorectal cancer risk, inflammatory bowel disorders (Crohn's Disease and Ulcerative Colitis) Hypertension (high blood pressure) and intestinal regularity. Recent human trials have reinforced the role of Prebiotics in preventing and possibly stopping early stage colon cancer. It has been argued that many of these health effects emanate from increased production of short-chain fatty acids (SCFA) by the stimulated beneficial bacteria. Thus food supplements specifically enhancing the growth of SCFA producing intestinal bacteria (such as clostridia and bacteroides species) are widely recognized to be beneficial.

While research does clearly demonstrate that prebiotics lead to increased production of these SCFA's, more research is required to establish a direct causal connection. It has been argued that prebiotics are beneficial to Crohn's Disease through production of SCFAs to nourish the colon walls, and beneficial to Ulcerative Colitis through reduction of Hydrogen Sulfide gas due to reduction of sulfate-reducing bacteria, which do not thrive in the slightly acidic environment SCFAs create.

The immediate addition of substantial quantities of prebiotics to the diet may result in a temporary increase in gas, bloating or bowel movement. It has been argued that chronically low consumption of prebiotic-containing foods in the typical Western diet may exaggerate this effect.

Human colonic bacteria substrates are relatively stable. Production of SCFA and fermentation quality are reduced during long-term diets of low fibre intake. Until bacterial flora are gradually established to habilitate or restore intestinal tone, nutrient absorption will be impaired and colonic transit time temporarily increased with an immediate addition of higher prebiotic intake.

2

Developments in Metagenomics for Accessing Novel Genes

Metagenomics is the study of metagenomes, genetic material recovered directly from environmental samples. The broad field may also be referred to as environmental genomics, ecogenomics or community genomics. Traditional microbiology and microbial genome sequencing rely upon cultivated clonal cultures. This relatively new field of genetic research enables studies of organisms that are not easily cultured in a laboratory as well as studies of organisms in their natural environment.

Early environmental gene sequencing cloned specific genes (often the 16S rRNA gene) to produce a profile of diversity in a natural sample. Such work revealed that the vast majority of microbial biodiversity had been missed by cultivation-based methods. Recent studies use "shotgun" Sanger sequencing or massively parallel pyrosequencing to get (mostly) unbiased samples of all genes from all members of sampled communities.

History

Origin of the Term

The term "metagenomics" was first used by Jo Handelsman, Jon Clardy, Robert M. Goodman, and others, and first appeared in publication in 1998. The term metagenome referenced the idea that a collection of genes sequenced from the environment could be analysed in a way analogous to the study of a single genome. The exploding interest in environmental genetics, along with the buzzword-like nature of the term, has resulted in the broader use of metagenomics to describe any sequencing of genetic material from environmental (i.e. uncultured) samples, even work that focuses on one organism or gene. Recently, Kevin Chen and Lior Pachter (researchers at the University of California, Berkeley) defined metagenomics as "the application of modern genomics techniques to the study of communities of microbial organisms directly in their natural environments, bypassing the need for isolation and lab cultivation of individual species."

Environmental Gene Surveys

Conventional sequencing begins with a culture of identical cells as a source of DNA. However, early metagenomic studies revealed that there are probably large groups of microorganisms in many environments that cannot be cultured and thus cannot be sequenced. These early studies focused on 16S ribosomal RNA sequences which are relatively short, often conserved within a species, and generally different between species. Many 16S rRNA sequences have been found which do not belong to any known cultured species, indicating that there are numerous non-isolated organisms out there.

Early molecular work in the field was conducted by Norman R. Pace and colleagues, who used PCR to explore the diversity of ribosomal RNA sequences. The insights gained from these breakthrough studies led Pace to propose the idea of cloning DNA directly from environmental samples as early as 1985. This led to the first report of isolating and cloning bulk DNA from an environmental sample, published by Pace and colleagues in 1991 while Pace was in the Department of Biology at Indiana University. Considerable efforts ensured that these were not PCR false positives and supported the existence of a complex community of unexplored species. Although this methodology was limited to exploring highly conserved, non-protein coding genes, it did support early microbial morphology-based observations that diversity was far more complex than was known by culturing methods.

Soon after that, Healy reported the metagenomic isolation of functional genes from "zoolibraries" constructed from a complex culture of environmental organisms grown in the laboratory on dried grasses in 1995. After leaving the Pace laboratory, Ed DeLong continued in the field and has published work that has largely laid the groundwork for environmental phylogenies based on signature 16S sequences, beginning with his group's construction of libraries from marine samples.

Longer Sequences from Environmental Samples

Recovery of DNA sequences longer than a few thousand base pairs from environmental samples was very difficult until recent advances in molecular biological techniques, particularly related to constructing libraries in bacterial artificial chromosomes (BACs), provided better vectors for molecular cloning.

Shotgun Metagenomics

Advances in bioinformatics, refinements of DNA amplification, and proliferation of computational power have greatly aided the analysis of DNA sequences recovered from environmental samples. These advances have enabled the adaptation of shotgun sequencing to metagenomic samples. The approach, used to sequence many cultured microorganisms as well as the human genome, randomly shears DNA, sequences many short sequences, and reconstructs them into a consensus sequence.

In 2002, Mya Breitbart, Forest Rohwer, and colleagues used environmental shotgun sequencing to show that 200 litres of seawater contains over 5000 different viruses. Subsequent studies showed that there are >1000 viral species in human stool and possibly a million different viruses per kilogram of marine sediment, including many bacteriophages. Essentially all of the viruses in these studies were new species. In 2004, Gene Tyson, Jill Banfield, and colleagues at the University of California, Berkeley and the Joint Genome Institute sequenced DNA extracted from an acid mine drainage system. This effort resulted in the complete, or nearly complete, genomes for a handful of bacteria and archaea that had previously resisted attempts to culture them. It was now possible to study entire genomes without the biases associated with laboratory cultures.

Global Ocean Sampling Expedition

Beginning in 2003, Craig Venter, leader of the privately-funded parallel of the Human Genome Project, has led the Global Ocean Sampling Expedition, circumnavigating the globe and collecting metagenomic samples throughout. All of these samples are sequenced using shotgun sequencing, in hopes that new genomes (and therefore new organisms) would be identified. The pilot project, conducted in the Sargasso Sea, found DNA from nearly 2000 different species, including 148 types of bacteria never before seen. As of 2009, Venter has circumnavigated the globe and thoroughly explored the West Coast of the United States, and is currently in the midst of a two-year expedition to explore the Baltic, Mediterranian and Black Seas.

Pyrosequencing

In 2006 Robert Edwards, Forest Rohwer, and colleagues at San Diego State University published the first sequences of environmental samples generated with so-called next generation sequencing, in this case chip-based pyrosequencing developed by 454 Life Sciences. This technique for sequencing DNA generates shorter fragments than conventional techniques, however this limitation is compensated for by the very large number of sequences generated. In addition, this technique does not require cloning the DNA before sequencing, removing one of the main biases in metagenomics.

MEGAN

In 2007, Daniel Huson and Stephan Schuster developed and published the first stand-alone metagenome analysis tool, MEGAN, which can be used to perform a first analysis of a metagenomic shotgun dataset. This tool was originally developed to analyse the metagenome of a mammoth sample. However in a recent study by Monzoorul et al. 2009, it was shown that adopting the LCA approach (of MEGAN) solely based on bit-score of the alignment leads to a number of false positive assignments especially in the context of metagenomic sequences originating from new organisms. This study proposed a new approach called SOrt-ITEMS which used several alignment parameters to increase the accuracy of assignments.

MG-RAST

In 2007, Folker Meyer and Robert Edwards and a team at Argonne National Laboratory and the University of Chicago released the Metagenomics RAST server (MG-RAST) a community resource for metagenome data set analysis. The SEED based free, public resource has so far (October 2009) been used for the analysis of over 4000 metagenome data sets. As of October 2009 100+ giga-basepairs of DNA have been analysed via MG-RAST, more than 350 public data sets are freely available for comparison within MG-RAST.

Applications

Metagenomics can improve strategies for monitoring the impact of pollutants on ecosystems and for cleaning up contaminated environments. Increased understanding of how microbial communities cope with pollutants is helping assess the potential of contaminated sites to recover from pollution and increase the chances of bioaugmentation or biostimulation trials to succeed.

Recent progress in mining the rich genetic resource of non-culturable microbes has led to the discovery of new genes, enzymes, and natural products. The impact of metagenomics is witnessed in the development of commodity and fine chemicals, agrochemicals and pharmaceuticals where the benefit of enzyme-catalysed chiral synthesis is increasingly recognized.

Metagenomic sequencing is being used to characterize the microbial communities from 15-18 body sites from at least 250 individuals. This is part of the Human Microbiome initiative with primary goals to determine if there is a core human microbiome, to understand the changes in the human microbiome that can be correlated with human health, and to develop new technological and bioinformatics tools to support these goals.

It is well known that the vast majority of microbes have not been cultivated. Functional metagenomics strategies are being used to explore the interactions between plants and microbes through cultivation-independent study of the microbial communities.

Microbial Diversity

Much of the interest in metagenomics comes from the discovery that the vast majority of microorganisms had previously gone unnoticed. Traditional microbiological methods relied upon laboratory cultures of organisms. Surveys of ribosomal RNA (rRNA) genes taken directly from the environment revealed that cultivation based methods find less than 1% of the bacteria and archaea species in a sample.

Gene Surveys

Shotgun sequencing and screens of clone libraries reveal genes present in environmental samples. This provides information both on which organisms are present and what metabolic processes are possible in the community. This can be helpful in understanding the ecology of a community, particularly if multiple samples are compared to each other.

Environmental Genomes

Shotgun metagenomics also is capable of sequencing nearly complete microbial genomes directly from the environment. Because the collection of DNA from an environment is largely uncontrolled, the most abundant organisms in an environmental sample are most highly represented in the resulting sequence data. To achieve the high coverage needed to fully resolve the genomes of underrepresented community members, large samples, often prohibitively so, are needed. On the other hand, the random nature of shotgun sequencing ensures that many of these organisms will be represented by at least some small sequence segments. Due to the limitations of microbial isolation methods, the vast majority of these organisms would go unnoticed using traditional culturing techniques.

Community Metabolism

Many bacterial communities show significant division of labour in metabolism. Waste products of some organisms are metabolites for others. Working together they turn raw resources into fully metabolized waste. Using comparative gene studies and expression experiments with microarrays or proteomics researchers can piece together a metabolic network that goes beyond species boundaries. Such studies require detailed knowledge about which versions of which proteins are coded by which species and even by which strains of which species. Therefore, community genomic information is another fundamental part (as metabolomics or proteomics) to be able to estimate how metabolites are possibly transferred and transformed through a community.

Pathogenomics

Pathogen infections are among the leading causes of infirmity and mortality among humans and animals in the world. Until recently, it has been difficult to compile information to understand the generation of pathogen virulence factors as well as pathogen behaviour in a host environment. The study of Pathogenomics attempts to utilize genomic and metagenomics data gathered from high through-put technologies (e.g. sequencing or DNA microarrays), to understand microbe diversity and interaction as well as host-microbe interactions involved in disease states. The bulk of pathogenomics research concerns itself with pathogens that affect human health; however, studies also exist for plant and animal infecting microbes.

History

In the early investigation of microbial genomics, it was difficult and costly to obtain sequence information for any pathogen. In 1995, the first pathogen genome, that of Haemophilus influenza was sequenced by traditional Sanger methods. Sanger methods, however, were slow and costly. The emergence of second generation high through-put sequencing technologies have allowed for microbial sequence information to be obtained much more quickly and at a considerably lower cost.. Largely thanks to second generation sequencing methods, hundreds of pathogen genomes have been sequenced since 1995. The emergence of second generation high through-put sequencing technologies have allowed for microbial sequence information to be obtained much more quickly and at a considerably lower cost. This influx of information is also due to the capacity of sequencing platforms to evaluate the sequences of many organisms in parallel.

With the sequences of many organisms available for analysis, scientists, through their investigations, began to challenge some of the earlier tenants of bacterial genome structure. Older paradigms of microbial genomics believed that only a few strains were sufficient to represent a specific bacterial species.

It was thought that bacterial genomes, like eukaryotes, were relatively stable. In 2001, however, the sequences of Escherichia coli 0157:H7 was obtained in a study by Perna and her colleges; the study showed that two members of the same bacterial species can differ as much as 30% in genomic content. It became evident that sequencing multiple strains for a species, rather than a few selectively chosen ones, was necessary to understand the diversity in a microbial species gene pool. It was also increasingly important to understand how to account for these differences in genomic content across a species strains and how it may contribute to pathogenic behaviour or prevent the formation of pathogens. More recently, the sequenced genomic data has been catalogued in databases and made publicly available online (there do also exist non-publicly available databases in the private sector). The availability and influx of this information presses upon those who conduct pathogenomics research to come up with a way of drawing meaningful conclusions from this data. In addition, the availability of such data on the internet encourages global collaboration of labs.

Microbe Analysis

Pathogens may be prokaryotic (archaea or bacteria), single-celled Eukarya or viruses. Prokaryotic genomes have typically been easier to sequence due to smaller genome size compared to Eukarya. Due to this, there is a bias in reporting pathogenic bacterial behaviour. More recently there have been increased efforts to sequence Eukarya genomes and more will be underway in the future. Regardless of this bias in reporting, many of the dynamic genomic events are similar across all the types of pathogen organisms. Pathogenomics does not focus exclusively on understanding pathogen-host interaction. Insight of individual or cooperative pathogen behaviour provides knowledge into the development or inheritance of pathogen virulence factors. Through a deeper understanding of the small subunits that cause infection, it may be able possible to develop novel therapeutics that are efficient and cost effective.

Analysing Individual Microbes

Dynamic genomes with high plasticity are necessary to allow pathogens, especially bacteria, to survive in changing environments. With the assistance of high throughput sequencing methods and in silico technologies, it is possible to detect, compare and catalogue many of these dynamic genomic events. Particular interest is in understanding how genomic events lead to pathogen development and how these events may be interrupted to prevent it.

Causes of Genomic Diversity

Three forces act in shaping the pathogen genome: gene gain, gene loss, and genome rearrangement. The knowledge and detection of these genomic dynamic events are necessary in the construction of useful therapeutic tools to combat pathogens.

Bacterial Genome Evolutionary Dynamics

Gene Loss/Genome Decay

Gene loss or genome decay occurs when a gene is no longer used by the microbe or when a microbe attempts to adapt to a new ecological niche.

Sequencing efforts and microarray analysis have exposed a large number of pseudo genes in some bacterial pathogen species. Mycobacterium leprae for example has been found to contain nearly as many pseudo genes as function genes. M. leprae is not the only microbe exhibiting such behaviour; in his article, Dr. Pallen reports similar properties from Yersinia pestis (the plague pathogen) and also Salmonella enterica. The inactivation of genes is typically associated with a change in the lifestyle of an organism, which can involve adapting to a new niche. The presence of extensive pseudogenes is contrary to another orthodox belief that all the genes in a bacterial genome are functional for some purpose.

It is possible to detect the presence of pseudogenes and the marks of genome decay through whole-genome sequencing in combination with comparative genomics Comparative genomics has helped to reveal that pathogens may favour losing genes in order to live in a host-associated niche and become endosymbionts. Sometimes the shedding of certain genes also renders a pathogen microbe harmless. The analysis of Listeria strains, for example, has shown that a reduced genome size has led to the generation of a non-pathogenic Listeria strain from a pathogenic precursor.

Gene Gain/Gene Duplication

One of the key forces driving gene gain is thought to be horizontal (lateral) gene transfer (LGT). It is of particular interest in microbial studies because these mobile genetic elements may introduce virulence factors into a new genome. An important comparative study conducted by Gill et al. in 2005 postulated that LGT may have been the cause for pathogen variations between Staphylococcus epidermidis and Staphylococcus aureus. There still, however, remains scepticism about the frequency of LGT, its identification, and its impact.

New and improved methodologies have been engaged, especially in the study of phylogenetics, to validate the presence and effect of LGT.

Gene gain and gene duplication events are balanced by gene loss, such that despite their dynamic nature, the genome of a bacterial species remains approximately the same size.

Genome Rearrangement

Mobile genetic insertion sequences can play a role in genome rearrangement activities. Pathogens that do not live in an isolated environment have been found to contain a large number of insertion sequence elements and various repetitive segments of DNA. The combination of these two genetic elements is thought help mediate homologous recombination. There are pathogens, such as Burkholderia mallei, and Burkholderia pseudomallei which have been shown to exhibit genome-wide rearrangements due to insertion sequences and repetitive DNA segments. At this time, no studies demonstrate genome-wide rearrangement events directly giving rise to pathogenic behaviour in a microbe. This does not mean it is not possible. Genome-wide rearrangements do, however, contribute to the plasticity of bacterial genome, which may prime the conditions for other factors to introduce, or lose, virulence factors.

Single Nucleotide Polymorphism

Single nucleotide polymorphisms(SNPs) are also a genomic variable that adds to the diversity of pathogen strains. Current efforts attempt to catalogue the various SNPs in pathogen strains.

Analysis of Genomic Diversity

There is a need to analyse more than a single genome sequence of a pathogen species to understand pathogen mechanisms. Comparative genomics is a powerful methodology that has gained more applicability with the recent increased amount of sequence information. There are several examples of successful comparative genomics studies, among them the analysis of Listeria. and Escherichia coli. The most important topic comparative genomics, in a pathogenomic context, attempts to address the difference between pathogenic and non-pathogenic microbes. This inquiry, though, proves to be very difficult to analyse, since a single bacterial species can have many strains and the genomic content of each of these strains can vary.

Pan-genomes and Core Genomes

The diversity within pathogen genomes makes it difficult to identify the total number of genes that are associated within all strains of a pathogen species. In fact, it has been thought that the total number of genes associated with a single pathogen species may be unlimited, although some groups are attempting to derive a more empirical value.. For this reason it was necessary to introduce the concept of pan-genomes and core genomes. Pan-genome and core genome literature also tends to have a bias towards reporting on prokaryotic pathogen organisms. Caution may need to be exercised when extending the definition of a pan-genome or a core-genome to the other pathogen organisms; this is because there is no formal evidence of the properties of these pan-genomes. Here, it will be assumed that the definitions may in fact extend, since all pathogen organisms share in the same dynamic genomic events and rely upon variability within strains as a mechanism of survival and virulence.

A core genome is the set of genes found across all strains of a pathogen species. A pan-genome is the entire gene pool for that pathogen species, and includes genes that are not shared by all strains. Pan-genomes may be open or closed depending on whether comparative analysis of multiple strains reveals no new genes (closed) or many new genes (open) compared to the core genome for that pathogen species. In the open pan-genome, genes may be further characterized as dispensable or strain specific. Dispensable genes are those found in more than one strain, but not in all strains, of a pathogen species. Strain specific genes are those found only in one strain of a pathogen species. The differences in pan-genomes are reflections of the life style of the organism. For example, Streptococcus agalactiae, which exists in diverse biological niches, has a broader pan-genome when compared with the more environmentally isolated Bacillus anthracis. Comparative genomics approaches are also being used to understand more about the pan-genome.

Mobile Genetic Elements that Encode Virulence Factors

Three genetic elements of human-affecting pathogens contribute to the transfer of virulence factors: plasmids, pathogenicity island, and prophages. Pathogencity islands and their detection are the focous of several bioinformatics efforts involved in pathogenomics.

Analysing Microbe-Microbe Interactions

Microbe-host interactions tend to overshadow the consideration of microbe-microbe interactions. Microbe-microbe interactions though can lead to chronic states of infirmity that are difficult to understand and treat.

Bioflims

Biofilms are an example of microbe-microbe interactions and are thought to be associated with up to 80% of human infections. Recently it has been shown that there are specific genes and cell surface proteins involved in the formation of biofilm. These genes and also surface proteins may be characterized through in silico methods to form an expression profile of biofilm interacting bacteria. This expression profile may be used in subsequent analysis of other microbes to predict biofilm microbe behaviour, or to understand how to dismantle biofilm formation.

Host Microbe Analysis

A microbe may be influenced by hosts to either adapt to their new environment or learn to evade it. An insight into these behaviours will provide beneficial insight for potential therapeutics. The most detailed outline of host-microbe interaction initiatives is outlined by the Pathogenomics European Research Agenda. Its report emphasises the following features:

- *Microarray analysis of host and microbe gene expression during infection.* This is important for identifying the expression of virulence factors that allow a pathogen to survive a host's defence mechanism. Pathogens tend to undergo an assortment of changed in order to subvert and hosts immune system, in some case favouring a hyper variable genome state. The genomic expression studies will be complimented with protein-protein interaction networks studies.
- *Using RNA interference (RNAi) to identify host cell functions in response to infections.* Infection depends on the balance between the characteristics of the host cell and the pathogen cell. In some cases, there can be an overactive host response to infection, such as in meningitis, which can overwhelm the host's body. Using RNA, it will be possible to more clearly identify how a host cell

defends itself during times of acute or chronic infection. This has also been applied successfully is Drosophila.

- *Not all microbe interactions in host environment are malicious.* Commensal flora, which exists in various environments in animals and humans may actually help combating microbial infections. The human flora, such as the gut for example, is home to a myriad of microbes.

The diverse community within the gut has been heralded to be vital for human health. There a number of projects under way to better understand the ecosystems of the gut. The sequence of commensal Escherichia coli strain SE11, for example, has already been determined from the faecal matter of a healthy human and promises to be the first of many studies. Through genomic analysis and also subsequent protein analysis a better understand of the beneficial properties of commensal flora will be investigated in hopes of understanding how to build a better therapeutic.

Eco-evo Perspective

The "eco-evo" perspective on pathogen-host interactions emphasizes the influences ecology and the environment on pathogen evolution. The dynamic genomic factors such as gene loss, gene gain and genome rearrangement are all strongly influenced by changes in the ecological niche that a particular microbial strain resides in. Microbes may switch from being pathogenic and non-pathogenic due to changing environments. Studies of the plague, Yersinia pestis, are prominent demonstration of how over time a microbe, in this case Yersinia pestis, evolves from a gastrointestinal pathogen to a very highly pathogenic microbe through dynamic genomic events. These flips between being pathogenic and non-pathogenic status and how they occurred with respect to ecological or environmental changes is important in novel therapeutic development to combat microbial infections.

Applications

Many of the future challenges of pathogenomics begin with handling and making sense of the large influx of data that now is available to the research community. Mining the data for useful information proves to be applicable to many facets of epidemiology. Bioinformatics approaches are providing much of the power for rapidly mining, organising, analysing, visualizing and annotating the data catalogued in databases.

Note that pathogenomics research could shed light for extensions of pathogens, or non-pathogens, that are not related to human, plant or animal health ; using microbes for bioremediation is one example. There is some but very little dialogue, however, concerning these extensions to pathogens and their relation to pathogenomics. It would be more suitable to categorize pathogen/non-pathogen applications that are unrelated to infection under the more general category of microbe genomics. Some general reviews speak extensively both about pathogen related and non-related applications in the same article. In the advent of new technology, it is easy to forget some of the basic things that prevent pathogen infections from starting and spreading.

While there of course exists more deadly and difficult to handle pathogen infections, there do also exist less dangerous ones. Historically, human health greatly improved with more emphasis on healthy life styles including better hygiene practices and access to clean recourses of water and nutrition. While pathogenomics can help provide insights into treatment and detection of some less potent pathogens, it must be kept in mind that there are many pathogens and only so much funding.

Reverse Vaccinology

The variability of genomes can make the developed of a vaccine very difficult, and antigen variation cannot match pathogen variation. Reverse vaccinology is a novel approach that may develop vaccines to combat pathogens more effectively. Reverse Vaccinology has already been successfully applied to Neisseria meningitides, Streptococcus pneumoniae and Chlamydia spp. Reverse Vaccinology applies not only to strain specific vaccines, but also the development of pan-genome vaccines. Lastly, comparative vaccinology attempts to compare the differences between pathogen and non-pathogen variants of a microbe to filter genes that are unique to the pathogen version. There are several vaccines developed through reverse Vaccinology that are currently in clinical trials.

Countering Bioterrorism

In 2005 the sequence of the 1918 Spanish influenza was completed. Accompanied with phylogenetic analysis, it was possible to supply a detailed account of the virus' evolution and behaviour, in particular it's adaptation to humans. Following the sequencing of the Spanish influenza, the pathogen was also reconstructed. When inserted into mice, the pathogen proved to be incredibly deadly.The 2001 anthrax attacks shed light on the possibility of the bioterrorism as being more of a real than imagined threat. Bioterrorism was anticipated in the Iraq war, with soldiers even being inoculated for an small pox attack. Using technologies and insight gained from reconstruction the Spanish influenza, it may be possible to prevent future deadly planted outbreaks of disease. There is a strong ethical concern however, as to whether the resurrection of old viruses is necessary and whether it is actually more harm than good.

Prospective Biotechnological Interventions of Manipulation of Rumen Bacteria: From Vision to Realism

The rumen, also known as a paunch, forms the larger part of the reticulorumen, which is the first chamber in the alimentary canal of ruminant animals. It serves as the primary site for microbial fermentation of ingested feed. The smaller part of the reticulorumen is the reticulum, which is fully continuous with the rumen, but differs from it with regard to the texture of its lining.

Brief Anatomy

The rumen is composed of several muscular sacs, the cranial sac, ventral sac, ventral blindsac, and reticulum.

The lining of the rumen wall is covered in small fingerlike projections called papillae, which are flattened, approximately 5 mm in length and 3 mm wide in cattle. The reticulum (derived from the Latin for net) is lined with ridges that form a hexagonal honeycomb pattern. The ridges are approximately 0.1- 0.2 mm wide and are raised 5 mm above the reticulum wall. The hexagons in the reticulum are approximately 2-5 cm wide in cattle. These features increase the surface area of the reticulorumen wall, facilitating the absorption of volatile fatty acids. Despite the differences in the texture of the lining of the two parts of the reticulorumen, it represents one functional space.

Stratification and Mixing of Digesta

Digesta in the rumen is not uniform, but rather is stratified into gas, liquid, and particles of different sizes, densities, and other physical characteristics. Additionally, digesta does not merely enter and exit the rumen without event, but it is subject to extensive mixing, and travels along complicated flow paths. Though they may seem trivial at first, these complicated stratification, mixing, and flow patterns of digesta are a key aspect of digestive activity in the ruminant and thus warrant detailed discussion. After being swallowed, ingesta travels down the oesophagus and is deposited in the dorsal part of the reticulum. Contractions of the reticulorumen propel and mix the recently ingested feed into the ruminal mat. The mat is a thick mass of digesta, consisting of partially degraded, long, fibrous material. Most material in the mat has been recently ingested, and as such, has considerable fermentable substrate remaining. Microbial fermentation proceeds rapidly in the mat, releasing many gases. Some of these gases are trapped in the mat, causing the mat to be buoyant. As fermentation proceeds, fermentable substrate is exhausted, gas production decreases, and particles lose buoyancy due to loss of entrapped gas. Digesta in the mat hence goes through a phase of increasing buoyancy followed by decreasing buoyancy. Simultaneously, the size of digesta particles–relatively large when ingested–is reduced by microbial fermentation and, later, rumination. At a certain point, particles are dense and small enough that they may "fall" through the rumen mat into the ventral sac below, or they may be swept out of the rumen mat into the reticulum by liquid gushing through the mat during ruminal contractions.

Once in the ventral sac, digesta continues to ferment at decreased rates, further losing buoyancy and decreasing in particle size. It is soon swept into the ventral reticulum by ruminal contractions.

In the ventral reticulum, less dense, larger digesta particles may be propelled up into the oesophagus and mouth during contractions of the reticulum. Digesta is chewed in the mouth in a process known as rumination, then expelled back down the oesophagus and deposited in the dorsal sac of the reticulum, to be lodged and mixed into the ruminal mat again. Denser, small particles stay in the ventral reticulum during reticular contraction, and then during the next contraction may be swept out of the reticulorumen with liquid through the reticulo-omasal orifice, which leads to the next chamber in the ruminant animal's alimentary canal, the omasum.

Water and saliva enter through the rumen to form a liquid pool. Liquid will ultimately escape from the reticulorumen from absorption through the wall, or through passing through the reticulo-omosal orifice, as digesta does. However, since liquid cannot be trapped in the mat as digesta can, liquid passes through the rumen much more quickly than digesta does. Liquid often acts as a carrier for very small digesta particles, such that the dynamics of small particles is similar to that of liquid.

The uppermost area of the rumen, the headspace, is filled with gases (such as methane, carbon dioxide, and, to a much lower degree, molecular hydrogen) released from fermentation and anaerobic respiration of food. These gases are regularly expelled from the reticulorumen through the mouth, in a process called eructation.

Digestion

Digestion in the reticulorumen is a complex process. Digestion occurs through fermentation by microbes in the reticulorumen rather than the animal *per se.* The reticulorumen is one of the few organs present in animals in which digestion of cellulose and other recalcitrant carbohydrates can proceed to any appreciable degree.

The main substrates of digestion in the reticulorumen are non-structural carbohydrates (starch, sugar, and pectin), structural carbohydrates (hemicellulose and cellulose), and nitrogen-containing compounds (protein, peptides, and amino acids). Both non-structural and structural carbohydrates are hydrolyzed to monosaccharides or disaccharides by microbial enzymes. The resulting mono- and disaccharides are transported into the microbes. Once within microbial cell walls, the mono- and disaccharides may be assimilated into microbial biomass or fermented to volatile fatty acids (VFAs) acetate, propionate, butyrate, lactate, valerate and other branched-chain VFAs via glycolysis and other biochemical pathways to yield energy for the microbial cell. Most VFAs are absorbed across the reticulorumen wall, directly into the blood stream, and are used by the ruminant as substrates for energy production and biosynthesis. Some branched chained VFAs are incorporated into the lipid membrane of rumen microbes. Protein is hydrolyzed to peptides and amino acids by microbial enzymes, which are subsequently transported across the microbial cell wall for assimilation into cell biomass, primarily. Peptides, amino acids, ammonia, and other sources of nitrogen originally present in the feed can also be used directly by microbes with little to no hydrolysis. Non-amino acid nitrogen is used for synthesis of microbial amino acids. In situations in which nitrogen for microbial growth is in excess, protein and its derivatives can also be fermented to produce energy, yielding ammonia.

Lipids, lignin, minerals, and vitamins play a less prominent role in digestion than carbohydrates and protein, but they are still critical in many ways. Lipids are partly hydrolysed and hydrogenated, and glycerol, if present in the lipid, is fermented. Lipids are otherwise inert in the rumen. Some carbon from carbohydrate or protein may be used for de novo synthesis of microbial lipid. High levels of lipid, particularly unsaturated lipid, in the rumen are thought to poison microbes and suppress fermentation activity. Lignin, a phenolic compound, is recalcitrant to digestion, through it can be solubilized by fungi. Lignin is thought to shield associated nutrients from digestion and hence limits degradation. Minerals are absorbed by microbes and are necessary to their growth. Microbes in turn synthesize many vitamins, such as cyanocobalamin, in great quantities—often great enough to sustain the ruminant even when vitamins are highly deficient in the diet.

Microbes in the Reticulorumen

Microbes in the reticulorumen include bacteria, protozoa, fungi, archaea, and viruses. Bacteria, along with protozoa, are the predominant microbes and by mass account for 40-60% of total microbial matter in the rumen. They are categorized into several functional groups, such as fibrolytic, amylolytic, and proteolytic types, which preferentially digest structural carbohydrates, non-structural carbohydrates, and protein, respectively. Protozoa (40-60% of microbial mass) derive their nutrients through phagocytosis of other microbes, and degrade and digest feed carbohydrates, especially starch and sugars, and protein. Although protozoa are not essential for rumen functioning, their presence has pronounced effects. Ruminal fungi make up only 5-10% of microbes and are absent on diets poor in fibre. Despite their low numbers, the fungi still occupy an important niche in the rumen because they hydrolyse some ester linkages between lignin and hemicellulose or cellulose, and help break down digesta particles. Rumen Archaea, approximately 3% of total microbes, are mostly autotrophic methanogens and produce methane through anaerobic respiration.

Most of the hydrogen produced by bacteria, protozoa and fungi is used by these methanogens to reduce carbon dioxide to methane. The maintenance of low partial pressure of hydrogen by methanogens is essential for proper functioning of the rumen. Viruses are present in unknown numbers and do not contribute to any fermentation or respiration activity. However, they do lyse microbes, releasing their contents for other microbes to assimilate and ferment in a process called microbial recycling, although recycling through the predatory activities of protozoa is quantitatively more important.

Microbes in the reticulorumen eventually flow out into the omasum and the remainder of the alimentary canal. Under normal fermentation conditions the environment in the reticulorumen is weakly acidic and is populated by microbes that are adapted to a pH between roughly 5.5 and 6.5; since the abomasum is strongly acidic (pH 2 to 4), it acts as a barrier that largely kills reticulorumen flora and fauna as they flow into it. Subsequently, microbial biomass is digested in the small intestine and smaller molecules (mainly amino acids) are absorbed and transported in the portal vein to the liver.

The digestion of these microbes in the small intestine is a major source of nutrition, as microbes usually supply some 60 to 90% of the total amount of amino acids absorbed. On starch-poor diets, they also provide the predominant source of glucose absorbed from the small intestinal contents. Under conditions of ruminal acidosis, when the environment of the reticulorumen has become too acidic (usually due to excessive fermentation of starches and sugars into VFA and lactate), microbes that favour a lower pH may start to dominate the ecosystem of the reticulorumen. This gives rise to rumen acidosis and often feed intake of the ruminant will drop.

- Bacteria, protozoa, and fungi exist together in the cow's rumen. Bacteria make up about half of the living organisms but do more than half of the rumen's digestive work.
- Rumen bacteria are classified into fibre digesters, starch and sugar digesters, lactate using bacteria, and hydrogen-using bacteria. They cooperate together and cross feed.

Bacteria

Bacteria make up about half of the living organisms inside of the rumen. However, they do more than half of the work in the rumen. The bacteria work together. Some breakdown certain carbohydrates and proteins which are then used by others. Some require certain growth factors, such as B-vitamins, which are made by others. Some bacteria help to clean up the rumen of others' end products, such as hydrogen ions, which could otherwise accumulate and become toxic to other organisms. This is called "cross-feeding".

Classification of Rumen Bacteria

Fibre–Digesting (or Cellulolytic) Bacteria:

The fibre-digesters are some of the "fussiest" bacteria in the rumen. They are very sensitive to acid. When a cow has acidosis ($pH<6.0$), the rumen produces a lower proportion of acetate to propionate because the fibre-digesters who primarily make acetate are not working well. Also, high levels of rumen available fat (generally over 5% of the diet) reduce the growth of the fibre-digesters. The exact reason for fat's negative effect on the fibre-digesters is not known. Some think that it reduces the microbe's ability to move nutrients into and out of its body. Others think that the fat coats fibre particles making it difficult for the fibre-digesting microbes to get in to do their work.

Bacterial species: Ruminococcus flavefacians, Ruminococcus albus, Bacteriodes succinogenes, Butyrivibrio fibrisolvens.

Growth Requirements: Cellulose, Hemicellulose, Pectin.

Many Also Require: Ammonia, Isoacids (Branched-Chain Amino Acids), Starch, B-vitamins- Fermentation Products: Acetate, Butyrate, Hydrogen (H_2), Carbon Dioxide (CO_2).

Rumen pH Requirement: High pH (above 6.0).

Fat Tolerance: Low.

Susceptibility to Ionophores (Bovatec and Rumensin): Some are susceptible.

Reproduction Speed: Slow.

Starch and Sugar-Digesting (or Amylolytic) Bacteria:

Starch and sugar-digesters make up a significant part of the rumen's bacterial population. Generally, high-producing dairy cows are fed diets containing more than 30% starches and sugars, so these bacteria are greatly needed. Even if a cow is on an all-straw diet, the fibre-digesters still never account for more than 25% of the rumen bacterial population. Starch and sugar-digesters are still present, cross-feeding off of the fibre-digesters' byproducts.

Bacterial Species

Bacteriodes ruminocola, Bacteriodes amylophilus, Selenomonas ruminantium, Streptococcus bovis, Succinomonas amylolytica.

Growth Requirements

Sugar, Starch, Peptides, Amino Acids.

Many Also Require: Ammonia, B-vitamins.

Fermentation Products

Propionate, Butyrate, Acetate, Lactate, Hydrogen (H_2), Carbon Dioxide (CO_2)

Rumen pH Requirement: Tolerate a lower (more acidic) pH (5.7).

Fat Tolerance

Higher than fibre digesters.

Susceptibility to Ionophors (Bovatec and Rumensin):

Most aren't susceptible.

Reproduction Speed

Faster than fibre digesters.

Streptococcus bovis, "The Rumen Weed"

Streptococcus bovisis present only when large amounts of starch or sugars are fed and pH is low. It produces lactic acid, a stronger acid than many of the other VFA's produced in the rumen. When conditions are favourable for Streptococcus bovis, it will grow explosively (doubling every 13 minutes). This type of growth produces rumen acidosis. Streptococcus bovis is controlled by ionophores. This is one of the major reasons for the favourable growth responses seen by the addition of ionophores to the diets of feedlot cattle.

Lactate-Using Bacteria

Includes: Megasphaera elsdenii, "The Rumen Maid".

As mentioned above, some bacteria, such as Streptococcus bovis, produce a strong acid called lactic acid. Megasphacra elsdenii uses lactic acid to grow. This helps to clean up the rumen a bit and raise rumen pH, aiding the growth of the acid-intolerant fibre-digesters.

Hydrogen-Using (or Methane) Bacteria: Under normal rumen conditions, hydrogen (H_2) does not accumulate in the rumen because it's used by hydrogen-using bacteria, such as Methanobacterium ruminantium.

Growth Requirements: Carbon dioxide and hydrogen

Fermentation products: Methane

The methane bacteria commonly produce methane in this way:

$$4H_2 + CO_2 \longrightarrow CH_4 + 2H_2O$$

Protozoa

As much as 50% of the microbial mass in the rumen can be made up of protozoa. However, their role, as compared to the rumen bacteria, is not as significant. The protozoa are actually predators to the bacteria in the rumen — they eat the bacteria for dinner! Protozoa are about 40 times the size of rumen bacteria.

The rumen protozoa produce fermentation end-products similar those made by the bacteria, particularly acetate, butyrate, and hydrogen. Rumen methane bacteria actually attach and live on the surface of rumen protozoa for immediate access to hydrogen.

Rumen protozoa eat large amounts of starch at one time and can store it in their bodies. This may help to slow down the production of acids that lower rumen pH, benefiting the rumen.

Rumen protozoa multiply very slowly in the rumen — over 15-24 hours – as opposed to the bacteria that may take as little as 13 minutes to multiply. For this reason, the rumen protozoa hide out in the slower moving fibre mat of the rumen so that they aren't washed out before they have a chance to multiply. Low roughage diets reduce the retention of fibre in the rumen and may decrease the number of protozoa in a cow's rumen.

Rumen Fungi

Fungi are known to exist in the rumen (up to 8% of the total mass) but they are poorly understood. They attach to feed particles and they reproduce very slowly. They may help out the fibre-digesting bacteria by doing some of the initial work of splitting fibrous material apart and making it more accessible for the bacteria. Higher numbers of fungi have been found in the rumens of cows fed very poorly digestible sub-tropical forages.

Towards a Post-materialist Understanding of Science –lessons Learnt form the Interface of Biodynamic Agriculture and Research

Alternative forms of practicing and understanding agriculture are becoming increasingly important on a global level. Among those in Europe biodynamic agriculture is representing the oldest form of 'alternative' agriculture, which from the beginning on was challenging theory and practice of conventional agriculture.

Biodynamic agriculture is based on non-materialist philosophy, known as 'anthroposophy', created by Rudolf Steiner. It represents a 'local' knowledge emerging at the beginning of the 20. Century and was further developed by close interaction of farmers willing to apply the new methods and scientists which were recognising the limitations of conventional science and the interpretations associated to it. Biodynamic

Agriculture was one of the most prominent opponents of conventinal agriculture until the rising of the environmental movement at the beginning of the 70-ies.

Biodynamic agriculture was participating in COMPAS from the beginning on because many roots of this farming concept are connected to wisdom coming from other cultures all along human evolution. Reencountering with traditional forms of agriculture could create room for mutual learning and synergies of all partners involved. The way in which Biodynamic Agriculture dealt and still deals with science as taught in Universities will be presented in this paper because we think that this is an important aspect for sharing with other forms of local knowledge.

Biodynamic agriculture studies natural phenomena on a non materialistic background, expressed in a theory of cognition as developed in anthroposophy. It starts by recognising that concepts always have a generic character, which by relating them to perceived phenomena (demand for concepts) become individualised making that phenomena which are at first incomprehensible become freed from their lack of context through putting the concepts back in their proper context. Only through this last step is reality attained. Reality is therefore the conceptually or spiritually permeated world of phenomena (Witzenmann, 1977/78). It follows from this that in reality there is no spiritless matter. But there is certainly a matterless spirituality in the form of thinking. Based on this a new understanding of plants and animals farmers and researchers had to create a specific form of biodynamic animal and plant breeding.

What these concepts mean in prac tice will be exemplified by presenting main stages and outcomes of the process through which local knowledge and experiences of a group of biodynamic farmers and researchers were creating their own understanding of biodynamic plant and animal breeding. The role played by the small group of scientists involved in this process allowed to identify main challenges that had to be faced when trying to transform materialist science into a new form of science which is post-materialist; in stead of giving the solution to the farmers it aims to support them in the systematisation of their own experiences and knowledge, based on a social learning process of all actors involved. Main lessons of this process will be presented.

Starting Points

The building of the group of farmers and researchers was motivated by the need for getting to a better understanding of the epistemological background which is framing the concretely defined standards for biodynamic farming that in turn are the basis for the yearly certification of the farms made by independent external specialists. The certification verifies the fulfilment of the standards on behalf of the farmers, which in turn guaranties high quality food products to the consumers.

This is rewarded by paying a higher price for food certifies with the "Demeter" label ranging from 10-30 % at an average (DEMETER, 2003).

Farmers and researches placed special interest in some basic questions on which the standards are stressing and that are representing mayor concerns for both farmers and researchers, because they signify additional investments in time, money and efforts for understanding the complex interrelations between cosmos, earth, animals, plants and humans. Among the most discussed issued we find the following aspects of biodynamic agriculture:

The stable type and the internal arrangement and fittings must meet the following requirements (cited according to DEMETER, 2003):

- The sleeping stalls for cattle are to have appropriate bedding.
- Fully slatted floors (more than 50%) are not permitted and the slatted area may not be calculated as resting-place.
- Cow trainers are not permitted.
- Sufficient area to be provided, and the herd managed to allow the expression of social behaviour and unhindered feeding.
- There must be at least as many feeding/sleeping stalls as there are animals in the stable. In stables with ad lib feeding, fewer feeding stations may be offered.

Dairy cattle and cows suckling calves are to have access to pasture during the summer half-year. Where this is not possible, access to the open air must be available all year round. Young stock (breeding replacements) have the same requirement for freedom of movement reasons. To tie up young replacement or fattening stock in housing all year round is not allowed. Cows should be given freedom of movement at calving. A calving bay should be provided for if stable renovations occur.

The horns of ruminants have significance for the development of life forces. They provide an opposing balance of forces to the intensive digestion an d absorption processes. They are a part of the total being of the cow. In comparison to other animal types, cattle manure has a particularly stimulating effect on soil fertility.

The horns also have a large significance as a sheath in the production of the Biodynamic preparations.

The fodder must be appropriate and contain as high a content of roughage (greenfeed e.g. pasture, hay, silage) as possible, but at least 60% DM throughout the entire year. The majority of summer feeds must be green material, preferably grazed from pasture.

The animals should be born and reared on a certified Biodynamic farm, if possible as part of a permanent herd. Poultry chickens should hatch after natural incubation. A principle of the Biodynamic method is the keeping of male sires on the farm, and is therefore highly recommended. Artificial insemination cannot fully replace the effect of the male influence in the farm herd, and is not recommended. It is not permitted to produce animals using genetic manipulation, or by the use of biotechnology (embryo transfer, sperm separation for sex determination).

All this in terms of farm management highly relevant practices are closely related to a specific understanding on what an animal is and how one must care for it. The objective of the group was therefore to engage in a social learning process that allows to deepen the understanding of what an animal is considering the experiences and knowledge of the practitioners and researchers as equally valid.

A Post-materialist Epistemology

In a first step the epistemological basis of biodynamic agriculture had to be clarified for all members of the group. The researchers engaged in the field of ethology, biology and agronomy supported this first phase of reflection.

The assumed subjectivity of thinking (KANT, 1877) contradicts the engineer's and also farmers experience who-by thinking and understanding and acting- can effectively influence the inorganic sphere of the world.

To overcome this contradiction, a cognitive theory is required that shows how man can understand the world.

This epistemology was successfully established and implemented by STEINER (1921). In trying to observe one's own cognition processes, one can easily apply and modify FURRER's (1980) arguments-. "it is not enough for us to think, act and perceive. Experiments can only be carried out if nature itself is thinkable, treatable and perceivable. Yet we can only find out by experiment whether nature has these qualities."

However, nature's essential quality is not its questionability but its mental conceivability and its functionality. One could even say that any science is the science of behaviour i.e. ethology, since all thinkable expectations of behaviour (hypotheses) are observed in the experiment as nature's actual behaviour.

The cognition process can be followed up by anyone who observes his or her own efforts to understand. The perceivable and questionable world phenomena,- which needs explanations- draws on human concepts created through thinking. Experiments show whether or not the expectations concerning behaviour are confirmed by world behaviour. If this is the case, the world of phenomena is no longer questionable, its functionality and its laws will be recognised and understood. In other words, the world becomes real; we can realise the world. By linking conception and perception our concepts-which always have a generic character-become individualised, e.g. a generic concept becomes an individual representation. The generic concept of an injury, for example, becomes individualised to a laceration on the femur.

By carrying out this cognition process in all the different spheres of conception-corresponding to the different spheres of being (inorganic, physiological, psychological and spiritual)-mankind recognizes the world and thus creates the different sciences (physics, biology, psychology and the humanities). Cognitive ethics derives from the links between scientific understanding of the world and of human nature.

Implementing these ethics by acting according to reason and understanding leads to cultural progress in the form of a metamorphosis of natural evolution.

Ethology of Farm Animals

By applying this cognition process to animal behaviour, it becomes apparent that animal behaviour cannot be understood without the concept of autonomous activity, because this behaviour is not simply the causal effect of the environment, but an expression of the animal actively coming to terms with its surroundings. The animal applies its species-typical behaviour in order to satisfy its wants and to avoid suffering. This has given rise to the "prerequisites to meet the animal's wants and to avoid suffering' theory. The satisfaction of wants and the avoidance of suffering leads to selforganisation and self-preservation of the animal physics, corresponding to the "theory of the prerequisites to meet needs and to avoid injuries" by TSCHANZ (1987). Housing systems that do not comply with the behaviour patterns of a species can impair or even prevent species-typical behaviour.

3

DNA Vaccination

DNA vaccination is a technique for protecting an organism against disease by injecting it with genetically engineered DNA to produce an immunological response. Nucleic acid vaccines are still experimental, and have been applied to a number of viral, bacterial and parasitic models of disease, as well as to several tumour models. DNA vaccines have a number of advantages over conventional vaccines, including the ability to induce a wider range of immune response types.

Vaccines are among the greatest achievements of modern medicine – in industrial nations, they have eliminated naturally-occurring cases of smallpox, and nearly eliminated polio, while other diseases, such as typhus, rotavirus, hepatitis A and B and others are well controlled. Conventional vaccines, however, only cover a small number of diseases, and infections that lack effective vaccines kill millions of people every year, with AIDS, hepatitis C and malaria being particularly common.

First generation vaccines are whole-organism vaccines – either live and weakened, or killed forms. Live, attenuated vaccines, such as smallpox and polio vaccines, are able to induce killer T-cell (T_C or CTL) responses, helper T-cell (T_H) responses and antibody immunity. However, there is a small risk that attenuated forms of a pathogen can revert to a dangerous form, and may still be able to cause disease in immunocompromised people (such as those with AIDS). While killed vaccines do not have this risk, they cannot generate specific killer T cell responses, and may not work at all for some diseases. In order to minimise these risks, so-called *second generation vaccines* were developed. These are subunit vaccines, consisting of defined protein antigens (such as tetanus or diphtheria toxoid) or recombinant protein components (such as the hepatitis B surface antigen). These, too, are able to generate T_H and antibody responses, but not killer T cell responses.

DNA vaccines are *third generation vaccines*, and are made up of a small, circular piece of bacterial DNA (called a plasmid) that has been genetically engineered to produce one or two specific proteins (antigens) from a pathogen. The vaccine DNA is injected into the cells of the body, where the "inner machinery" of the host cells "reads" the DNA and converts it into pathogenic proteins. Because these proteins are recognised as foreign, when they are processed by the host cells and displayed on their surface, the immune system is alerted, which then triggers a range of immune responses. These DNA vaccines developed from "failed" gene therapy experiments. The first demonstration of a plasmid-induced immune response was when mice inoculated with a plasmid expressing human growth hormone elicited antibodies instead of altering growth.

Current Use

Thus far, few experimental trials have evoked a response sufficiently strong enough to protect against disease, and the usefulness of the technique, while tantalizing, remains to be conclusively proven in human trials. However, in June 2006 positive results were announced for a bird flu DNA vaccine and a veterinary DNA vaccine to protect horses from West Nile virus has been approved. In August 2007, a preliminary study in DNA vaccination against multiple sclerosis was reported as being effective.

Plasmid Vectors for Use in Vaccination

Vector Design

DNA vaccines elicit the best immune response when highly active expression vectors are used. These are plasmids which usually consist of a strong viral promoter to drive the in vivo transcription and translation of the gene (or complementary DNA) of interest. Intron A may sometimes be included to improve mRNA stability and hence increase protein expression. Plasmids also include a strong polyadenylation/ transcriptional termination signal, such as bovine growth hormone or rabbit beta-globulin polyadenylation sequences. Multicistronic vectors are sometimes constructed to express more than one immunogen, or to express an immunogen and an immunostimulatory protein. Because the plasmid is the "vehicle" from which the immunogen is expressed, optimising vector design for maximal protein expression is essential. One way of enhancing protein expression is by optimising the codon usage of pathogenic mRNAs for eukaryotic cells. Pathogens often have different AT contents than the species being immunized, so altering the gene sequence of the immunogen to reflect the codons more commonly used in the target species may improve its expression.

Another consideration is the choice of promoter. The SV40 promoter was conventionally used until research showed that vectors driven by the Rous Sarcoma Virus (RSV) promoter had much higher expression rates. More recently, expression rates have been further increased by the use of the cytomegalovirus (CMV) immediate early promoter. Inclusion of the Mason-Pfizer monkey virus (MPV)-CTE with/without rev increased envelope expression. Furthermore the CTE+rev construct was significantly more immunogenic then CTE alone vector. Additional modifications to improve expression rates have included the insertion of enhancer sequences, synthetic introns, adenovirus tripartite leader (TPL) sequences and modifications to the polyadenylation and transcriptional termination sequences. An example of DNA vaccine plasmid is pVAC, it uses SV40 promoter.

Vaccine Insert Design

Immunogens can be targeted to various cellular compartments in order to improve antibody or cytotoxic T-cell responses. Secreted or plasma membrane-bound antigens are more effective at inducing antibody responses than cytosolic antigens, while cytotoxic T-cell responses can be improved by targeting antigens for cytoplasmic degradation and subsequent entry into the major histocompatibility complex (MHC) class I pathway. This is usually accomplished by the addition of N-terminal ubiquitin signals.

The conformation of the protein can also have an effect on antibody responses, with "ordered" structures (like viral particles) being more effective than unordered structures. Strings of minigenes (or MHC class I epitopes) from different pathogens are able to raise cytotoxic T-cell responses to a number of pathogens, especially if a TH epitope is also included.

Delivery Methods

DNA vaccines have been introduced into animal tissues by a number of different methods. The two most popular approaches are injection of DNA in saline, using a standard hypodermic needle, and gene gun delivery. A schematic outline of the construction of a DNA vaccine plasmid and its subsequent delivery by these two methods into a host is illustrated at Scientific American. Injection in saline is normally conducted intramuscularly (IM) in skeletal muscle, or intradermally (ID), with DNA being delivered to the extracellular spaces. This can be assisted by electroporation; by temporarily damaging muscle fibres with myotoxins such as bupivacaine; or by using hypertonic solutions of saline or sucrose. Immune responses to this method of delivery can be affected by many factors, including needle type, needle alignment, speed of injection, volume of injection, muscle type, and age, sex and physiological condition of the animal being injected.

Gene gun delivery, the other commonly used method of delivery, ballistically accelerates plasmid DNA (pDNA) that has been adsorbed onto gold or tungsten microparticles into the target cells, using compressed helium as an accelerant.

Alternative delivery methods have included aerosol instillation of naked DNA on mucosal surfaces, such as the nasal and lung mucosa, and topical administration of pDNA to the eye and vaginal mucosa. Mucosal surface delivery has also been achieved using cationic liposome-DNA preparations, biodegradable microspheres, attenuated *Shigella* or *Listeria* vectors for oral administration to the intestinal mucosa, and recombinant adenovirus vectors.

The method of delivery determines the dose of DNA required to raise an effective immune response. Saline injections require variable amounts of DNA, from 10 μg 1 mg, whereas gene gun deliveries require 100 to 1000 times less DNA than intramuscular saline injection to raise an effective immune response. Generally, 0.2 μg – 20 μg are required, although quantities as low as 16 ng have been reported. These quantities vary from species to species, with mice, for example, requiring approximately

10 times less DNA than primates. Saline injections require more DNA because the DNA is delivered to the extracellular spaces of the target tissue (normally muscle), where it has to overcome physical barriers (such as the basal lamina and large amounts of connective tissue, to mention a few) before it is taken up by the cells, while gene gun deliveries bombard DNA directly into the cells, resulting in less "wastage".

Another approach to DNA vaccination is expression library immunization (ELI). Using this technique, potentially all the genes from a pathogen can be delivered at one time, which may be useful for pathogens which are difficult to attenuate or culture. ELI can be used to identify which of the pathogen's genes induce a protective response. This has been tested with *Mycoplasma pulmonis*, a murine lung pathogen with a relatively small genome, and it was found that even partial expression libraries can induce protection from subsequent challenge.

Immune Response Raised by DNA Vaccines

Helper T-Cell Responses

Antigen presentation stimulates T cells to become either "cytotoxic" CD8+ cells or "helper" CD4+ cells. Cytotoxic cells directly attack other cells carrying certain foreign or abnormal molecules on their surfaces. Helper T cells, or Th cells, coordinate immune responses by communicating with other cells. In most cases, T cells only recognize an antigen if it is carried on the surface of a cell by one of the body's own MHC, or major histocompatibility complex, molecules.

DNA immunization is able to raise a range of T_H responses, including lymphoproliferation and the generation of a variety of cytokine profiles. A major advantage of DNA vaccines is the ease with which they can be manipulated to bias the type of T-cell help towards a TH1 or TH2 response. Each type of response has distinctive patterns of lymphokine and chemokine expression, specific types of immunoglobulins expressed, patterns of lymphocyte trafficking, and types of innate immune responses generated.

Raising of Different Types of T-cell Help

The type of T-cell help raised is influenced by the method of delivery and the type of immunogen expressed, as well as the targeting of different lymphoid compartments. Generally, saline needle injections (either IM or ID) tend to induce TH1 responses, while gene gun delivery raises TH2 responses. This is true for intracellular and plasma membrane-bound antigens, but not for secreted antigens, which seem to generate TH2 responses, regardless of the method of delivery.

Generally the type of T-cell help raised is stable over time, and does not change when challenged or after subsequent immunizations which would normally have raised the opposite type of response in a naïve animal. However, Mor *et al.*. (1995) immunized and boosted mice with pDNA encoding the circumsporozoite protein of the mouse malarial parasite *Plasmodium yoelii* (PyCSP) and found that the initial TH2 response changed, after boosting, to a TH1 response.

Mechanistic Basis for Different Types of T-Cell Help

It is not understood how these different methods of DNA immunization, or the forms of antigen expressed, raise different profiles of T-cell help. It was thought that the relatively large amounts of DNA used in IM injection were responsible for the induction of TH1 responses. However, evidence has shown no differences in TH type due to dose. It has been postulated that the type of T-cell help raised is determined by the differentiated state of antigen presenting cells. Dendritic cells can differentiate to secrete IL-12 (which supports TH1 cell development) or IL-4 (which supports TH2 responses). pDNA injected by needle is endocytosed into the dendritic cell, which is then stimulated to differentiate for TH1 cytokine production, while the gene gun bombards the DNA directly into the cell, thus bypassing TH1 stimulation.

Practical Uses of Polarised T-Cell Help

This polarisation in T-cell help is useful in influencing allergic responses and autoimmune diseases. In autoimmune diseases, the goal would be to shift the self-destructive TH1 response (with its associated cytotoxic T cell activity) to a non-destructive TH2 response. This has been successfully applied in predisease priming for the desired type of response in preclinical models and somewhat successful in shifting the response for an already established disease.

Cytotoxic T-cell Responses

One of the greatest advantages of DNA vaccines is that they are able to induce cytotoxic T lymphocytes (CTL) without the inherent risk associated with live vaccines. CTL responses can be raised against immunodominant and immunorecessive CTL epitopes, as well as subdominant CTL epitopes, in a manner which appears to mimic natural infection. This may prove to be a useful tool in assessing CTL epitopes of an antigen, and their role in providing immunity.

Cytotoxic T-cells recognise small peptides (8-10 amino acids) complexed to MHC class I molecules (Restifo et al., 1995). These peptides are derived from endogenous cytosolic proteins which are degraded and delivered to the nascent MHC class I molecule within the endoplasmic reticulum (ER). Targeting gene products directly to the ER (by the addition of an amino-terminal insertion sequence) should thus enhance CTL responses. This has been successfully demonstrated using recombinant vaccinia viruses expressing influenza proteins, but the principle should be applicable to DNA vaccines too. Targeting antigens for intracellular degradation (and thus entry into the MHC class I pathway) by the addition of ubiquitin signal sequences, or mutation of other signal sequences, has also been shown to be effective at increasing CTL responses.

CTL responses can also be enhanced by co-inoculation with co-stimulatory molecules such as B7-1 or B7-2 for DNA vaccines against influenza nucleoprotein, or GM-CSF for DNA vaccines against the murine malaria model *P. yoelii*. Co-inoculation with plasmids encoding co-stimulatory molecules IL-12 and TCA3 have also been shown to increase CTL activity against HIV-1 and influenza nucleoprotein antigens.

Humoral (Antibody) Response

Antibody responses elicited by DNA vaccinations are influenced by a number of variables, including type of antigen encoded; location of expressed antigen (i.e. intracellular vs. secreted); number, frequency and dose of immunizations; site and method of antigen delivery, to name a few.

Kinetics of Antibody Response

Humoral responses after a single DNA injection can be much longer-lived than after a single injection with a recombinant protein. Antibody responses against hepatitis B virus (HBV) envelope protein (HBsAg) have been sustained for up to 74 weeks without boost, while life-long maintenance of protective response to influenza haemagglutinin has been demonstrated in mice after gene gun delivery. Antibody-secreting cells migrate to the bone marrow and spleen for long-term antibody production, and are generally localised there after one year.

DNA-raised antibody responses rise much more slowly than when natural infection or recombinant protein immunization occurs. It can take as long as 12 weeks to reach peak titres in mice, although boosting can increase the rate of antibody production. This slow response is probably due to the low levels of antigen expressed over several weeks, which supports both primary and secondary phases of antibody response.

Additionally, the titres of specific antibodies raised by DNA vaccination are lower than those obtained after vaccination with a recombinant protein. However, DNA immunization-induced antibodies show greater affinity to native epitopes than recombinant protein-induced antibodies. In other words, DNA immunization induces a qualitatively superior response. Antibody can be induced after just one vaccination with DNA, whereas recombinant protein vaccinations generally require a boost. As mentioned previously, DNA immunization can be used to bias the TH profile of the immune response, and thus the antibody isotype, which is not possible with either natural infection or recombinant protein immunization. Antibody responses generated by DNA are useful not just in vaccination but as a preparative tool, too. For example, polyclonal and monoclonal antibodies can be generated for use as reagents.

Mechanistic Basis for DNA Raised Immune Responses

DNA Uptake Mechanism

When DNA uptake and subsequent expression was first demonstrated *in vivo* in muscle cells, it was thought that these cells were unique in this ability because of their extensive network of T-tubules. Using electron microscopy, it was proposed that DNA uptake was facilitated by caveolae (or, non-clathrin coated pits). However, subsequent research revealed that other cells (such as keratinocytes, fibroblasts and epithelial Langerhans cells) could also internalize DNA. This phenomenon has not been the subject of much research, so the actual mechanism of DNA uptake is not known.

Two theories are currently popular – that *in vivo* uptake of DNA occurs non-specifically, in a method similar to phago- or pinocytosis, or through specific receptors. These might include a 30kDa surface receptor, or macrophage scavenger receptors. The 30kDa surface receptor binds very specifically to 4500-bp genomic DNA fragments (which are then internalised) and is found on professional APCs and T-cells. Macrophage scavenger receptors bind to a variety of macromolecules, including polyribonucleotides, and are thus also candidates for DNA uptake. Receptor mediated DNA uptake could be facilitated by the presence of polyguanylate sequences. Further research into this mechanism might seem pointless, considering that gene gun delivery systems, cationic liposome packaging, and other delivery methods bypass this entry method, but understanding it might be useful in reducing costs (e.g. by reducing the requirement for cytofectins), which will be important in the food animals industry.

Antigen Presentation by Bone Marrow-derived Cells

Studies using chimeric mice have shown that antigen is presented by bone-marrow derived cells, which include dendritic cells, macrophages and specialised B-cells called professional antigen presenting cells (APC) Iwasaki et al., 1997). After gene gun inoculation to the skin, transfected Langerhans cells migrate to the draining lymph node to present antigen. After IM and ID injections, dendritic cells have also been found to present antigen in the draining lymph node and transfected macrophages have been found in the peripheral blood. Besides direct transfection of dendritic cells or macrophages, cross priming is also known to occur following IM, ID and gene gun deliveries of DNA. Cross priming occurs when a bone marrow-derived cell presents peptides from proteins synthesised in another cell in the context of MHC class 1. This can prime cytotoxic T-cell responses and seems to be important for a full primary immune response.

Role of the Target Site

IM and ID delivery of DNA initiate immune responses differently. In the skin, keratinocytes, fibroblasts and Langerhans cells take up and express antigen, and are responsible for inducing a primary antibody response. Transfected Langerhans cells migrate out of the skin (within 12 hours) to the draining lymph node where they prime secondary B- and T-cell responses. In skeletal muscle, on the other hand, striated muscle cells are most frequently transfected, but seem to be unimportant in mounting an immune response. Instead, IM inoculated DNA "washes" into the draining lymph node within minutes, where distal dendritic cells are transfected and then initiate an immune response. Transfected myocytes seem to act as a "reservoir" of antigen for trafficking professional APCs.

Maintenance of Immune Response

DNA vaccination generates an effective immune memory via the display of antigen-antibody complexes on follicular dendritic cells (FDC), which are potent B-cell stimulators. T-cells can be stimulated by similar, germinal centre dendritic cells. FDC are able to generate an immune memory because antibodies production "overlaps" long-term expression of antigen, allowing antigen-antibody immunocomplexes to form and be displayed by FDC.

Interferons

Both helper and cytotoxic T-cells can control viral infections by secreting interferons. Cytotoxic T cells usually kill virally infected cells. However, they can also be stimulated to secrete antiviral cytokines such as INF-γ and TNF-α, which don't kill the cell but place severe limitations on viral infection by down-regulating the expression of viral components. DNA vaccinations can thus be used to curb viral infections by non-destructive IFN-mediated control. This has been demonstrated for the hepatitis B virus. IFN-γ is also critically important in controlling malaria infections, and should be taken into consideration when developing anti-malarial DNA vaccines.

Modulation of the Immune Response

Cytokine Modulation

For a vaccine to be effective, it must induce an appropriate immune response for a given pathogen, and the ability of DNA vaccines to polarise T-cell help towards TH1 or TH2 profiles, and generate CTL and/or antibody when required, is a great advantage in this regard. This can be accomplished by modifications to the form of antigen expressed (i.e. intracellular vs. secreted), the method and route of delivery, and the dose of DNA delivered. However, it can also be accomplished by the co-administration of plasmid DNA encoding immune regulatory molecules, i.e. cytokines, lymphokines or co-stimulatory molecules. These "genetic adjuvants" can be administered a number of ways:

- as a mixture of 2 separate plasmids, one encoding the immunogen and the other encoding the cytokine;
- as a single bi- or polycistronic vector, separated by spacer regions; or
- as a plasmid-encoded chimera, or fusion protein.

In general, co-administration of pro-inflammatory agents (such as various interleukins, tumour necrosis factor, and GM-CSF) plus TH2 inducing cytokines increase antibody responses, whereas pro-inflammatory agents and TH1 inducing cytokines decrease humoral responses and increase cytotoxic responses (which is more important in viral protection, for example). Co-stimulatory molecules like B7-1, B7-2 and CD40L are also sometimes used.

This concept has been successfully applied in topical administration of pDNA encoding IL-10. Plasmid encoded B7-1 (a ligand on APCs) has successfully enhanced the immune response in anti-tumour models, and mixing plasmids encoding GM-CSF and the circumsporozoite protein of P. yoelii (PyCSP) has enhanced protection against subsequent challenge (whereas plasmid-encoded PyCSP alone did not). It was proposed that GM-CSF may cause dendritic cells to present antigen more efficiently, and enhance IL-2 production and TH cell activation, thus driving the increased immune response. This can be further enhanced by first priming with a pPyCSP and pGM-CSF mixture, and later boosting with a recombinant poxvirus expressing PyCSP. However, co-injection of plasmids encoding GM-CSF (or IFN-γ, or IL-2) and a fusion protein of *P. chabaudi* merozoite surface protein 1 (C-terminus)-hepatitis B virus surface protein (PcMSP1-HBs) actually abolished protection against challenge, compared to protection acquired by delivery of pPcMSP1-HBs alone.

The advantages of using genetic adjuvants are their low cost and simplicity of administration, as well as avoidance of unstable recombinant cytokines and potentially toxic, "conventional" adjuvants (such as alum, calcium phosphate, monophosphoryl lipid A, cholera toxin, cationic and mannan-coated liposomes, QS21, carboxymethylcellulose and ubenimix). However, the potential toxicity of prolonged cytokine expression has not been established, and in many commercially important animal species, cytokine genes still need to be identified and isolated. In addition, various plasmid encoded cytokines modulate the immune system differently according to the time of delivery. For example, some cytokine plasmid DNAs are best delivered after the immunogen pDNA, because pre- or co-delivery can actually decrease specific responses, and increase non-specific responses.

Immunostimulatory CpG Motifs

Plasmid DNA itself appears to have an adjuvant effect on the immune system. Bacterially derived DNA has been found to trigger innate immune defence mechanisms, the activation of dendritic cells, and the production of TH1 cytokines. This is due to recognition of certain CpG dinucleotide sequences which are immunostimulatory. CpG stimulatory (CpG-S) sequences occur twenty times more frequently in bacterially derived DNA than in eukaryotes. This is because eukaryotes exhibit "CpG suppression" – i.e. CpG dinucleotide pairs occur much less frequently than expected. Additionally, CpG-S sequences are hypomethylated. This occurs frequently in bacterial DNA, while CpG motifs occurring in eukaryotes are all methylated at the cytosine nucleotide. In contrast, nucleotide sequences which inhibit the activation of an immune response (termed CpG neutralising, or CpG-N) are over represented in eukaryotic genomes. The optimal immunostimulatory sequence has been found to be an unmethylated CpG dinucleotide flanked by two 5' purines and two 3' pyrimidines. Additionally, flanking regions outside this immunostimulatory hexamer must be guanine-rich to ensure binding and uptake into target cells.

The innate system works synergistically with the adaptive immune system to mount a response against the DNA encoded protein. CpG-S sequences induce polyclonal B-cell activation and the upregulation of cytokine expression and secretion. Stimulated macrophages secrete IL-12, IL-18, TNF-α, IFN-α, IFN-β and IFN-γ, while stimulated B-cells secrete IL-6 and some IL-12.

Manipulation of CpG-S and CpG-N sequences in the plasmid backbone of DNA vaccines can ensure the success of the immune response to the encoded antigen, and drive the immune response toward a TH1 phenotype. This is useful if a pathogen requires a TH response for protection. CpG-S sequences have also been used as external adjuvants for both DNA and recombinant protein vaccination with variable success rates. Other organisms with hypomethylated CpG motifs have also demonstrated the stimulation of polyclonal B-cell expansion. However, the mechanism behind this may be more complicated than simple methylation – hypomethylated murine DNA has not been found to mount an immune response.

Most of the evidence for the existence of immunostimulatory CpG sequences comes from murine studies. Clearly, extrapolation of this data to other species should be done with caution – different species may require different flanking sequences, as binding specificities of scavenger receptors differ between species. Additionally, species such as ruminants may be insensitive to immunostimulatory sequences due to the large gastrointestinal load they exhibit. Further research may be useful in the optimisation of DNA vaccination, especially in the food animal industry.

Alternative Boosts

DNA-primed immune responses can be boosted by the administration of recombinant protein or recombinant poxviruses. "Prime-boost" strategies with recombinant protein have successfully increased both neutralising antibody titre, and antibody avidity and persistence, for weak immunogens, such as HIV-1 envelope protein. Recombinant virus boosts have been shown to be very efficient at boosting DNA-primed CTL responses. Priming with DNA focuses the immune response on the required immunogen, while boosting with the recombinant virus provides a larger amount of expressed antigen, leading to a large increase in specific CTL responses.

Prime-boost strategies have been successful in inducing protection against malarial challenge in a number of studies. Primed mice with plasmid DNA encoding *Plasmodium yoelii* circumsporozoite surface protein (PyCSP), then boosted with a recombinant vaccinia virus expressing the same protein had significantly higher levels of antibody, CTL activity and IFN-γ, and hence higher levels of protection, than mice immunized and boosted with plasmid DNA alone. This can be further enhanced by priming with a mixture of plasmids encoding PyCSP and murine GM-CSF, before boosting with recombinant vaccinia virus. An effective prime-boost strategy for the simian malarial model *P. knowlesi* has also been demonstrated. Rhesus monkeys were primed with a multicomponent, multistage DNA vaccine encoding two liver-stage antigens-the circumsporozoite surface protein (PkCSP) and sporozoite surface protein 2 (PkSSP2)- and two blood stage antigens-the apical merozoite surface protein 1 (PkAMA1) and merozoite surface protein 1 (PkMSP1p42). They were then boosted with a recombinant canarypox virus encoding all four antigens (ALVAC-4). Immunized monkeys developed antibodies against sporozoites and infected erythrocytes, and IFN-γ-secreting T-cell responses against peptides from PkCSP. Partial protection against sporozoite challenge was achieved, and mean parasitemia was significantly reduced, compared to control monkeys. These models, while not ideal for extrapolation to *P. falciparum* in humans, will be important in pre-clinical trials.

Additional Methods of Enhancing DNA-Raised Immune Responses

Formulations of DNA

The efficiency of DNA immunization can be improved by stabilising DNA against degradation, and increasing the efficiency of delivery of DNA into antigen presenting cells. This has been demonstrated by coating biodegradable cationic microparticles (such as poly(lactide-co-glycolide) formulated with cetyltrimethylammonium bromide) with DNA. Such DNA-coated microparticles can be as effective at raising CTL as recombinant vaccinia viruses, especially when mixed with alum. Particles 300 nm in diameter appear to be most efficient for uptake by antigen presenting cells.

Alphavirus Vectors

Recombinant alphavirus-based vectors have also been used to improve DNA vaccination efficiency. The gene encoding the antigen of interest is inserted into the alphavirus replicon, replacing structural genes but leaving non-structural replicase genes intact. The Sindbis virus and Semliki Forest virus have been used to build recombinant alphavirus replicons. Unlike conventional DNA vaccinations, however, alphavirus vectors kill transfected cells, and are only transiently expressed. Also, alphavirus replicase genes are expressed in addition to the vaccine insert. It is not clear how alphavirus replicons raise an immune response, but it is thought that this may be due to the high levels of protein expressed by this vector, replicon-induced cytokine responses, or replicon-induced apoptosis leading to enhanced antigen uptake by dendritic cells.

Vector DNA

A vector is a DNA molecule used as a vehicle to transfer foreign genetic material into another cell. The four major types of vectors are plasmids, bacteriophages and other viruses, cosmids, and artificial chromosomes. Common to all engineered vectors are an origin of replication, a multicloning site, and a selectable marker.

The vector itself is generally a DNA sequence that consists of an insert (transgene) and a larger sequence that serves as the "backbone" of the vector. The purpose of a vector which transfers genetic information to another cell is typically to isolate, multiply, or express the insert in the target cell. Vectors called expression vectors (expression constructs) specifically are for the expression of the transgene in the target cell, and generally have a promoter sequence that drives expression of the transgene. Simpler vectors called transcription vectors are only capable of being transcribed but not translated: they can be replicated in a target cell but not expressed, unlike expression vectors. Transcription vectors are used to amplify their insert.

Insertion of a vector into the target cell is usually called transformation for bacterial cells, transfection for eukaryotic cells, although insertion of a viral vector is often called transduction.

Characteristics

Two common vectors are plasmids and viral vectors.

Plasmids

Plasmids are double-stranded generally circular DNA sequences that are capable of automatically replicating in a host cell. Plasmid vectors minimalistically consist of an origin of replication that allows for semi-independent replication of the plasmid in the host and also the transgene insert. Modern plasmids generally have many more features, notably including a "multiple cloning site" which includes nucleotide overhangs for insertion of an insert, and multiple restriction enzyme consensus sites to either side of the insert. In the case of plasmids utilized as transcription vectors, incubating bacteria with plasmids generates hundreds or thousands of copies of the vector within the bacteria in hours, and the vectors can be extracted from the bacteria, and the multiple cloning site can be cut by restriction enzymes to excise the hundredfold or thousandfold amplified insert. These plasmid transcription vectors characteristically lack crucial sequences that code for polyadenylation sequences and translation termination sequences in translated mRNAs, making protein expression from transcription vectors impossible. plasmids may be-conjugative/transmissible and non conjugative. conjugative-mediate DNA transfer through conjugation and therefore spread rapidly among the bacterial ceels of a population. e.g- F plasmid, many R and some col plasmids. nonconjugative- do not mediate DNA through conjugation. e.g.- many R and col plasmids.

Viral Vectors

Viral vectors are generally genetically-engineered viruses carrying modified viral DNA or RNA that has been rendered noninfectious, but still contain viral promoters and also the transgene, thus allowing for translation of the transgene through a viral promoter. However, because viral vectors frequently are lacking infectious sequences, they require helper viruses or packaging lines for large-scale transfection. Viral vectors are often designed for permanent incorporation of the insert into the host genome, and thus leave distinct genetic markers in the host genome after incorporating the transgene. For example, retroviruses leave a characteristic retroviral integration pattern after insertion that is detectable and indicates that the viral vector has incorporated into the host genome.

Transcription

Transcription is a necessary component in all vectors: the premise of a vector is to multiply the insert (although expression vectors later also drive the translation of the multiplied insert). Thus, even stable expression is determined by stable transcription, which generally depends on promoters in the vector. However, expression vectors have a variety of expression patterns: constitutive (consistent expression) or inducible (expression only under certain conditions or chemicals). This expression is based on different promoter activities, not post-transcriptional activities. Thus, these two different types of expression vectors depend on different types of promoters.

Viral promoters are often used for constitutive expression in plasmids and in viral vectors because they normally reliably force constant transcription in many cell lines and types.

Inducible expression depends on promoters that respond to the induction conditions: for example, the murine mammary tumour virus promoter only initiates transcription after dexamethasone application and the *Drosophilia* heat shock promoter only initiates after high temperatures. transcription is the synthesis of mRNA. Genetic information is copied from DNA to RNA.

Expression

Expression vectors require translation of the vector's insert, thus requiring more components than simpler transcription-only vectors. Expression vectors require sequences that encode for:

- Polyadenylation tail: Creates a polyadenylation tail at the end of the transcribed pre-mRNA that protects the mRNA from exonucleases and ensures transcriptional and translational termination: stabilizes mRNA production.
- Minimal UTR length: UTRs contain specific characteristics that may impede transcription or translation, and thus the shortest UTRs or none at all are encoded for in optimal expression vectors.
- Kozak sequence: Vectors should encode for a Kozak sequence in the mRNA, which assembles the ribosome for translation of the mRNA.

Features

Modern vectors may encompass additional features besides the transgene insert and a backbone:

- Promoter: Necessary component for all vectors: used to drive transcription of the vector's transgene.
- Genetic markers: Genetic markers for viral vectors allow for confirmation that the vector has integrated with the host genomic DNA.
- Antibiotic resistance: Vectors with antibiotic-resistance open reading frames allow for survival of cells that have taken up the vector in growth media containing antibiotics through antibiotic selection.
- Epitope: Vector contains a sequence for a specific epitope that is incorporated into the expressed protein. Allows for antibody identification of cells expressing the target protein.
- β-galactosidase: Some vectors contain a sequence for β-galactosidase, an enzyme that digests galactose, within which a multiple cloning site, the region in which a gene may be inserted, is located. An insert successfully ligated into the vector will disrupt the β-galactosidase gene and thus unable to digest galactose. Cells containing vector with an insert may be identified using blue/white selection by growing cells in media containing an analogue of galactose (X-gal). Cells expressing β-galactosidase (therefore doesn't contain an insert) appear as blue colonies. White colonies would be selected as those that may contain an insert. Other proteins which may function similarly as a reporter include green fluorescent protein and luciferase.
- Targeting sequence: Expression vectors may include encoding for a targeting sequence in the finished protein that directs the expressed protein to a specific

organelle in the cell or specific location such as the periplasmic space of bacteria.

- Protein purification tags: Some expression vectors include proteins or peptide sequences that allows for easier purification of the expressed protein. Examples include polyhistidine-tag, glutathione-S-transferase, and maltose binding protein. Some of these tags may also allow for increased solubility of the target protein. The target protein is fused to the protein tag, but a protease cleavage site positioned in the polypeptide linker region between the protein and the tag allows the tag to be removed later.

HIV Vaccine

An HIV vaccine is the theoretical vaccine which would be given to persons without HIV in order to vaccinate them against getting HIV, the virus which causes AIDS. No effective vaccine against HIV exists. As there is not a known cure for AIDS, the search for a vaccine has become part of medical approaches against the disease.

It has been known for many years that HIV is an extremely difficult virus to render harmless, and no cure presently exists. Research into a vaccine is one of several strategies to reduce the worldwide harm from AIDS, with other approaches based upon antiviral treatments such as highly active antiretroviral therapy (HAART), and social approaches such as safe sex.

There is evidence that a vaccine may be possible. Work with monoclonal antibodies (MAb) has proven that the human body can defend itself against HIV, and certain individuals remain asymptomatic for decades after HIV infection. More recently in 2009, a number of potential candidates for antibodies and early stage results from clinical trials have been announced by various teams. However these are early results, and have either not been developed to the point of human testing, or not fully peer reviewed and replicated by other teams, at this time.

Overview

The urgency of the search for a vaccine against HIV stems from the AIDS-related death toll of over 25 million people since 1981. Indeed, in 2002, AIDS became the primary cause of mortality due to an infectious agent in Africa.

Alternative medical treatments to a vaccine do exist. Highly active antiretroviral therapy (HAART) has been highly beneficial to many HIV-infected individuals since its introduction in 1996 when the protease inhibitor-based HAART initially became available. HAART allows the stabilization of the patient's symptoms and viremia, but they do not cure the patient of HIV, nor of the symptoms of AIDS. And, importantly, HAART does nothing to prevent the spread of HIV through people with undiagnosed HIV infections. Safer sex measures have also proven insufficient to halt the spread of AIDS in the worst affected countries, despite some success in reducing infection rates. Therefore, an HIV vaccine is generally considered as the most likely, and perhaps the only way by which the AIDS pandemic can be halted. However, after over 20 years of research, HIV-1 remains a difficult target for a vaccine.

Difficulties in Developing an HIV Vaccine

In 1984, after the confirmation of the etiological agent of AIDS by scientists at the U.S. National Institutes of Health and the Pasteur Institute, the United States Health and Human Services Secretary Margaret Heckler declared that a vaccine would be available within two years. However, the classical vaccination approaches that have been successful in the control of various viral diseases by priming the adaptive immunity to recognize the viral envelope proteins have failed in the case of HIV-1. Some have stated that an HIV vaccine may not be possible without significant theoretical advances.

There are a number of factors that cause development of an HIV vaccine to differ from the development of other classic vaccines:

- Classic vaccines mimic natural immunity against reinfection generally seen in individuals recovered from infection; there are almost no recovered AIDS patients.
- Most vaccines protect against disease, not against infection; HIV infection may remain latent for long periods before causing AIDS.
- Most effective vaccines are whole-killed or live-attenuated organisms; killed HIV-1 does not retain antigenicity and the use of a live retrovirus vaccine raises safety issues.
- Most vaccines protect against infections that are infrequently encountered; HIV may be encountered daily by individuals at high risk.
- Most vaccines protect against infections through mucosal surfaces of the respiratory or gastrointestinal tract; the great majority of HIV infection is through the genital tract.

HIV Structure

The epitopes of the viral envelope are more variable than those of many other viruses. Furthermore, the functionally important epitopes of the gp120 protein are masked by glycosylation, trimerisation and receptor-induced conformational changes making it difficult to block with neutralising antibodies. The ineffectiveness of previously developed vaccines primarily stems from two related factors.

- First, HIV is highly mutable. Because of the virus' ability to rapidly respond to selective pressures imposed by the immune system, the population of virus in an infected individual typically evolves so that it can evade the two major arms of the adaptive immune system; humoral (antibody-mediated) and cellular (mediated by T cells) immunity.
- Second, HIV isolates are themselves highly variable. HIV can be categorized into multiple clades and subtypes with a high degree of genetic divergence. Therefore, the immune responses raised by any vaccine need to be broad enough to account for this variability. Any vaccine that lacks this breadth is unlikely to be effective.

The difficulties in stimulating a reliable antibody response has led to the attempts to develop a vaccine that stimulates a response by cytotoxic T-lymphocytes.

Another response to the challenge has been to create a single peptide that contains the least variable components of all the known HIV strains.

Animal Model

The typical animal model for vaccine research is the monkey, often the macaque. Monkeys can be infected with SIV or the chimeric SHIV for research purposes. However, the well-proven route of trying to induce neutralizing antibodies by vaccination has stalled because of the great difficulty in stimulating antibodies that neutralise heterologous primary HIV isolates. Some vaccines based on the virus envelope have protected chimpanzees or macaques from homologous virus challenge, but in clinical trials, individuals who were immunised with similar constructs became infected after later exposure to HIV-1.

There are some differences between SIV and HIV that may introduce challenges in the use of an animal model.

As published on 27 November 2009 in Journal of Biology, there is new animal model strongly resembling that of HIV in humans. Generalized immune activation as a direct result of activated CD4+ T cell killing- performed in mice allows new ways of testing HIV behaviour.

Clinical Trials to Date

Several vaccine candidates are in varying phases of clinical trials.

Phase I

Most initial approaches have focused on the HIV envelope protein. At least thirteen different gp120 and gp160 envelope candidates have been evaluated, in the US predominantly through the AIDS Vaccine Evaluation Group. Most research focused on gp120 rather than gp41/gp160, as the latter are generally more difficult to produce and did not initially offer any clear advantage over gp120 forms. Overall, they have been safe and immunogenic in diverse populations, have induced neutralizing antibody in nearly 100% recipients, but rarely induced CD8+ cytotoxic T lymphocytes (CTL). Mammalian derived envelope preparations have been better inducers of neutralizing antibody than candidates produced in yeast and bacteria. Although the vaccination process involved many repeated "booster" injections, it was very difficult to induce and maintain the high anti-gp120 antibody titers necessary to have any hope of neutralizing an HIV exposure.

The availability of several recombinant canarypox vectors has provided interesting results that may prove to be generalizable to other viral vectors. Increasing the complexity of the canarypox vectors by inclusion of more genes/epitopes has increased the percent of volunteers that have detectable CTL to a greater extent than did increasing the dose of the viral vector. Importantly, CTLs from volunteers were able to kill peripheral blood mononuclear cells infected with primary isolates of HIV, suggesting that induced CTLs could have biological significance. In addition, cells from at least some volunteers were able to kill cells infected with HIV from other clades, though the pattern of recognition was not uniform among volunteers. Canarypox is the first candidate HIV vaccine that has induced cross-clade functional CTL responses. The first phase I trial of the candidate vaccine in Africa was launched early in 1999 with Ugandan volunteers. The study determined the extent to which Ugandan volunteers have CTL that are active against the subtypes of HIV prevalent in Uganda, A and D.

Other strategies that have progressed to phase I trials in uninfected persons include peptides, lipopeptides, DNA, an attenuated Salmonella vector, lipopeptides, p24, etc. Specifically, candidate vaccines that induce one or more of the following are being sought:

- neutralizing antibodies active against a broad range of HIV primary isolates;
- cytotoxic T cell responses in a vast majority of recipients;
- strong mucosal immune responses.

Phase II

On December 13, 2004, the HIV Vaccine Trials Network (HVTN) began recruiting for the STEP study, a 3,000-participant phase II clinical trial of a novel HIV vaccine, at sites in North America, South America, the Caribbean and Australia. The trial was co-funded by the National Institute of Allergy and Infectious Diseases (NIAID), which is a division of the National Institutes of Health (NIH), and the pharmaceutical company Merck & Co. Merck developed the experimental vaccine called V520 to stimulate HIV-specific cellular immunity, which prompts the body to produce T cells that kill HIV-infected cells. In previous smaller trials, this vaccine was found to be safe, because of the lack of adverse effects on the patients. The vaccine showed induced cellular immune responses against HIV in more than half of volunteers.

V520 contains a weakened adenovirus that serves as a carrier for three subtype B HIV genes. Subtype B is the most prevalent HIV subtype in the regions of the study sites. Adenoviruses are among the main causes of upper respiratory tract ailments such as the common cold. Because the vaccine contains only three HIV genes housed in a weakened adenovirus, study participants cannot become infected with HIV or get a respiratory infection from the vaccine. It was announced in September 2007 that the trial for V520 would be discontinued after it determined that the vaccination was ineffective. The foremost issue facing the rAd5 adenovirus that was used is the high prevalence of the adenovirus-specific antibodies as a result of prior exposure to the virus. Adenovirus vectors and many other viral vectors currently used in HIV vaccines, will induce a rapid memory immune response against the vector. This results in an impediment to the development of a T cell response against the inserted antigen (HIV antigens) Additionally, it appears that V520 may have made some recipients more receptive to infection by HIV-1.

The HVTN expected to finish the study in 2009, but ceased further treatment administration and declared the vaccine ineffective at preventing HIV-infection in September 2007. The results of the trial have caused some to call for a reexamination of vaccine development strategies.

Phase III

In February 2003, VaxGen announced that their AIDSVAX vaccine was a failure in North America as there was not a statistically significant reduction of HIV infection within the study population. This same vaccine was retested in Thailand within a vaccine regimen called RV 144 beginning in 2003, with positive results. In both cases the vaccines targeted gp120 and were specific for the geographical regions. The Thai trial was the largest AIDS vaccine trial to date when it started.

In October 2009, the results of the RV 144 trial were published. Initial results, released in September 2009 prior to publication of complete results, were encouraging for scientists in search of a vaccine. The study involved 16,395 participants who did not have HIV infection, 8197 of whom were given treatment consisting of two experimental vaccines targeting HIV types B and E that are prevalent in Thailand, while 8198 were given a placebo. The participants were tested for HIV every six months for three years. After three years, the vaccine group saw HIV infection rates reduced by more than 30% compared with those in the placebo group. However, after taking into account the seven people who had HIV infections at the time of their vaccination (two in the placebo group, five in the vaccine group) the percentage dropped to 26%.

Planned Clinical Trials

Novel approaches, including modified vaccinia Ankara (MVA), adeno-associated virus, Venezuelan equine encephalitis (VEE) replicons, and codon-optimized DNA have proven to be strong inducers of CTL in macaque models, and have provided at least partial protection in some models. Most of these approaches are in, or will soon enter, clinical studies.

Economics of Vaccine Development

A June 2005 study estimates that $682 million is spent on AIDS vaccine research annually. Economic issues with developing an AIDS vaccine include the need for advance purchase commitment (or advance market commitments) because after an AIDS vaccine has been developed, governments and NGOs may be able to bid the price down to marginal cost.

Classification of all Theoretically Possible HIV Vaccines

Any theoretically possible HIV vaccines must inhibit or stop the HIV virion replication cycle.So, the targets of the vaccine are the following phases of the HIV virion cycle:

- Phase I. Free state
- Phase II. Attachment

- Phase III. Penetration
- Phase IV. Uncoating
- Phase V. Replication
- Phase VI. Assembling
- Phase VII. Releasing.

So, the possible approaches for the HIV vaccine are the following (in the bracket specified the *Phases* were it is possible to do).

Filtering of the Virions from Blood (Phase I)

- Biological approach for removing the HIV virions from the blood.
- Chemical approach for removing the HIV virions from the blood.
- Physical approach for removing the HIV virions from the blood.

Different Approaches to Catch the Virion (Phase I-III, VI, VII)

- Phagocytosis of the HIV virions.
- Chemical or organic based capture (creation of any skin or additional membrane around the virion) of HIV virions
- Chemical or organic attachments to the virion.

Different Approaches to Destroy or Damage the Virion or its Parts (Phase I-VII)

Here, "damage" means inhibiting or stopping the ability of virion to process any of the *Phase II-VII*. Here are the different classification of methods:

- By nature of method:
  - o Physical methods (*Phase I-VII*)
  - o Chemical and biological methods (*Phase I-VII*)
- By damaging target of the HIV virion structure:
  - o Damaging the Docking Glycoprotein gp120 (*Phase I-III, VI, VII*)
  - o Damaging the Transmembrane Glycoprotein gp41 (*Phase I-III, VI, VII*)
  - o Damaging the virion matrix (*Phase I-III, VI, VII*)
  - o Damaging the virion Capsid (*Phase I-III, VI, VII*)
  - o Damaging the Reverse Transcriptase (*Phase I-VII*)
  - o Damaging the RNA (*Phase I-VII*).

Blocking the Replication (Phase I)

- Insertion into blood chemical or organic compounds which binds to the gp120. Hypothetically, it can be pieces of the CD4 cell membranes with receptors. Any chemical and organic alternative (with ability to bind the gp120) of this receptors also can be used.
- Insertion into blood chemical or organic compounds which binds to the receptors of the CD4 cells.

Inhibiting Process of Phases (Drugs Already used for this Approach)

- Biological, chemical or physical approach to inhibit the *Attachment*
- Biological, chemical or physical approach to inhibit the *Penetration*
- Biological, chemical or physical approach to inhibit the *Uncoating* including introducing the mutation into the HIV
- Biological, chemical or physical approach to inhibit the *Replication* including introducing the mutation into the HIV
- Biological, chemical or physical approach to inhibit the *Assembling* including introducing the mutation into the HIV
- Biological, chemical or physical approach to inhibit (capping) the *Releasing.*

Methods of the Inhibiting of the Functionality of the Infected Cell (Phase VI-VII)

Inhibiting the life functions of the infected cell:

- Inhibiting the metabolism of the infected cell
- Inhibiting the energy exchange of the infected cell.

Future Work

According to Gary J. Nabel of the Vaccine Research Centre in Bethesda, Maryland, several hurdles must be overcome before scientific research will culminate in a definitive AIDS vaccine. First, greater translation between animal models and human trials must be established. Second, new, more effective, and more easily produced vectors must be identified. Finally, and most importantly, there must arise a robust understanding of the immune response to potential vaccine candidates. Emerging technologies that enable the identification of T-cell-receptor specificities and cytokine profiles will prove invaluable in hastening this process.

4

Qualitative Analysis and Antimicrobial Activity

English common names include Spanish cherry,, Medlar, and Bullet wood. Vernacular names include Pikul in Thai, Munamal in Sinhalese, Magizham in Tamil, Elenji in Malayalam, Tanjong in Malay, Vakula in Sanskrit, Bakul in Hindi, Bengali, in Marathi, Kha-Yay in Myanmar.

Tree Description

Bullet wood is an evergreen tree reaching a height of about 16 m. It flowers in April, and fruiting occurs in June. It is a medium-sized large tree growing on tropical forests in South Asia, Southeast Asia & Northern Australia. Leaves are glossy, dark green, oval shaped, 5–14 cm long and 2.5–6 cm wide. Flowers are cream, hairy and scented. Bark is thick and appears dark brownish black or grayish black in colour. It may reach up to a height of 30–60 ft with about 1 m in circumference.

Ayurvedic Uses

The bark, flowers, fruits and seeds are astringent, cooling, anthelmintic, tonic, and febrifuge. It is mainly used in dental ailments like bleeding gums, pyorrhea, dental caries and loose teeth. Extract of flowers used against heart diseases, leucorrhoea, menorrhagia and act as antiduretic in polyuria and antitoxin. The snuff made from the dried and powdered flowers used in a disease called Ahwa in which strong fever, headache and pain in the neck, shoulders and other parts of the body occurs. Ripened fruits facilitates in burning urination. The ripe fruit pounded and mixed with water is given to promote delivery in childbirth. Powder of dried flowers is a brain tonic and useful as a snuff to relieve cephalalgia. Decoration of bark is used to wash the wounds.

Other Uses

- The edible fruit is softly hairy becoming smooth, ovoid, bright red-orange when ripe.

- *Mimusops elengi* wood is luxurious wood. It is extremely hard, strong & tough timber. Heart wood is sharply defined from sapwood. Rich deep red in colour.Works easily & takes beautiful polish. Weight is 63 lbs per cubic foot.

The stem bark is dark gry in colour with striations & few cracks on the surface.

Antifungal Activity of a known Medicinal Plant Mimusops Elengi

L. Against Grain Moulds

Plants have been formed the basis of natural pesticides, that make excellent leads for new pesticide development. The potential of higher plants as a source of new drugs is still largely unexplored.

Hence, last decade witnessed an increase in the investigation on plants as a source of new biomolecules for human disease management. Traditionally plants have been well exploited by man for the treatment of human diseases, Ayurveda is a good example, but not much information is available on the exploitation of plant wealth for the management of plant diseases, especially against phytopathogenic fungi.

Fungi cause severe damage to stored food commodities. Among different species of fungi *Aspergillus* sp., *Fusarium* sp. and *Penicillium* sp. are associated with heavy loss of grains, fruits, vegetables and other plant products during picking, transit and storage rendering them unfit for human consumption by producing mycotoxins and affecting their nutritive value. Many seed borne fungi, which cause severe damage to stored food commodities, were generally managed by synthetic chemicals, which were considered both efficient and effective. The continuous use of these synthetic fungicides started unraveling nonbiodegradability and known to have residual toxicity to cause pollution. Pesticide pollution of soil and water bodies is well documented. Hence in recent time application of plant metabolites for plant disease management has become important viable component of Integrated Pest Management, as plant metabolites are eco-friendly.

Mimusops elengi is a wild plant distributed in tropical and subtropical region belonging to the family sapotaceae. Earlier report reveals that the fruits are used in chronic dysentery, constipations; flowers are used as snuff to relive headache, lotion for wounds and ulcers. Barks are used to increase fertility in women and known to have antiulcer activity. They are rich source of tannin, saponin, alkaloids, glucoside, ursolic acid (Anonymous, 1969).

A pentacylcic triterpene 3β, 6β, 19α, 23-tetrahydroxy-urs-12-ene reported from bark recorded moderate inhibiting activity against β– glucuronidase enzyme associated with gastric ulcers. Seeds of *Mimusops* is known to contain several saponins such as mimusin Misaponin A and 16α-hydroxy Mi-saponin A (Sahu *et al.*, 1997), taxifolin, α- spinasterol glucoside, Mi-glycoside 1, mimusopside A and B. Seed kernel also reported to have saponins (Lavaud *et al.*, 1996). A scientific and systematic phytochemical investigation of leaves with regard to the various biological activities in general and antifungal activity against phytopathogenic fungi in particular of this plant is lacking, hence the present study. Considering these, higher plants are routinely screened in our laboratory; during this routine screening *Mimusops elengi* recorded highly significant antifungal activity.

Thus detailed investigations were conducted to test the efficacy of *M. elengi* against important wide array of seed borne phytopathogenic fungi.

Materials and Methods

Collection of plant materials: Fresh leaves of Mimusops elengi free from diseases were collected from Mysore (India), washed thoroughly 2-3 times with running tap water and once with sterile water, shade-dried, powdered and used for extraction. A voucher specimen of the plant is deposited in the herbarium of Department of Studies in Botany, University of Mysore, Mysore. Karnataka, India.

Preparation of Extracts

Aqueous Extract : Samples (50 g) of shade dried, powder of leaves of *M. elengi* was macerated with 100 ml of sterile distilled water in a Waring blender (Waring International, new Hartford, CT, USA) for 10 min. The macerate was first filtered through double layer muslin cloth and then centrifuged at 4000 *g* for 30 min. The supernatant was filtered through Whatman No. 1 filter paper and heat sterilized at 120 °C for 30 min (Satish *et al.,* 1999). The extract was preserved aseptically in a brown bottle at 5 °C until further use.

Solvent Extracts : Fifty gram of shade dried, powder of *M. elengi* was filled in the thimble and extracted successively with petroleum ether, benzene, chloroform, methanol and ethanol using a Soxhlet extractor for 48 h. All the extracts were concentrated using rotary flash evaporator and preserved at 5 °C in airtight bottles until further use. All the extracts were subjected to antifungal activity assay.

Phytochemical Analysis

Phytochemical analysis of the evaporated methanol extract was conducted following the procedure of Anonymous (1985) and Harborne (1998). Methanol extract was fractioned in to different fractions as acidic, phenolic, alkaloid and neutral following the procedure of Roberts *et al.* (1981). All the fractions were again subjected to antifungal activity.

Isolation of Important Phytopathogenic Fungi

Seed samples of sorghum, paddy rice and maize were collected directly from farmer fields, regulated markets, warehouse and retail shops to isolate the important pathogenic fungi associated with these seeds. The collected seed samples were subjected to standard blotter method (ISTA, 1996) and incubated in alternative cycles of dark and light. On the seventh day of incubation, samples were screened for seed mycoflora with the help of stereobinocular microscope and compound microscope. Associated fungi were identified based on growth characteristic, mycelial morphology, spore morphology and other important characters using standard manuals. The fungi, which were frequently associated in higher percentage in sorghum were selected which served as tested fungi.

Anti-fungal Activity Assay

Different concentrations (10, 20, 30, 40 and 50%) of aqueous extract, all the solvent extracts and isolated constituents (Fraction I to IV) of *Mimusops elengi* were subjected to antifungal activity assay by poisoned food technique. In Czepak Dox Agar (CDA) medium, the extracts were added to the medium to achieve the desired concentrations in the medium, autoclaved and poured into Petridishes (20 ml each) and allowed to cool. After complete solidification of the medium, 5 mm disc of 7-day-old culture of the tested fungi were transferred. Four replicates were maintained for each concentration. The CDA media devoid of the extract served as control. The plates were incubated at 26±1°C for seven days.

The fungitoxicity of the extract in terms of percentage inhibition of mycelial growth was calculated by using the formula:

%inhibition = dc – dt x 100/dc

Where dc=Average increase in mycelial growth in control, dt=Average increase in mycelial growth in treatment (Singh and Tripathi, 1999). Synthetic fungicides, viz., Blitox, Captan, Dithane M-45 and Thiram were also tested at their recommended dosage (2000 ppm) for antifungal activity by poisoned food technique for comparative studies.

Solvent Extracts

Among different solvent extracts tested, methanol and ethanol extracts recorded highly significant antifungal activities. Whereas, activity was not observed in petroleum ether, benzene and chloroform extracts. Even for the methanol extract, all tested fungi showed over 50% of susceptibility among which, *D. halodes, D. tetramera, F. equiseti, F. graminearum, F.* moniliformae, F. proliferatum, F. solani, A. flavus, A. fumigtus, A. versicolor, *P. chrysogenum* and *P. griseofulvum* which recorded over 80% of mycelial growth inhibition at 2 mg/ml concentration. *A. niger* was the least susceptible and *A. flavus* was the most susceptible among the 24 tested fungi. Antifungal activities were varied among different tested pathogenic fungi against ethanol extract, which recorded significant activities next to methanol extract.

Phytochemical Analysis

Phytochemical analysis of methanol extract revealed the presence of carbohydrates and glycosides, proteins and amino acids, alkaloids, phenolic compounds and tannin. Phytosterols, saponins, oils, gum and mucilage were found absent in methanol extract.

Rauvolfia Serpentine

Hindus used this plant for centuries as a febrifuge and as an antidote to the bites of poisonous reptiles like snakes. It was also used to treat dysentery and other painful affections of the intestinal canal. Some believed it caused uterine contraction and promoted the expulsion of the fetus. It was also mentioned as a stomachic that cures fever. In the western parts of India the root was used as a remedy for painful affections of the bowels. With Aristolochia it was given to cure cholera; with *Holorrhena antidysenterica* and *Jatropha curcus* given in colic; with *Andrographis paniculata* it was given for fever. In the last decade its medicinal value has been accepted by the allopathic system.

Habitat It grows in India, Pakistan, Sri Lanka, Burma and Thailand. In India, it is widely distributed in the sub-Himalayan tract from Punjab to Nepal, Sikkim and Bhutan. It is also found in the lower hills of Gangetic plains, eastern and western Ghats and Andamans. It is mostly found in moist deciduous forests at altitudes ranging from sea level to an altitude of 1,200 m high. In the Deccan, it is associated with bamboo forests.

Morphology Description (Habit) It is an evergreen, perennial, glabrous and errect undershrub grows up to a height of 60 cm (rarely more than it). Roots are tuberous with pale brown cork. Leaves are in whorls of three, elliptic to lanceolate or obovate, bright green above, pale green below, tip acute or acuminate, base tapering and slender. Petioles long. Flowers are in many flowered irregular corymbose cymes. Peduncles long but pedicels stout. Flowers white, often has violet coloured tinge. Calyx glabrous, bright red and lanceolate. Corolla is longer than calyx, tube slender, swollen a little above the middle, lobes 3, and elliptic-oblong. Disc is cup shaped. Drupes are slightly connate, obliquely ovoid and purplish black in colour.

Principal Constituents Reserpine is the most important alkaloid present in root, stem and leaves of the plant. It varies from 1.7 to 3.0 %. The root barks has more than 90% of the total alkaloids in roots. The % of the alkaloid depends on the geographical place from where the plant is collected and also the season of collection. Generally samples from Assam have a higher % of alkaloids (2.57 %) than the other parts of India and December is the best month for the collection for getting more % of alkaloids. But, age of the plant has no effect on the % of alkaloid content (up to 4 years). Ajmalicine, ajmaline, isoajmaline, ajmalinine, chandrine, rauwolfinine, renoxidine, rescin-namine, reserpiline, reserpin, reserpinine, sarpagine, serpentine, serpentinine, tetraphyllicine, yohimbine, 3-epi-a-yohimbine are the minor alkaloids identified from samples collected from India.

Pharmacology

Reserpine has a highly complex pattern of activity. Besides the amine concentration in brain, it is also reported to influence the concentration of glycogen, acetyl choline, g-amino butyric acid, nucleic acids and anti-diuretic hormone. The effects of reserpine include respiratory inhibition, stimulation of peristalsis, myosis, relaxation of nictating membranes, and influence on the temperature regulating centre. It increases the volume and free acidity of gastric secretion. Reserpine reduces glycaemia in some cases but the effect is short-lived. In some patients it has a stimulating effect on prothrombin activity. Reserpine also favours permeation of blood into areas rendered ischemic by burns. It produces sedation and a lowering of blood pressure. If administered orally, in hypertension, the effects of reserpine are slow, seldom appearing before 3-6 days of administration and continuing for some time after withdrawal of the drug and have a cumulative effect. It is most valuable in young patients with mild labile hypertension associated with tachycardia. In long established hypertension, it is best used in conjunction with more potent hypertensive drugs such as hexamethonium or hydralazine. Combined with polythiazide, it is a useful hypotensive in mild to moderate thiazide, it is a useful hypotensive in mild to moderate conditions. The response to reserpine varies in patients and the dosage must be adjusted to individual requirements. In severe hypertension, it may be given by intravenous or intramuscular injection when the effect begins within a few hours. Parenteral therapy of reserpine is indicated in the treatment of hypertension only when oral administration is impracticable.

Deserpidine is almost as active as reserpine in its hypotensive and sedative activity. reduces hypotension and inhibits intestinal movements. Given with an equal amount of reserpine it was more hypotensive than either of the drugs in equivalent doses. It possesses anti-fibrillar activity. Serpentina causes marked inhibition of succinate dehydrogenase in brain and liver tissues. It produces a systemic and pulmonary hypotension due to a decrease in cardiac output; there is no change in coronary flow, but coronary vascular resistance is decreased and myocardial oxygen consumption is unaffected. Ajmaline has been reported to stimulate respirartion and intestinal movements. The action of ajmaline on systemic and pulmonary blood pressure is similar to that of serpentine Rauwolfinine has hypertensive properties on the autolysis of rat brain and liver tissue, but to a lesser extent than reserpine. In contrast to reserpine, the total extract of *R. serpentina* inhibited the acetylcholine-induced contraction of the enervated dorsal leech muscle. The whole crude drug is reported to contain some principles which bring about undesirable side effects such as purgation and sexual debility.

Toxicology In patients with cardiac arrhythmia, myocardial infarction or severe cardiac damage, bronchitis, asthma or gastric ulcer, reserpine has a relatively low toxicity, but even the minimum therapeutic doses may give rise to nasal congestion, lethargy, drowsiness, peculiar dreams, vertigo and gastro-intestinal upsets; sometimes dyspnea and urticarial rash may occur. Higher doses may cause flushing, injection of conjunctivae, insomnia, bradycardia, occasionally parkinsonism, and severe mental depression which may lead to suicide. Cases of asthenia and edema have also been reported. Side effects of reserpine are usually transient and quickly disappear on reducing the dosage or discontinuing treatment. Tolerance to reserpine does not develop and it does not appear to be habit-forming. Prolonged previous use of reserpine may cause disturbances in blood pressure during operation under general anesthesia, while some patients may be highly susceptible to a small parenteral dosage. When give to nursing mothers to increase the secretion of milk, it is excreted with milk but the amount is not therapeutically harmful3. Minimum therapeutic doses may give rise to nasal congestion, lethargy drowsiness, peculiar dreams, vertigo and gastro-intestinal upsets; sometimes dyspnea and urticarial rash may occur. Higher doses may cause flushing, injection of conjunctivae, insomnia, bradycardia, occasionally parkinsonism, and severe mental depression which may lead to suicide. Cases of asthenia and edema have also been reported. Side effects of reserpine are usually transient and quickly disappear on reducing the dosage or discontinuing treatment. Serpentine is more toxic than ajmaline or serpentinine.

Indications *Rauwolfia* has been employed for centuries for the relief of various central nervous system disorders, both psychic and motor, including anxiety states, excitement, maniacal behaviour associated with psychosis, schizophrenia, insanity insomnia and epilepsy. Extracts of the roots are valued for the treatment of intestinal disorders, particularly diarrhoea and dysentery and also as anthelmintic. Mixed with other plant extracts, they have been used in the treatment of cholera, colic and fever. The root was believed to stimulate uterine contration and recommended for use in child-birth in difficult cases. The juice of the leaves has been used as a remedy for opacity of the cornea.

Capparis Zeylanica

Capparis is a flowering plant genus in the family Capparaceae which is included in the Brassicaceae in the unrevised APG II system. These plants are shrubs or lianas and are collectively known as caper shrubs or caperbushes. *Capparis* species occur over a wide range of habitat in the subtropical and tropical zones.

Uses and Ecology

Caperbushes are mainly used by humans for their fruit, which are rich in micronutrients. *C. spinosa*, simply known as caper, yields fruit and more importantly flower buds, which are widely used pickled as a vegetable condiment. The fruit of other species, such as *karir* (*C. decidua*), are also used for cooking; *C. mitchellii* and the Wild passionfruit (the local subspecies of *C. spinosa*) are well-known bush tucker in Australia. Mabinlang seeds (*C. masaikai*) are eaten as sweets. Mabinlang is also used in Traditional Chinese Medicine. *Aspalathos*, the root of a shrub contained for example in the sacred Ancient Egyptian incense *kp.t* (kyphi), is sometimes considered to be *C. spinosa*. Other species have also recorded uses in herbalism and folk medicine; dedicated research is largely lacking however. Mabinlins are sweet-tasting proteins found in Mabinlang seed (and possibly in other *Capparis* species); at least one of them is highly resistant to heat. The market for mabinlins is presently not large, but this is mainly due to insufficient supply rather than to lack of demand.

Caperbushes from arid regions- chiefly *C. decidua*- are highly useful in landscape gardening, afforestation and reforestation. They can stop soil erosion and preserve agricultural land. Any large-flowered species can be used to attract butterflies. The Crimson Rose (*Atrophaneura hector*), a spectacular swallowtail butterfly of South Asia, likes to visit flowers of *C. spinosa* in the winter months for example.

The free radicals species contain unpaired electrons. The oxygen radicals, icluding superoxide radical (O2-), hydroxyl radical (OH-) and non-free radical species, such as H2O2 and singlet oxygen (1O2), are various forms of activated oxygen, generated in many redox processes. These radicals are trapped and destroyed by specific enzymes, such as superoxide dismutase, catalase and glutathione peroxidase.

Overproduction of free radicals, together with A, C and E avitaminosis and a reduced level of the above mentioned enzymes, is considered to be the main contributor to oxidative stress.

Besides, excessive generation of ROS, induced by various stimuli and which exceed the antioxidant capacity of the organism, leads to a variety of pathophysiological processes such as inflammation, diabetes, genotoxicity, and cancer.

Pandanus odorus Ridl. (Fragrant screw pine, Thai name: Toei-hom) belongs to the family Pandanaceae is an upright shrub, approximately 0.5–1.0 m high, consisting of stem and nominal support roots. Various parts of Toei-hom are used in food and traditional medicine. Patients and medical practitioners believe that the root and rhizome to be effective against diabetes. The decoction of the *P. odorus* root and rhizome has been traditionally used in treating diabetic patients without much scientific evidence.

It was reported that the root and rhizome of *Pandanus odorus* exert both hypoglycemic and hypolipidemic activity. At the same time isolation of 4-hydroxybenzoic acid from the root of *P. odorus* having hypoglycemic activity has been confirmed by Penchom *et al,* 1998.

As a part of our continuous effort to explore the antioxidant potential of the plants of Bangladesh, the aim of this research is to analyse the antioxidant activity of different parts of the plant *P. odorus.*

Materials and Methods

Plant materials: Different parts of the test plants were collected during the month of January, 2010 from Ramnagar, Comilla, Bangladesh and identified from the Bangladesh National Herbarium, Dhaka where a voucher specimen was deposited having the accession no. 34478.

Preparation of Crude Plant Extract: About 200 g of dried, ground separate parts of the plant were soaked in 1.5 L of 98% methanol for 5-7 days, stirring every 18 h using a sterilized glass rod, separately. The final extracts were passed through No. 1 Whatman filter paper (Whatman Ltd., UK) that is followed by solvent-solvent partitioning with petroleum ether, chloroform and ethyl acetate. The filtrates obtained were concentrated under vacuum in a rotary evaporator at 40 °C and stored at 4°C for further use.

Screening of Antioxidant Activity: Antioxidant activities of different parts of the test plant were determined on the basis of their scavenging potential of the stable DPPH free radical in both qualitative and quantitative assay.

Qualitative assay: A suitably diluted stock solutions were spotted on pre-coated silica gel TLC plates and the plates were developed in solvent systems of different polarities (polar, medium polar and nonpolar) to resolve polar and non-polar components of the extracts. The plates were dried at room temperature and were sprayed with 0.02% DPPH in ethanol. Bleaching of DPPH by the resolved band was observed for 10 minutes and the colour changes (yellow on purple background) were noted (Sadhu *et al.*, 2003).

Quantitative assay: The antioxidant activity of different parts of the plants *Pandanus odorus* were determined using the 1, 1-diphenyl-2-picrylhydrazyl (DPPH) free radical scavenging assay. The free radical scavenging capacity of the petroleum ether, chloroform and ethyl acetate fractionated extracts of different parts of the plant were determined using DPPH. DPPH solution (0.004% w/v) was prepared in 95% methanol. The crude extracts of different parts of the plant *Pandanus odorus* were mixed with 95% methanol to prepare the stock solution (10mg/100mL). The concentration of extract of different parts of *Pandanus odorus* solution was 10 mg/ 100 ml or 100μg/ml. From stock solution 2 ml, 4 ml, 6 ml, 8 ml & 10 ml of this solution were taken in five test tubes & by serial dilution with methanol and was made the final volume of each test tube up to 10 ml whose concentration was then 20μg/ml, 40μg/ml, 60μg/ml, 80μg/ml & 100μg/ml respectively.

Freshly prepared DPPH solution (0.004% w/v) was added in each of these test tubes containing extract of different parts of *Pandanus odorus* (20μg/ml, 40μg/ml, 60μg/ml, 80μg/ml, and 100μg/ml) and after 10 min, the absorbance was taken at 517 nm using a spectrophotometer.

Ascorbic acid was used as a reference standard and dissolved in distilled water to make the stock solution with the same concentration (10mg/100mL or 100μg/ml) of extract of different parts of *P.odorus*. Control sample was prepared containing the same volume without any extract and reference ascorbic acid. 95% methanol was used as blank.

Percent scavenging of the DPPH free radical was measured using the following equation—

% DPPH radical-scavenging = [(Absorbance of Control- Absorbance of test Sample) (Absorbance of Control)] x 100

Then % inhibitions were plotted against respective concentrations used and from the graph IC50 was calculated.

Results

In case of leaves, the chloroform fraction showed highest radical scavenging activity having IC50 value of 1.66 μg/ml followed by petroleum ether and ethyl acetate fraction having IC 50 value of 4.69 μg/ml and 26.21 μg/ml respectively. In case of root, the petroleum ether fraction showed highest radical scavenging activity having IC50 value of 1.42 μg/ml followed by ethyl acetate and chloroform fraction having IC50 value of 3.76 μg/ml and 7.01 μg/ml respectively. On the other hand the antioxidant activity of standard (Ascorbic Acid) was with IC50 value of 43.04μg/ml.

Phytomedicines to Combat Mycotic Infections Due to Dermatophytes

Dermatophytes (name based on the Greek for 'skin plants') are a common label for a group of three types of fungus that commonly causes skin disease in animals and humans. These anamorphic (asexual or imperfect fungi) genera are: *Microsporum*, *Epidermophyton* and *Trichophyton*. There are about 40 species in these three genera.

Species capable of reproducing sexually belong in the teleomorphic genus Arthroderma, of the Ascomycota. Dermatophytes cause infections of the skin, hair and nails due to their ability to obtain nutrients from keratinized material. The organisms colonize the keratin tissues and inflammation is caused by host response to metabolic by-products.

They are usually restricted to the nonliving cornified layer of the epidermis because of their inability to penetrate viable tissue of an immunocompetent host. Invasion does elicit a host response ranging from mild to severe.

Acid proteinases, elastase, keratinases, and other proteinases reportedly act as virulence factors. The development of cell-mediated immunity correlated with delayed hypersensitivity and an inflammatory response is associated with clinical cure, whereas the lack of or a defective cell-mediated immunity predisposes the host to chronic or recurrent dermatophyte infection.

Some of these infections are known as ringworm or tinea. Toenail and fingernail infections are referred to as onychomycosis. Dermatophytes usually do not invade living tissues, but colonize the outer layer of the skin. Occasionally the organisms do invade subcutaneous tissues, resulting in kerion development.

Types of Dermatophyte Infections

Athlete's Foot or Tinea Pedis

Contrary to the name, it does not affect just athletes. Tinea pedis affect men more than women. Frequently affects the webs between the toes first, before spreading to the sole of the foot in a "moccasin" pattern.

Jock Itch or Tinea Cruris

Frequently, the feet are also involved. The theory is that the feet get infected first from contact with the ground. The fungus spores are carried to the groin from scratching, from putting on underclothing or pants. Frequently extend from the groin to the perianal skin and gluteal cleft.

Ringworm of the Body or Tinea Corpora

Facial Ringworm or Tinea Faciei

Can be misdiagnosed for other conditions like psoriasis, discoid lupus, etc. Can be aggravated by treatment with topical steroid or immunosuppressive creams.

Blackdot Ringworm or Tinea Capitis

Infected hair shafts are broken off just at the base, leaving a black dot just under the surface of the skin. Scraping these residual black dot will yield the best diagnostic scrapings for microscopic exam. Numerous green arthrospores will be seen under the microscope inside the stubbles of broken hair shafts at 400x. Tinea capitis can not be treated topically, and must be treated systemically with antifungals.

Scalp Ringworm or Tinea Capitis

Ringworm of the Hands or Tinea Manuum

In most cases of tinea manuum, only one hand is involved. Frequently both feet are involved concurrently, thus the saying "one hand, two feet".

Diagnosis and Identification

Rapid in office testing can be done with scraping of the nail, skin, or scalp. Characteristic hyphae can be seen interspersed among the epithelial cells. Trichophyton tonsurans, the causative agent of tinea capitis (scalp infection) can be seen as solidly packed arthrospores within the broken hairshafts scraped from the plugged black dots of the scalp.

Fungal culture medium is used for positive identification of the species. Usually fungal growth is noted in 5 to 14 days. Microscopic morphology of the micro and macroconidia is the most reliable identification character, but a good slide preparation is needed, and also needed is the stimulation of sporulation in some strains.

Culture characteristics such as surface texture, topography and pigmentation are variable so they are the least reliable criteria for identification. Clinical information such as the appearance of the lesion, site, geographic location, travel history, animal contacts and race is also important, especially in identifying rare non-sporulating species like Trichophyton concentricum, Microsporum audouinii and Trichophyton schoenleinii.

A special media called Dermatophyte Test Medium (DTM) has been formulated to grow and identify dermatophytes. Without having to look at the colony, the hyphae, or macroconidia- one can identify the dermatophyte by a simple colour test.

The specimen (scraping from skin, nail, or hair) is embedded in the DTM culture medium. It is incubated at room temperature for 10 to 14 days. If the fungus is a dermatophyte, the medium will turn bright red. If the fungus is not a dermatophyte, no colour change will be noted. If kept beyond 14 days, false positive can result even with non-dermatophytes. Specimen from the DTM can be sent for species identification if desired.

Transmission

Dermatophytes are transmitted by direct contact with infected host (human or animal) or by direct or indirect contact with infected exfoliated skin or hair in clothing, combs, hair brushes, theatre seats, caps, furniture, bed linens, shoes, socks, towels, hotel rugs, sauna, bathhouse, and locker room floors. Depending on the species the organism may be viable in the environment for up to 15 months.

There is an increased susceptibility to infection when there is a preexisting injury to the skin such as scars, burns, excessive temperature and humidity. Adaptation to growth on humans by most geophilic species resulted in diminished loss of sporulation, sexuality, and other soil-associated characteristics.

Classification

Dermatophytes are classified as anthropophilic (humans), zoophilic (animals) or geophilic (soil) according to their normal habitat.

- Anthropophilic dermatophytes are restricted to human hosts and produce a mild, chronic inflammation.
- Zoophilic organisms are found primarily in animals and cause marked inflammatory reactions in humans who have contact with infected cats, dogs, cattle, horses, birds, or other animals. This is followed by a rapid termination of the infection.
- Geophilic species are usually recovered from the soil but occasionally infect humans and animals. They cause a marked inflammatory reaction, which limits the spread of the infection and may lead to a spontaneous cure but may also leave scars.

Frequency of Species

- About 76% of the dermatophyte species isolated are Trichophyton rubrum
- 27% are Trichophyton mentagrophytes
- 7% are Trichophyton verrucosum
- 3% are Trichophyton tonsurans
- Infrequently isolated (less than 1%) are Epidermophyton floccosum, Microsporum audouinii, Microsporum canis, Microsporum equinum, Microsporum nanum, Microsporum versicolor, Trichophyton equinum, Trichophyton kanei, Trichophyton raubitschekii, and Trichophyton violaceum.

Medications

- Topical medications like clotrimazole, butenafine, miconazole, and terbinafine.
- Systemic medications (oral) like fluconazole, griseofulvin, terbinafine, and itraconizole.
- Tea tree oil.

Treatment

Tinea corpora (body), tinea manus (hands), tinea cruris (groin), tinea pedis (foot) and tinea facie (face) can be treated topically. Tinea unguum (nails) usually will require oral treatment with terbinafine, itraconizole, or griseofulvin. Griseofulvin is usually not as effective as terbinafine or itraconizole. A lacquer (Penlac) can be used daily, but is ineffective unless combined with aggressive debridement of the affected nail. Tinea capitis (scalp) must be treated orally, as the medication must be present deep in the hair follicles to eradicate the fungus. Usually griseofulvin is given orally for 2 to 3 months. Clinically dosage up to twice the recommended dose might be used due to relative resistance of some strains of dermatophytes.

Tinea pedis is usually treated with topical medicines, like ketoconazole or terbinafine, and pills, or with medicines that contains miconazole, clotrimazole, or tolnaftate. Antibiotics may be necessary to treat secondary bacterial infections that occur in addition to the fungus (for example, from scratching).

Mycotic Infections

Organism

- Genus/Species: There are a large number of different genera and species of fungi that cause human diseases. Only a few of these specific agents will be presented and discussed on this page.

General Concepts

- The fungi represent a diverse, heterogeneous group of eukaryotes. Most of these organisms are plant pathogens and relatively few cause disease in humans.
- In nature, fungi generally grow by secreting enzymes that digest tissues but some are actually predacious.
- The growth of the fungi generally involves two phases; vegetative and reproductive.
    - o In the vegetative phase, the cells are haploid and divide mitotically. Most fungi exist as molds with hyphae but some fungi exist as unicellular yeast cells. Some fungi can change their morphology and are termed dimorphic. For example, *Candida* is found in the yeast form at 37°C but changes to the mold form at 25°C.
    - o In the reproductive phase, fungi may undergo either asexual or sexual reproduction. Asexual reproduction involves the generation of spores; sexual reproduction requires specific cellular structures that are used for taxonomic differentiation.
- The fungi are classified based on the characteristics of their sexual phase. For the kingdom Fungi, there are two phyla; Zygomycota and Dikaryomycota. The phylum Dikaryomycota is further divided into two subphyla; Ascomycotina and Basidiomycotina. A third group of the fungi for which a sexual phase has not been observed is termed Deuteromycotina. The table outlines this classification and gives representatives of each.

Group	*Representative Genera*
Phylum Zygomycota	*Rhizopus, Absidia, Mucor*
Phylum Dikaryomycota	
·Subphylum Ascomycotina	*Trichophyton, Histoplasma, Blastomyces*
·Subphylum Basidiomycotina	*Cryptococcus*
Form-class Deuteromycotina	*Candida, Epidermophyton, Coccidioides*

- Fungal membranes contain ergosterol rather than cholesterol and this provides a target for chemotherapy (azole derivatives interfere with ergosterol synthesis). Nonetheless, most anti-fungal antibiotics are still relatively toxic to the human host.

Distinctive Properties

- Mycotic infections are classified by the tissue levels that are colonized.
  - Superficial infections are generally limited to the outer layers of the skin and hair.
  - Cutaneous infections are located deeper in the epidermis, hair and nails.
  - Subcutaneous infections involve the dermis, subcutaneous tissues and muscle.
- In addition, mycotic infections may be systemic, generally originating in the lungs.
- Finally, some mycoses are termed opportunistic, and these may involve a variety of body sites.

Pathogenesis

- Mycotic disease is often a consequence of predisposing factors including age, stress or other pathologic conditions (e.g. cancer, diabetes, AIDS).
- Only the dermatophytes (*Trichophyton*, *Microsporum*) and *Candida* are communicable from human to human. The other agents are acquired from the environment (plants, soil, etc.).
- Fungi generally cause one of three distinct tissue responses;
  1. chronic inflammation (scarring, accumulation of lymphocytes)
  2. granulomatous inflammation (collections of modified epithelial cells, lymphocytes)
  3. acute suppurative inflammation (vascular congestion, exudation of plasma, accumulation of PMNs).
- Some of the tissue responses may be due to mycotoxins, which are fungal metabolites that are toxic to the host. Some agents produce LPS-like endotoxins or hemolysins or steroid-like toxins that affect the nervous system
- *Aspergillus* produces a toxin called aflatoxin that has a strong association with liver cancer. For example, in Thailand, where people generally consume about 25-times more aflatoxin in their diets, the incidence of cancer is about 10-fold greater.
- Systemic mycoses are generally asymptomatic but may have generalized symptoms including low grade fever, shaking chills, night sweats, malaise or appetite loss.

Host Defences

- Host defences against the fungi include nonspecific and specific factors:

- Nonspecific defences include the skin (lipids, fatty acids, normal flora), internal factors (mucous membranes, ciliated cells, macrophages), blood components, temperature, genetic and hormonal factors. In other words, both physical and chemical factors and phagocytic defences play major roles in prevention and control of mycotic disease.
- Specific defences include both humoral and cell-mediated. The role of humoral defences is somewhat controversial, since certain antibodies are not protective. It is possible that high titers of certain antibodies actually suppress the cell mediated defences. Nevertheless, some antibodies may be protective (e.g. antitoxins or opsonins). Generally, however, the cell-mediated defences are probably more important. Acquired resistance is usually T-cell mediated and persons with compromised cell-mediated defences generally show more disseminated disease.

Epidemiology

- Dermatophytes may be communicated from person to person by combs, towels, etc. These infections (termed "tineas" when affecting the skin) include ringworm, athlete's foot, jock itch, etc. *Candida* is a member of the normal vaginal flora; candidiasis is often associated with diabetes.
- In some cases of mycosis, occupation seems an important contributor. For example, *Sporothrix* is normally found in woody plants; hence, agricultural workers acquire disease more often. Similarly, *Histoplasma* is often found in bird or bat excreta; hence spelunkers (caves) or persons involved in community clean up may acquire more often.
- Most of the systemic diseases are geographically distributed; histoplasmosis and blastomycosis predominate in the Ohio and Mississippi Rivers valleys, paracoccidioidomycosis is found primarily in Brazil, coccidioidomycosis predominates in the Southwestern United States while cryptococcosis has a worldwide distribution.

Diagnosis

- Clinical: For the dermatophytes, appearance of the lesions is usually diagnostic. For systemic mycoses, the epidemiology and symptomology are useful clues.
- Laboratory: Treatment of skin scrapings with 10% potassium hydroxide can reveal hyphae or spores. Most fungi can be grown on Sabouraud's dextrose agar but they are often very difficult to speciate. Some fungi show a yellow fluorescence under 365 nm ultraviolet light. Skin testing for a delayed hypersensitivity response is useful for epidemiologic purposes but often not for diagnosis.

Control

- Sanitary: Control by sanitary means is difficult, but the incidence of communicable disease can be reduced by good hygiene.

- Immunological: No vaccines are currently available.
- Chemotherapeutic: Many antifungals are available but some are very toxic to the host and must be used with caution. Topical powders and creams often contain tolnaftate or azole derivatives (miconazole, clotrimazole, econazole) and are useful against superficial dermatophytes. Hair or nail disease may be treated with oral griseofulvin, but it is rather toxic. Sporotrichosis may be treated using potassium iodide or amphotericin B (oral). Systemic infections are generally treated by amphotericin B, 5-flourocytosine, miconazole or ketoconazole.

Phytomedicines as a New Crop Opportunity

Phytomedicines, simply defined, are a special category of plant drugs. They are standardized, which means that certain compounds in the plant material are quantified and elucidated so as to have a replicable final product, batch after batch. One of the criticisms traditionally leveled against natural medicines is the lack of standard levels of biological materials from the natural plants.

In many parts of the world, the United States is considered a third world country when it comes to phytomedicines. Unfortunately, our European colleagues are far ahead of us; they have a multi-billion dollar industry in phytomedicines, whereas ours is only a two hundred million dollar industry. Even so, there are a lot of exciting developments in the United States. Over the last five years, I estimate that the American medicinal herb industry has grown at a rate exceeding 20% per year and this trend continues today.

Phytomedicines represent new crop opportunities for several reasons. First, the companies who use these products are very keen to get high quality, sanitary material. This is becoming quite a serious problem. One example we have seen quite recently relates to problems in Eastern Europe, the source of many of these plants. Would you buy plant material which has been growing next to a steel plant in Eastern Europe? Most people would not. The United States is a good growers market because it has a stable political environment, has many different growing climates and conditions, excellent choice of growers, and the ability to grow according to industrial specifications. The Appalachian region, in particular, is very rich in medicinal flora.

Of growing concern is the extinction of medicinal plant species in this country. Given the growing demand for these botanicals and current problems in over-collecting, it is quite likely that some of the last populations of these plants will be collected over the next five to ten years if we don't begin to cultivate them commercially. Finally, these specialty crops could provide profitable growing opportunities for small farmers. Some of these crops appear to be ideally suited to the small farmer.

Important Medicinal Plants

Ginkgo biloba : The nuts of this tree are quite tasty when properly roasted. An extract (Funfgeld 1988) derived from the leaf of this tree is the single largest selling drug in Germany and France today (Foster 1991). The extract is used for peripheral circulation and has the unique property of making red blood cells more elastic and selectively dilating capillaries. Consequently, this extract is used in Europe for treating conditions of tinnitus, vertigo, cold hands, cold feet, macular degeneration, and dementia. In order to meet the demand for the leaf, there is a 400 hectare farm in South Carolina which produces over a million kilograms of dried leaves a year. There are other plantations now being developed to meet the international demand for this product.

American ginseng (Panax quinquifolium) : In Marathon County, Wisconsin, alone, this crop represents a one hundred million dollar a year crop (raw and finished product) which is primarily exported to China where it is widely used and appreciated. Oddly enough, we buy *Panax ginseng* from China because Americans prefer their material and they ours.

Saw palmetto (Serenoa repens) : The berries of the Saw Palmetto are quite useful for benign prostatic hypertrophy. This is a very big pharmaceutical in Germany and France. There is a well developed industry in Florida which produces berries for export to Europe.

Goldenseal (Hydrastis canadensis) : This is one of the fastest growing products in the United States in the natural products industry and used to be in the United States Pharmacopoeia until about 40 years ago. It is native to Appalachia and is one of the crops that is most threatened by extirpation because of growing demand. Ginseng growers in Wisconsin have been contacted to see if they could grow Goldenseal. To date, the project is showing signs of success, and it seems likely that a large percentage of this product will come from cultivated plots rather than naturally occurring populations.

Bloodroot (Sanguanaria canadensis) : This is very popular and is the active ingredient in a toothpaste product called Viadent. It is used to control plaque and gingivitis.

Siberian ginseng (Eleutherococcus senticosus) : This is a very popular and interesting product. It is called an adaptogen. It was developed in Russia by Dr. Breckman. Adaptogens are body regulators which promote proper balance. Many medical doctors question this hypothesis, but there is a significant amount of empirical and clinical data from Russia to support this.

Echinacca (Echinacea purpurea) : This is a popular nonspecific immunostimulant in Germany and much of Europe. There are hundreds of hectares of cultivated Echinacea in Europe. There is also a large organic farm, Trout Lake Farm in Washington State, which grows significant amounts of this product. Echinacca is used as a preventative for colds and flues and is quite effective when used in this way. It is one of Europe's most popular natural products.

Milk thistle (Silybum marianum) : This is used in Europe for liver conditions, treatment of acute mushroom poisoning, and other hepatoxic compounds.

Black cohosh (Cimicifuga racemosa) : This is an interesting native American plant. It has a rich tradition among Native Americans and by people who live in the Appalachian region. It is used by women for regulation of hormonal cycles.

Valerian (Valeriana officinalus) : This is one of the more popular sleep aids in Europe and has been widely used for hundreds of years. There is no reason why this cannot be grown in the United States.

Feverfew (Tanacetum parthenium) : Research in England has shown promising results in the treatment of migraine. A small amount of feverfew when taken orally can reduce the frequency and severity of migraine. It has great medical potential and could most definitely be grown in this country.

St. John's wort (Hypericum perforatum) : This is a popular European product. It has antiviral properties and also antidepressant properties.

Catnip (Nepeta cataria) : Not only is it my cat's favorite play thing, but it is also used as a mild sedative. This is currently being grown in the United States as a sedative, although not commercially.

Pacific yew (Taxus brevifolia) : At the moment, this is a most controversial and interesting phytomedicine. The pacific yew tree is the source of taxol which shows great promise for the treatment of ovarian and other cancers. However, there are not enough trees. These trees grow very slowly, and if harvested for taxol, the natural stands of the yew tree would be exhausted. The ability to obtain taxol from ornamental sources of *Taxus* as well as planting nurseries of the western yew for taxol extraction are underway.

Ginger (Zingiber officinale) : It is not only a well known spice, but it has been proven to be as effective as dramamine in reducing nausea and motion sickness. It has also been quite useful for pregnant women who are suffering nausea in early pregnancy and is quite safe.

Conclusions

In the United States, the policies of the Food and Drug Administration (FDA), historically, have been unhelpful to natural products. These plant products are polypharmaceutics, meaning they have multiple compounds, and FDA is not presently prepared to review products containing more than one compound. Demanding evidence that each individual component in an extract is safe and effective is a matter of scientific curiosity, but has little to do with the inherent questions of safety and effectiveness of the extract. Unfortunately, FDA still struggles with this concept.

The natural healthcare market is a very fast growing one. It is projected that at current inflation rates for health care, by the year 2030, health care costs could consume 100% of the gross national product. Something must be done. People are turning more and more to prevention and wellness programs, including natural medicines. As this trend develops, phytomedicines could become an important new alternative crop in the United States.

5

Biological Warfare : A Serious Threat in the 21st Century

Biological warfare (BW), also known as germ warfare, is the deliberate use of disease causing biological agents such as protozoa, fungi, bacteria, protists, or viruses, to kill or incapacitate humans, other animals or plants. Biological weapons (often referred to as *bioweapons*) are living organisms or replicating entities (virus) that reproduce or replicate within their host victims.

Biological weapons may be employed in various ways to gain a strategic or tactical advantage over an adversary, either by threat or by actual deployment. Like some of the chemical weapons, biological weapons may also be useful as area denial weapons. These agents may be lethal or non-lethal, and may be targeted against a single individual, a group of people, or even an entire population. They may be developed, acquired, stockpiled or deployed by nation states or by non-national groups. In the latter case, or if a nation-state uses it clandestinely, it may also be considered bioterrorism.

There is an overlap between biological warfare and chemical warfare, as the use of toxins produced by living organisms is considered under the provisions of both the Biological Weapons Convention and the Chemical Weapons Convention. Toxins and Psychochemical weapons are often referred to as *midspectrum agents*. Unlike bioweapons, these midspectrum agents do not reproduce in their host and are typically characterized by shorter incubation periods.

Overview

Offensive biological warfare, including mass production, stockpiling and use of biological weapons, was outlawed by the 1972 Biological Weapons Convention (BWC). The rationale behind this treaty, which has been ratified or acceded to by 163 countries as of 2009, is to prevent a biological attack which could conceivably result in large numbers of civilian fatalities and cause severe disruption to economic and societal infrastructure. Many countries, including signatories of the BWC, currently pursue research into the defence or protection against BW, which is not prohibited by the BWC.

A nation or group that can pose a credible threat of mass casualty has the ability to alter the terms on which other nations or groups interact with it. Biological weapons allow for the potential to create a level of destruction and loss of life far in excess of nuclear, chemical or conventional weapons, relative to their mass and cost of development and storage. Therefore, biological agents may be useful as strategic deterrents in addition to their utility as offensive weapons on the battlefield.

As a tactical weapon for military use, a significant problem with a BW attack is that it would take days to be effective, and therefore might not immediately stop an opposing force. Some biological agents (especially smallpox, plague, and tularemia) have the capability of person-to-person transmission via aerosolized respiratory droplets. This feature can be undesirable, as the agent(s) may be transmitted by this mechanism to unintended populations, including neutral or even friendly forces. While containment of BW transmission is less of a concern for certain criminal or terrorist organizations, it remains a significant concern for the military and civilian populations of virtually all nations.

Characteristics of Biological Weapons

Anti-personnel

Ideal characteristics of a biological agent to be used as a weapon against humans are high infectivity, high virulence, non-availability of vaccines, and availability of an effective and efficient delivery system. Stability of the weaponized agent (ability of the agent to retain its infectivity and virulence after a prolonged period of storage) may also be desirable, particularly for military applications.

The primary difficulty is not the production of the biological agent, as many biological agents used in weapons can often be manufactured relatively quickly, cheaply and easily. Rather, it is the weaponization, storage and delivery in an effective vehicle to a vulnerable target that pose significant problems.

For example, *Bacillus anthracis* is considered an effective agent for several reasons. First, it forms hardy spores, perfect for dispersal aerosols. Second, this organism is not considered transmissible from person to person, and thus rarely if ever causes secondary infections. A pulmonary anthrax infection starts with ordinary influenza-like symptoms and progresses to a lethal hemorrhagic mediastinitis within 3–7 days, with a fatality rate that is 90% or higher in untreated patients. Finally, friendly personnel can be protected with suitable antibiotics.

A large-scale attack using anthrax would require the creation of aerosol particles of 1.5 to 5 microns. Too large and the particles would not reach the lower respiratory tract. Too small and the particles would be exhaled back out into the atmosphere. At this size, conductive powders tend to aggregate because of electrostatic charges, hindering dispersion. So the material must be treated to insulate and neutralize the charges. The weaponized agent must be resistant to degradation by rain and ultraviolet radiation from sunlight, while retaining the ability to efficiently infect the human lung. There are other technological difficulties as well, chiefly relating to storage of the weaponized agent.

Agents considered for weaponization, or known to be weaponized, include bacteria such as *Bacillus anthracis, Brucella spp., Burkholderia mallei, Burkholderia pseudomallei, Chlamydophila psittaci, Coxiella burnetii, Francisella tularensis,* some of the Rickettsiaceae (especially *Rickettsia prowazekii* and *Rickettsia rickettsii*), *Shigella spp., Vibrio cholerae,* and *Yersinia pestis.* Many viral agents have been studied and/or weaponized, including some of the Bunyaviridae (especially Rift Valley fever virus), Ebolavirus, many of the Flaviviridae (especially Japanese encephalitis virus), Machupo virus, Marburg virus, Variola virus, and Yellow fever virus. Fungal agents that have been studied include *Coccidioides spp.*.

Toxins that can be used as weapons include ricin, staphylococcal enterotoxin B, botulinum toxin, saxitoxin, and many mycotoxins. These toxins and the organisms that produce them are sometimes referred to as select agents. In the United States, their possession, use, and transfer are regulated by the Centres for Disease Control and Prevention's Select Agent Program.

Anti-agriculture

Biological warfare can also specifically target plants to destroy crops or defoliate vegetation. The United States and Britain discovered plant growth regulators (i.e., herbicides) during the Second World War, and initiated an herbicidal warfare program that was eventually used in Malaya and Vietnam in counter insurgency. Though herbicides are chemicals, they are often grouped with biological warfare as bioregulators in a similar manner as biotoxins. Scorched earth tactics or destroying livestock and farmland were carried out in the Vietnam war (cf. Agent Orange) and Eelam War in Sri Lanka.

The United States developed an anti-crop capability during the Cold War that used plant diseases (bioherbicides, or mycoherbicides) for destroying enemy agriculture. It was believed that destruction of enemy agriculture on a strategic scale could thwart Sino-Soviet aggression in a general war. Diseases such as wheat blast and rice blast were weaponized in aerial spray tanks and cluster bombs for delivery to enemy watersheds in agricultural regions to initiate epiphytotics (epidemics among plants). When the United States renounced its offensive biological warfare program in 1969 and 1970, the vast majority of its biological arsenal was composed of these plant diseases.

In 1980s Soviet Ministry of Agriculture had successfully developed variants of foot-and-mouth disease, and rinderpest against cows, African swine fever for pigs, and psittacosis to kill chicken. These agents were prepared to spray them down from tanks attached to airplanes over hundreds of miles. The secret program was code-named "Ecology".

Attacking animals is another area of biological warfare intended to eliminate animal resources for transportation and food. In the First World War, German agents were arrested attempting to inoculate draft animals with anthrax, and they were believed to be responsible for outbreaks of glanders in horses and mules. The British tainted small feed cakes with anthrax in the Second World War as a potential means of attacking German cattle for food denial, but never employed the weapon. In the 1950s, the United States had a field trial with hog cholera. During the Mau Mau Uprising in 1952, the poisonous latex of the African milk bush was used to kill cattle.

Unconnected with inter-human wars, humans have deliberately introduced the rabbit disease Myxomatosis, originating in South America, to Australia and Europe, with the intention of reducing the rabbit population- which had devastating but temporary results, with wild rabbit populations reduced to a fraction of their former size but survivors developing immunity and increasing again.

Biodefence

Biodefence refers to short term, local, usually military measures to restore biosecurity to a given group of persons in a given area — in the civilian terminology, it is a very robust biohazard response. It is technically possible to apply biodefence measures to protect animals or plants, but this is generally uneconomic. However, protection of water supplies and food supplies are often a critical part of biodefence. Various definitions of biosafety emerged in different professions to guarantee non-human health. Biodefence is most often discussed in the context of biowar or bioterrorism, and is generally considered a military or emergency response term.

Biodefence applies to two distinct target populations: civilian non-combatant and military combatant (troops in the field).

Biodefence of Troops in the Field

The Department of Defence (or "DoD") has focused since at least 1998 on the development and application of vaccine-based biodefences. In a July 2001 report commissioned by the DoD, the "DoD-critical products" were stated as vaccines against anthrax (AVA and Next Generation), smallpox, plague, tularemia, botulinum, ricin, and equine encephalitis. Note that two of these targets are toxins (botulinum and ricin) while the remainder are infectious agents.

Role of Public Health Departments and Disease Surveillance

It is important to note that all of the classical and modern biological weapons organisms are animal diseases, the only exception being smallpox. Thus, in any use of biological weapons, it is highly likely that animals will become ill either simultaneously with, or perhaps earlier than humans. Indeed, in the largest biological weapons accident known– the anthrax outbreak in Sverdlovsk (now Yekaterinburg) in the Soviet Union in 1979, sheep became ill with anthrax as far as 200 kilometres from the release point of the organism from a military facility in the southeastern portion of the city. Thus, a robust surveillance system involving human clinicians and veterinarians may identify a bioweapons attack early in the course of an epidemic, permitting the prophylaxis of disease in the vast majority of people (and/or animals) exposed but not yet ill.

For example in the case of anthrax, it is likely that by 24 – 36 hours after an attack, some small percentage of individuals (those with compromised immune system or who had received a large dose of the organism due to proximity to the release point) will become ill with classical symptoms and signs (including a virtually unique chest X-ray finding, often recognized by public health officials if they receive timely reports). By making these data available to local public health officials in real time, most models of anthrax epidemics indicate that more than 80% of an exposed population can receive antibiotic treatment before becoming symptomatic, and thus avoid the moderately high mortality of the disease.

Identification of Bioweapons

The goal of biodefence is to integrate the sustained efforts of the national and homeland security, medical, public health, intelligence, diplomatic, and law enforcement communities. Health care providers and public health officers are among the first lines of defence. In some countries private, local, and provincial (state) capabilities are being augmented by and coordinated with federal assets, to provide layered defences against biological weapons attacks. During the first Gulf War the United Nations activated a biological and chemical response team, Task Force Scorpio, to respond to any potential use of weapons of mass destruction on civilians.

The traditional approach toward protecting agriculture, food, and water: focusing on the natural or unintentional introduction of a disease is being strengthened by focused efforts to address current and anticipated future biological weapons threats that may be deliberate, multiple, and repetitive. The growing threat of biowarfare agents and bioterrorism has led to the development of specific field tools that perform on-the-spot analysis and identification of encountered suspect materials. One such technology, being developed by researchers from the Lawrence Livermore National Laboratory (LLNL), employs a "sandwich immunoassay", in which fluorescent dye-labelled antibodies aimed at specific pathogens are attached to silver and gold nanowires.

In the Netherlands, the company TNO has designed Bioaerosol Single Particle Recognition eQuipment (BiosparQ). This system would be implemented into the national response plan for bioweapons attacks in the Netherlands.

Researchers at Ben Gurion University in Israel are developing a different device called the BioPen, essentially a "Lab-in-a-Pen", which can detect known biological agents in under 20 minutes using an adaptation of the ELISA, a similar widely employed immunological technique, that in this case incorporates fibre optics.

History

Biological warfare has been practiced repeatedly throughout history. Before the 20th century, the use of biological agents took three major forms:

- Deliberate poisoning of food and water with infectious material
- Use of microorganisms, toxins or animals, living or dead, in a weapon system
- Use of biologically inoculated fabrics.

Antiquity

The earliest documented incident of the intention to use biological weapons is recorded in Hittite texts of 1500–1200 B.C, in which victims of plague were driven into enemy lands. Although the Assyrians knew of ergot, a parasitic fungus of rye which produces ergotism when ingested, there is no evidence that they poisoned enemy wells with the fungus, as has been claimed.

According to Homer's epic poems about the legendary Trojan War, the *Iliad* and the *Odyssey,* spears and arrows were tipped with poison. During the First Sacred War in Greece, in about 590 BC, Athens and the Amphictionic League poisoned the water supply of the besieged town of Kirrha (near Delphi) with the toxic plant hellebore. The Roman commander Manius Aquillus poisoned the wells of besieged enemy cities in about 130 BC.

During the 4th century BC Scythian archers tipped their arrow tips with snake venom, human blood, and animal feces to cause wounds to become infected. There are numerous other instances of the use of plant toxins, venoms, and other poisonous substances to create biological weapons in antiquity.

In 184 B.C, Hannibal of Carthage had clay pots filled with venomous snakes and instructed his soldiers to throw the pots onto the decks of Pergamene ships. In about AD 198, the Parthian city of Hatra (near Mosul, Iraq) repulsed the Roman army led by Septimius Severus by hurling clay pots filled with live scorpions at them.

Middle Ages

The Mongol Empire established commercial and political connections between the Eastern and Western areas of the world, but it did so through the most mobile army ever seen. The armies, being the most rapidly moving travelers who had ever moved between the steppes of East Asia (where bubonic plague was and remains endemic among small rodents) and managed to keep the chain of infection without a break until they reached, and infected, peoples and rodents who had never encountered it. The ensuing Black Death may have killed almost half of the population of Europe in the next decades, changing the course of Asian and European history.

During the Middle Ages, victims of the bubonic plague were used for biological attacks, often by flinging the infected corpses (formites)and excrement over castle walls using catapults. In 1346, during the siege of Kafa (now Feodossia Ukraine) the attacking Tartar Forces which were subjugated by the Mongol empire under Genghis Khan, used the bodies of Mongol warriors of the Golden Horde who had died of plague, as weapons. The dead warriors were thrown over the walls of the besieged city of Crimean. An outbreak of plague followed and the defending forces retreated, followed by the conquest of the city by the Mongol army. It has been speculated that this operation may have been responsible for the advent of the Black Death in Europe.

At the siege of Thun-l'Eveque in 1340, during the Hundred Years' War, the attackers catapulted decomposing animals into the besieged area.

In 1422, during the siege of Karlstein Castle in Bohemia, Hussite attackers used catapults to throw dead (but not plague-infected) bodies and 2000 carriage-loads of dung over the walls.

The last known incident of using plague corpses for biological warfare occurred in 1710, when Russian forces attacked the Swedes by flinging plague-infected corpses over the city walls of Reval (Tallinn). However, during the 1785 siege of La Calle, Tunisian forces flung diseased clothing into the city.

18th Century

The Native American population was devastated after contact with the Old World due to the introduction of many different fatal diseases. There are two documented cases of alleged and attempted germ warfare. The first, during a parley at Fort Pitt on June 24, 1763, Ecuyer gave representatives of the besieging Delawares two blankets and a handkerchief that had been exposed to smallpox, hoping to spread the disease to the Natives in order to end the siege. William Trent, the militia commander, left records that clearly indicated that the purpose of giving the blankets was "to Convey the Smallpox to the Indians."

British commander Lord Jeffrey Amherst and Swiss-British officer Colonel Henry Bouquet certainly discussed this, in the course of Pontiac's Rebellion; there still exists correspondence referencing the idea of giving smallpox-infected blankets to enemy Indians. Historian Francis Parkman verifies four letters from June 29, July 13, 16 and 26th, 1763. Excerpts: Commander Lord Jeffrey Amherst writes July 16, 1763, "P.S. You will Do well to try to Inocculate the Indians by means of Blankets, as well as to try Every other method that can serve to Extirpate this Execrable Race. I should be very glad your Scheme for Hunting them Down by Dogs could take Effect,..." Colonel Henry Bouquet replies July 26, 1763, "I received yesterday your Excellency's letters of 16th with their Inclosures. The signal for Indian Messengers, and all your directions will be observed."

While the intent of carrying out biological warfare is clear, there is debate among historians as to whether this actually took place despite Bouquet's affirmative reply to Amherst, and the continuing correspondence on the point. Smallpox is highly infectious and does not require contaminated blankets to spread uncontrollably, and together with measles, influenza, chicken pox, etc. had been doing so since the arrival of Europeans and their animals. Historians have been unable to establish whether or not the Amherst plan was implemented, particularly in light of the fact that smallpox was already present in the region, and that scientific knowledge of disease at that time had yet to develop an understanding of infection vectors, nor in the case of smallpox a full acknowledgment of the protective effect of a cowpox infection.

Regardless of whether the plan was carried out, trade and combat provided ample opportunity for transmission of the disease.

The diseases which depopulated the New World can be traced to Eurasia where people had long lived with them and developed some immunological ability to survive their presence. Without similarly long ancestral exposure, indigenous Americans were immunologically naive and extremely vulnerable.

19th Century

In 1834, Massachusetts diarist Richard Henry Dana visited San Francisco on a merchant ship. His ship traded many items including blankets with Mexicans and Russians who had established outposts on the northern side of the San Francisco Bay. Local histories document that the California plague epidemic began at the Russian fort soon after they left. It is possible that the blankets were the source of the contamination (hidden fleas, or rats, perhaps), but another possible source was a Chinese ship making port in San Francisco at the same time. Plague became established in California and has since become endemic throughout much of the North American West. Native rodents have suffered a severe population decline, only partly due to human eradication action. During the American Civil War, General Sherman reported that Confederate forces shot farm animals in ponds upon which the Union troops depended for drinking water. This would have made the water unpleasant to drink, though perhaps the death caused might not have been that desired. A Confederate doctor planned and may have carried out a bacteriological attack on Northern populations across the Canadian border.

Jack London, in his story "Yah! Yah! Yah!"", described a punitive European expedition to a South Pacific island deliberately exposing the Polynesian population to measles, of which many of them died. While much of the material for London's *South Sea Tales* is derived from his personal experience in the region, it is not known whether this particular incident is historical.

20th Century

During the First World War, the Empire of Germany pursued an ambitious biological warfare program. Using diplomatic pouches and couriers, the German General Staff supplied small teams of saboteurs in the Russian Duchy of Finland, and in the then-neutral countries of Romania, the United States, and Argentina.

In Finland, saboteurs mounted on reindeer placed ampoules of anthrax in stables of Russian horses in 1916. Anthrax was also supplied to the German military attache in Bucharest, as was glanders, which was employed against livestock destined for Allied service.

German intelligence officer and US citizen Dr. Anton Casimir Dilger established a secret lab in the basement of his sister's home in Chevy Chase, Maryland, that produced glanders which was used to infect livestock in ports and inland collection points including, at least, Newport News, Norfolk, Baltimore, and New York, and probably St. Louis and Covington, Kentucky. In Argentina, German agents also employed glanders in the port of Buenos Aires and also tried to ruin wheat harvests with a destructive fungus.

The Geneva Protocol of 1925 prohibited the use of chemical weapons and biological weapons, but said nothing about experimentation, production, storage, or transfer; later treaties did cover these aspects. Twentieth-century advances in microbiology enabled the first pure-culture biological agents to be developed by World War II.

The interwar period was a period of development by many nations, most notably the Empire of Japan. Secret Imperial Japanese Army Unit 731, based primarily at Pingfan in Manchuria commanded by Lieutenant General Shirô Ishii, did research on BW, conducted often fatal human experiments on prisoners, and produced biological weapons for combat use during the Second Sino-Japanese War.

Biological experiments, often using twins with one subject to the procedure and the other as a control, were carried out by Nazi Germany on concentration camp inmates, particularly by Joseph Mengele.

1937–1945

During the Sino-Japanese War (1937–1945) and World War II, the Imperial Japanese Army made use of biological weapons against both Chinese soldiers and civilians in several military campaigns. Three veterans of Unit 731 testified, in a 1989 interview to the Asahi Shimbun, that they were part of a mission to contaminate the Horustein river with typhoid near the Soviet troops during the Battle of Khalkhin Gol. In 1940, the Imperial Japanese Army Air Force bombed Ningbo with ceramic bombs full of fleas carrying the bubonic plague. A film showing this operation was seen by the imperial princes Tsuneyoshi Takeda and Takahito Mikasa during a screening made by mastermind Shiro Ishii. During the Khabarovsk War Crime Trials the accused, such as Major General Kiyashi Kawashima, testified that as early as 1941 some 40 members of Unit 731 air-dropped plague-contaminated fleas on Changde. These operations caused epidemic plague outbreaks.

Many operations were ineffective due to inefficient delivery systems, using disease-bearing insects rather than dispersing the agent as an bioaerosol cloud. Nevertheless, some modern Chinese historians estimate that 400,000 Chinese died as a direct result of Japanese field testing and operational use of biological weapons.

In response to biological weapons development in Japan, and at the time suspected in Nazi Germany, the United States, United Kingdom, and Canada initiated a BW development programs in 1941 that resulted in the weaponization of tularemia, anthrax, brucellosis, and botulism toxin.

The centre for United States military BW research was Fort Detrick, Maryland, where USAMRIID is currently based; the first director was pharmaceutical executive George W. Merck. Some biological and chemical weapons research and testing was also conducted at Dugway Proving Grounds in Utah, at a munition manufacturing complex in Terre Haute, Indiana, and at a tract on Horn Island, Mississippi.

Much of the British work was carried out at Porton Down. Field testing carried out in the United Kingdom during World War II left Gruinard island in Scotland contaminated with anthrax for the next 48 years.

1946 to 1972

During the 1948 Israel War of Independence, International Red Cross reports raised suspicion that the Jewish Haganah militia had released Salmonella typhi bacteria into the water supply for the city of Acre, causing an outbreak of typhoid among the inhabitants. Egyptian troops later claimed to have captured disguised Haganah soldiers near wells in Gaza, whom they executed for allegedly attempting another attack. Israel denies these allegations.

During the Cold War, US conscientious objectors were used as consenting test subjects for biological agents in a program known as Operation Whitecoat. There were also many unpublicized tests carried out on the public during the Cold War.

Considerable research on the topic was performed by the United States, the Soviet Union, and probably other major nations throughout the Cold War era, though it is generally believed that biological weapons were never used after World War II. This view was challenged by China and North Korea, who accused the United States of germ warfare in the Korean War (1950–1953).

Cuba has also accused the United States of spreading human and animal disease on their island nation.

At the time of the Korean War the United States had only weaponized one agent, brucellosis ("Agent US"), which is caused by *Brucella suis*. The original weaponized form used the M114 bursting bomblet in M33 cluster bombs.

While the specific form of the biological bomb was classified until some years after the Korean War, in the various exhibits of biological weapons that Korea alleged were dropped on their country nothing resembled an M114 bomblet. There were ceramic containers that had some similarity to Japanese weapons used against the Chinese in World War II, developed by Unit 731.

Some of the Unit 731 personnel were imprisoned by the Soviets, and would have been a potential source of information on Japanese weaponization. The head of Unit 731, Lieutenant General Shiro Ishii, was granted immunity from war crimes prosecution in exchange for providing information to the United States on the Unit's activities.

The Korean War allegations also stressed the use of disease vectors, such as fleas, which, again, were probably a legacy of Japanese biological warfare efforts. The United States initiated its weaponization efforts with disease vectors in 1953, focused on Plague-fleas, EEE-mosquitoes, and yellow fever- mosquitoes (OJ-AP). However, US medical scientists in occupied Japan undertook extensive research on insect vectors, with the assistance of former Unit 731 staff, as early as 1946.

The United States Air Force was not satisfied with the operational qualities of the M114/US and labelled it an interim item until the United States Army Chemical Corps could deliver a superior weapon. The Air Force also changed its plans and wanted lethal biologicals.

The Chemical Corps then initiated a crash program to weaponize anthrax (N) in the E61 1/2-lb hour-glass bomblet. Though the program was successful in meeting its development goals, the lack of validation on the infectivity of anthrax stalled standardization.

Around 1950 the Chemical Corps also initiated a program to weaponize tularemia (UL). Shortly after the E61/N failed to make standardization, tularemia was standardized in the 3.4" M143 bursting spherical bomblet. This was intended for delivery by the MGM-29 Sergeant missile warhead and could produce 50% infection over a 7-square-mile (18 km^2) area.

Unlike anthrax, tularemia had a demonstrated infectivity with human volunteers (Operation Whitecoat). Furthermore, although tularemia is treatable by antibiotics, treatment does not shorten the course of the disease.

In addition to the use of bursting bomblets for creating biological aerosols, the Chemical Corps started investigating aerosol-generating bomblets in the 1950s. The E99 was the first workable design, but was too complex to be manufactured. By the late 1950s the 4.5" E120 spraying spherical bomblet was developed; a B-47 bomber with a SUU-24/A dispenser could infect 50% or more of the population of a 16-square-mile (41 km^2) area with tularemia with the E120. The E120 was later superseded by dry-type agents.

Dry-type biologicals resemble talcum powder, and can be disseminated as aerosols using gas expulsion devices instead of a burster or complex sprayer. The Chemical Corps developed Flettner rotor bomblets and later triangular bomblets for wider coverage due to improved glide angles over Magnus-lift spherical bomblets. Weapons of this type were in advanced development by the time the program ended.

United States President Richard Nixon signed an executive order on November 1969, which stopped production of biological weapons in the United States and allowed only scientific research of lethal biological agents and defensive measures such as immunization and biosafety. The biological munition stockpiles were destroyed, and approximately 2,200 researchers became redundant.

United States Special Forces and the CIA also had an interest in biological warfare, and a series of special munitions was created for their operations. The covert weapons developed for the military (M1, M2, M4, M5, and M32- or Big Five Weapons) were destroyed in accordance with Nixon's executive order to end the offensive program. The CIA maintained its collection of biologicals well into 1975 when it became the subject of the senate Church Committee.

The Biological Weapons Convention

In 1972, the United States signed the Biological and Toxic Weapons Convention, which banned the "development, production and stockpiling of microbes or their poisonous products except in amounts necessary for protective and peaceful research." By 1996, 137 countries had signed the treaty; however it is believed that since the signing of the Convention the number of countries capable of producing such weapons has increased.

The Soviet Union continued research and production of offensive biological weapons in a program called Biopreparat, despite having signed the convention. The United States was unaware of the program until Dr. Vladimir Pasechnik defected in 1989, and Dr. Kanatjan Alibekov, the first deputy director of Biopreparat defected in 1992.

After the 1991 Persian Gulf War, Iraq admitted to the United Nations inspection team to having produced 19,000 litres of concentrated botulinum toxin, of which approximately 10,000 L were loaded into military weapons; the 19,000 litres have never been fully accounted for. This is approximately three times the amount needed to kill the entire current human population by inhalation, although in practice it would be impossible to distribute it so efficiently, and, unless it is protected from oxygen, it deteriorates in storage.

On September 18, 2001 and for a few days after several letters were received by members of the United States Congress and media outlets containing anthrax spores: the attack killed five people. The identity of the perpetrator remained unknown until 2008, when a primary suspect was named.

List of BW Institutions and Programs by Country

According to the United States Office of Technology Assessment, since disbanded, seventeen countries were believed to possess biological weapons in 1995: Libya, North Korea, South Korea, Iraq, Taiwan, Syria, Israel, Iran, China, Egypt, Vietnam, Laos, Cuba, Bulgaria, India, South Africa, and Russia.

United States

The United States biological weapons program officially began in the spring 1943 on orders from U.S. President Franklin Roosevelt. Research continued following World War II as the U.S. built up a large stockpile of biological agents and weapons. Throughout its history the program was secret. It became controversial when it was later revealed that laboratory and field testing (some of the latter using simulants on non-consenting individuals) had been common. The official policy of the United States was first to deter the use of bio-weapons against U.S. forces and secondarily to retaliate if deterrence failed. There exists no evidence that the U.S. ever used biological agents against an enemy in the field.

In 1969, President Richard Nixon ended all offensive (i.e., non-defensive) aspects of the U.S. bio-weapons program. In 1975 the U.S. ratified both the 1925 Geneva Protocol and the 1972 Biological Weapons Convention (BWC) — these are international treaties outlawing biological warfare. Recent U.S. biodefence programs, however, have raised concerns that the US may be pursuing research that is outlawed by the BWC.

History

Early History (1918-41)

The United States' first interest in any form of biological warfare came at the close of World War I. The only agent the U.S. tested was the toxin ricin. The U.S. conducted tests concerning two methods of ricin dissemination, the first, involved adhering the toxin to shrapnel for delivery by artillery shell, which was successful. The other method, delivering an aerosol cloud of ricin, proved less successful. Neither delivery method was perfected before the war in Europe ended.

In the early 1920s suggestions that the U.S. begin a biological weapons program were coming from within the Chemical Warfare Service (CWS). Chief of the CWS, Amos Fries, decided that such a program would not be "profitable" for the U.S. Japan's Shiro Ishii began promoting biological weapons during the 1920s and toured biological research facilities worldwide, including in the United States. Though Ishii concluded that the U.S. was developing a bio-weapons programs he was incorrect. In fact, Ishii concluded that all major powers he visited was developing a bio-weapons program. As the interwar period continued, the United States did not emphasize biological weapons development or research. While the U.S. was spending very little time on biological weapons research, its future allies and enemies in the upcoming second World War were researching the potential of biological weapons as early as 1933.

World War II (1941-45)

Despite the World War I-era interest in ricin, as World War II erupted the United States Army still maintained the position that biological weapons was, for the most part, impractical. Other nations, notably France, Japan and the United Kingdom, thought otherwise and had begun their own biological weapons programs. Thus, as late as 1942 the U.S. had no biological weapons capabilities. Initial interest in biological weapons by the Chemical Warfare Service began in 1941. That fall, U.S. Secretary of War Henry L. Stimson requested that the National Academy of Sciences (NAS) undertake consideration of U.S. biological warfare. He wrote to Dr. Frank B. Jewett, then president of the NAS:

Because of the dangers that might confront this country from potential enemies employing what may be broadly described as biological warfare, it seems advisable that investigations be initiated to survey the present situation and the future possibilities. I am therefore, asking if you will undertake the appointment of an appropriate committee to survey all phases of this matter. Your organization already has before it a request from The Surgeon General for the appointment of a committee by the Division of Medical Sciences of the National Research Council to examine one phase of the matter.

In response the NAS formed a committee, the War Bureau of Consultants (WBC), which issued a report on the subject in February 1942. The report, among other items, recommended the research and development of an offensive biological weapons program.

The British, and the research undertaken by the WBC, pressured the U.S. to begin biological weapons research and development and in November 1942 U.S. President Franklin Roosevelt officially approved an American biological weapons program. In response to the information provided by the WBC, Roosevelt ordered Stimson to form the War Research Service (WRS). Established within the Federal Security Agency, the WRS' stated purpose was to promote "public security and health", but, in reality, the WRS was tasked with coordinating and supervising the U.S. biological warfare program. In the spring of 1943 the U.S. Army Biological Warfare Laboratories were established at Fort (then Camp) Detrick in Maryland.

Though initially, under George Merck, the WRS contracted several universities to participate in the U.S. biological weapons program, the program became large quickly and before long it was under the full control of the CWS. By November 1943 the biological weapons facility at Detrick was completed, in addition, the United States constructed three other facilities- a biological agent production plant at Vigo County near Terre Haute, Indiana, a field-testing site on Horn Island in Mississippi, and another field site near Granite Peak in Utah. According to an official history of the period, "the elaborate security precautions taken [at Camp Detrick] were so effective that it was not until January 1946, 4 months after VJ Day, that the public learned of the wartime research in biological weapons".

Cold War (1946-69)

Immediately following World War II, production of U.S. biological warfare agents went from "factory-level to laboratory-level". Meanwhile, work on biological weapons delivery systems increased. By 1950 the principal U.S. bio-weapons facility was located at Camp Detrick in Maryland under the auspices of the Research and Engineering Division of the U.S. Army Chemical Corps. The U.S. also maintained bio-warfare facilities at Fort Terry, an animal research facility on Plum Island. From the end of World War II through the Korean War, the U.S. Army, the Chemical Corps and the U.S. Air Force all made great strides in their biological warfare programs, especially concerning delivery systems.

The U.S. biological program expanded significantly during the Korean War. From 1952-1954 the Chemical Corps maintained a biological weapons research and development facility at Fort Terry on Plum Island, New York. The Fort Terry facility's focus was on anti-animal biological weapon research and development; the facility researched more than a dozen potential BW agents. A facility was opened in Pine Bluff, Arkansas, Pine Bluff Arsenal and by 1954 the production of weapons-grade agents began.

End of the Program (1969-73)

President Richard M. Nixon issued his "Statement on Chemical and Biological Defence Policies and Programs" on November 25, 1969 in a speech from Fort Detrick. The statement ended, unconditionally, all U.S. offensive biological weapons programs. Nixon noted that biological weapons were unreliable and stated:

The United States shall renounce the use of lethal biological agents and weapons, and all other methods of biological warfare. The United States will confine its biological research to defensive measures such as immunization and safety measures.

In his speech Nixon called his move "unprecedented"; and it was in fact the first review of the U.S. biological warfare program since 1954. Despite the lack of review, the biological warfare program had increased in cost and size since 1961; when Nixon ended the program the budget was $300 million annually. Nixon's statement confined all biological weapons research to defensive-only and ordered the destruction of the existing U.S. biological arsenal.

U.S. biological weapons stocks were destroyed over the next few years. A $12 million disposal plan was undertaken at Pine Bluff Arsenal, where all U.S. anti-personnel biological agents were stored. That plan was completed in May 1972 and included decontamination of facilities at Pine Bluff. Other agents, including anti-crop agents such as wheat stem rust, were stored at Beale Air Force Base and Rocky Mountain Arsenal. These anti-crop agents, along with agents at Fort Detrick used for research purposes were destroyed in March 1973.

Budget History

From the onset of the U.S. biological weapons program in 1943 through the end of World War II the United States spent $400 million on biological weapons, mostly on research and development. The budget for fiscal year 1966 was $38 million. When Nixon ended the U.S. bio-weapons program it represented the first review of the U.S. biological warfare program since 1954. Despite the lack of review, the biological warfare program had increased in cost and size since 1961; when Nixon ended the program the budget was $300 million annually.

Geneva Protocol and BWC

The 1925 Geneva Protocol, ratified by most major powers in the 1920s and 30s, had still not been ratified by the United States at the dawn of World War II. Among the Protocol's provisions, was a ban on bacteriological warfare. The Geneva Protocol had encountered opposition in the U.S. Senate, in part due to strong lobbying against it by the Chemical Warfare Service, and it was never brought to the floor for a vote when originally introduced. Regardless, on June 8, 1943 President Roosevelt affirmed a no-first-use policy for the United States concerning biological weapons. Even with Roosevelt's declaration opposition to the Protocol remained strong; in 1949 the Protocol was among several old treaties returned to President Harry S. Truman unratified.

When Nixon ended the U.S. bio-weapons program in 1969 he also announced that he would resubmit the Geneva Protocol to the U.S. Senate. This was a move Nixon was considering as early as July 1969. The announcement included language that indicated the Nixon administration was moving toward an international agreement on an outright ban on bio-weapons. Thus, the Nixon administration became the world's leading anti-biological weapons voice calling for an international treaty. The Eighteen Nation Disarmament Committee was discussing a British draft of a biological weapons treaty which the United Nations General Assembly approved in 1968 and that NATO supported. These arms control talks would eventually lead to the Biological Weapons Convention, the international treaty outlawing biological warfare. Prior, to the Nixon announcement only Canada supported the British draft. Beginning in 1972, the Soviet Union, United States and more than 100 other countries signed the BWC. The United States ratified the Geneva Protocol in 1975.

Post-1969 bio-defence Program

Both the U.S. bio-weapons ban and the Biological Weapons Convention restricted any work in the area of biological warfare to defensive in nature. In reality, this gives BWC member-states wide latitude to conduct biological weapons research because the BWC contains no provisions for monitoring of enforcement. The treaty, essentially, is a gentlemen's agreement amongst members backed by the long-prevailing thought that biological warfare should not be used in battle.

After Nixon declared an end to the U.S. bio-weapons program debate in the Army centered around whether or not toxin weapons were included in the president's declaration. Following Nixon's November 1969 order, scientists at Fort Detrick worked on one toxin, *Staphylococcus* enterotoxin type B (SEB), for several more months. Nixon ended the debate when he added toxins to the bio-weapons ban in February 1970. The U.S. also ran a series of experiments with anthrax, code named Project Baçchus, Project Clear Vision and Project Jefferson in the late 1990s and early 2000s.

In recent years certain critics have claimed the U.S. stance on biological warfare and the use of biological agents has differed from historical interpretations of the BWC. For example, it is said that the U.S. now maintains that the Article I of the BWC (which explicitly bans bio-weapons), does not apply to "non-lethal" biological agents. Previous interpretation was stated to be in line with a definition laid out in Public Law 101-298, the Biological Weapons Anti-Terrorism Act of 1989. That law defined a biological agent as:

any micro-organism, virus, infectious substance, or biological product that may be engineered as a result of biotechnology, or any naturally occurring or bioengineered component of any such microorganism, virus, infectious substance, or biological product, capable of causing death, disease, or other biological malfunction in a human, an animal, a plant, or another living organism; deterioration of food, water, equipment, supplies, or material of any kind...

According to the Federation of American Scientists, U.S. work on non-lethal agents exceeds limitations in the BWC.

Agents and Weapons

When the U.S. biological warfare program ended in 1969 it had developed seven mass-produced, battle-ready biological weapons in the form of agents that cause: anthrax, tularemia, brucellosis, Q-fever, VEE, and botulism. In addition Staphylococcal Enterotoxin B was produced as an incapacitating agent. In addition to the agents that were ready to be used the U.S. program conducted research into the weaponization of more than 20 other agents. They included: smallpox, EEE and WEE, AHF, Hantavirus, BHF, Lassa fever, glanders, melioidosis, plague, yellow fever, psittacosis, typhus, dengue fever, Rift Valley fever (RVF), CHIKV, late blight of potato, rinderpest, Newcastle disease, bird flu, and the toxin ricin.

Besides the numerous pathogens that afflict human beings, the U.S. had developed an arsenal of anti-agriculture biological agents. These included rye stem rust spores (stored at Edgewood Arsenal, 1951–1957), wheat stem rust spores (stored at the same facility 1962- 1969), and the causative agent of rice blast.

A U.S. facility at Fort Terry primarily focused on anti-animal biological agents. The first agent that was a candidate for development was foot and mouth disease (FMD). Besides FMD, five other top secret biological weapons projects were commissioned on Plum Island. The other four programs researched included RVF, rinderpest, African swine fever, plus eleven miscellaneous exotic animal diseases. The eleven miscellaneous pathogens were: Blue tongue virus, bovine influenza, bovine virus diarrhea (BVD), fowl plague, goat pneumonitis, mycobacteria, "N" virus, Newcastle disease, sheep pox, Teschers disease, and vesicular stomatitis.

Work on delivery systems for the U.S. bio-weapons arsenal led to the first mass-produced biological weapon in 1952, the M33 cluster bomb. The M33's sub-munition, the pipe bomb like, cylindrical M114 bomb, was also completed and battle-ready by 1952. Other delivery systems researched and at least partially developed during the 1950s included the E77 balloon bomb and the E86 cluster bomb. The peak of U.S. biological weapons delivery system development came during the 1960s. Production of cluster bomb sub-muntions began to shift from the cylindrical bomblets to spherical bomblets, which had a larger coverage area. Development of the spherical E120 bomblet took place in the early 1960s as did development of the M143 bomblet, similar to the chemical M139 bomblet. The experimental Flettner rotor bomblet was also developed during this time period. The Flettner rotor was called, "probably one of the better devices for disseminating microorganisms", by William C. Patrick III.

Alleged Uses

Cuba

It has been rumored that the U.S. employed biological weapons against the Communist island nation of Cuba. Noam Chomsky claimed that evidence exists implicating the U.S. in biological warfare in Cuba, but his evidence was disputed. Allegations in 1962 held that CIA operatives had contaminated a shipment of sugar while it was in storage in Cuba. Again, in 1962, a Canadian agricultural technician assisting the Cuban government claimed he was paid \$5,000 to infect Cuban turkeys with the deadly Newcastle disease. Though the technician later claimed he had just pocketed the money, many Cubans and some Americans believed a clandestinely administered biological weapons agent was responsible for a subsequent outbreak of the disease in Cuban turkeys. In 1971 the first serious outbreak of swine flu in the Western Hemisphere occurred in Cuba, and Cubans alleged that U.S. covert biological warfare was responsible for this outbreak, which led to the preemptive slaughter of 500,000 pigs. Evidence linking these incidents to biological warfare has not been confirmed.

Accusations have continued to come out of Havana charging U.S. use of bio-weapons on the island. The Cuban government blamed the U.S. for a 1981 outbreak of dengue fever that sickened more than 300,000. Dengue, a vector-borne disease usually carried by mosquitoes, killed 158 people that year in Cuba, including 101 children under 15. Poor relations between Cuba and the United States coupled with confirmed U.S. research into entomological warfare during the 1950s made these charges seem not implausible. However, dengue fever occurs naturally in the region of the world where Cuba is located.

Korean War

North Korean and Chinese officials leveled accusations that during the Korean War the United States engaged in biological warfare in North Korea. The claim is dated to the period of the war, and has been thoroughly denied by the U.S. In 1998, Stephen Endicott and Edward Hagermann claimed that the accusations were true in their book, *The United States and Biological Warfare: Secrets from the Early Cold War and Korea* The book received mixed reviews, some called it "bad history" and "appalling", while other praised the case the authors made.

In 1952 the Chinese and North Koreans insinuated that mysterious outbreaks of disease in North Korea and China were due to U.S. biological attacks. Despite assertions that this did not occur from the International Red Cross and World Health Organization, whom the Chinese denounced as Western-biased, the Chinese government pursued an investigation by the World Peace Council. A committee led by Joseph Needham gathered evidence for a report that included eyewitness testimony, and testimony from doctors as well as four American Korean War prisoners who confirmed use of biological weapons by the U.S. The U.S. government denied the accusations and their denial was generally supported by top scientists in the West. In Eastern Europe, and China, North Korea it was widely believed that the accusations were true.

The same year Endicotts' book was published Kathryn Weathersby and Milton Leitenberg of the Cold War International History Project at the Woodrow Wilson Centre in Washington released a cache of Soviet and Chinese documents which revealed the North Korean claim was an elaborate disinformation campaign. In addition, a Japanese journalist claims to have seen similar evidence of a Soviet disinformation campaign and that the evidence supporting its occurrence was faked.

Others have revived these claims more recently. In March 2010, the allegations were investigated by the Al Jazeera English news programme People & Power. In this program, Professor Mori Masataka investigated historical artifacts in the form of bomb casings from US biological weapons, contemporary documentary evidence and eye witness testimonies. From the evidence he collected, Professor Mori concluded that the United States did in fact test biological weapons on North Korea during the Korean War.

Experimentation and Testing

Entomological Testing

The United States seriously researched the potential of entomological warfare (EW) during the Cold War. EW is a specific type of biological warfare which aims to use insects as weapon, either directly or through their potential to act as vectors. During the 1950s the United States conducted a series of field tests using entomological weapons. Operation Big Itch, in 1954, was designed to test munitions loaded with uninfected fleas (*Xenopsylla cheopis*). In May 1955 over 300,000 yellow fever mosquitoes (*Aedes aegypti*) were dropped over parts of the U.S. state of Georgia to determine if the air-dropped mosquitoes could survive to take meals from humans. The mosquito tests were known as Operation Big Buzz. The U.S. engaged in at least two other EW testing programs, Operation Drop Kick and Operation May Day. A 1981 Army report outlined these tests as well as multiple cost-associated issues that occurred with EW.

Experiments on Consenting Individuals

Operation Whitecoat involved the controlled testing of many serious agents on military personnel consented to experimentation, and understood the risks involved. No deaths are known to have resulted from this program.

Experiments on Non-consenting Individuals

Testing on Unwitting Military Personnel

In August 1949 a U.S. Army Special Operations Division, operating out of Fort Detrick in Maryland, set up its first test at The Pentagon in Washington, D.C. Operatives sprayed harmless bacteria into the building's air conditioning system and observed as the microbes spread throughout the Pentagon.

The US military acknowledges that it tested several chemical and biological weapons on US military personnel in the desert facility, including the East Demilitarization Area near Deseret Chemical Depot/Deseret Chemical Test Centre at Fort Douglas, Utah, but takes the position that the tests have contributed to long-term illnesses in only a handful of exposed personnel. Veterans who took part believe they were also exposed to Agent Orange. The Department of Veterans Affairs denies almost all claims for care and compensation made by veterans who believe they got sick as a result of the tests. The US military for decades remained silent about "Project 112" and its victims, a slew of tests overseen by the Army's Deseret Test Centre in Salt Lake City. Project 112 starting in the 1960s tested chemical and biological agents, including VX, sarin and e. Coli, on military personnel who did not know they were being tested. After the Defence Department finally acknowledged conducting the tests on unwitting human subjects, it agreed to help the Veterans' Affairs Department track down those who were exposed, but a General Accountability Office report in 2008 scolded the military for ceasing the effort.

Testing on Unwitting Civilians

Medical experiments were conducted on a large scale on civilians who had not consented to participate. Often, these experiments took place in urban areas in order to test dispersion methods. Questions were raised about detrimental health effects after experiments in San Francisco, California, were followed by a spike in hospital visits; however, in 1977 the Centres for Disease Control and Prevention determined that there was no association between the testing and the occurrence of pneumonia or influenza. The San Francisco test involved a U.S. Navy ship that sprayed Serratia marcescens from the bay; it traveled more than 30 miles. One dispersion test involved laboratory personnel disguised as passengers spraying harmless bacteria in Ronald Reagan Washington National Airport.

Scientists tested biological pathogens, including *Bacillus globigii*, which were thought to be harmless, at public places such as subways. A light bulb containing *Bacillus globigii* was dropped on New York City's subway system; the result was strong enough to affect people prone to illness (also known as Subway Experiment). Based on the circulation measurements, thousands of people would have been killed if a dangerous microbe was released in the same manner. A jet aircraft released material over Victoria, Texas, that was monitored in the Florida Keys.

GAO Report

In February, 2008, the Government Accountability Office (GAO) released report GAO-08-366 titled, "Chemical and Biological Defence, DOD and VA Need to Improve Efforts to Identify and Notify Individuals Potentially Exposed during Chemical and Biological Tests." The report stated that tens of thousands of military personnel and civilians may have been exposed to biological and chemical substances through DOD tests. In 2003, the DOD reported it had identified 5,842 military personnel and estimated 350 civilians as being potentially exposed during the testing, known as Project 112.

The GAO scolded the U.S. Department of Defence's (DOD) 2003 decision to stop searching for people affected by the tests was premature. The GAO report also found that the DoD made no effort to inform civilians of exposure, and that the United States Department of Veterans Affairs (VA) is failing to use available resources to inform veterans of possible exposure or to determine if they were deceased. After the DoD halted efforts to find those who may have been affected by the tests, veteran health activisits and others identified approximately 600 additional individuals who were potentially exposed during Project 112. Some of the individuals were identified after the GAO reviewed records stored at the Dugway Proving Ground, others were identified by the Institute of Medicine. Many of the newly identified suffer from long term illnesses that may have been caused by the biological or chemical testing.

- Fort Detrick, Maryland :
  - o U.S. Army Biological Warfare Laboratories (1943–69)
    - * Building 470
    - * One-Million-Liter Test Sphere
    - * Operation Whitecoat
  - o United States Army Medical Unit (1954–69)
  - o U.S. Army Medical Research Institute of Infectious Diseases (USAMRIID; 1969–present)
  - o National Biodefence Analysis and Countermeasures Centre (NBACC; Projected: 2008)
- Project Bacchus
- Project Clear Vision
- Project SHAD.

Project SHAD stands for Project Shipboard Hazard and Defence, a series of Cold War-era tests by the United States Department of Defence of biological weapons and chemical weapons. Exposures of uninformed and unwilling humans during the testing to the test substances, particularly the exposure to United States military personnel then in service, has added controversy to recent revelations of the project.

History

Project SHAD was part of a larger effort by the Department of Defence called Project 112. The Project began in 1962 during John F. Kennedy's administration, and it is largely believed that neither Kennedy nor subsequent Presidents knew of Project 112 or SHAD. However, Robert McNamara, Kennedy's Secretary of Defence, did know of and approved these tests. There is also some evidence that demonstrates local governments were involved with these tests, though it is unclear how exactly they aided with Project SHAD. The official statement on Project SHAD's purpose was "...to identify U.S. war ships vulnerabilities to attacks with biological or chemical warfare agents and to develop procedures to respond to such attacks while maintaining a warfighting capability." 134 tests were planned initially, but only 46 tests were actually completed. In these tests, chemical and biological agents were introduced to military personnel, who were at the time ignorant that they were involved in such an experiment. Nerve agents and chemicals include, but are not limited to, VX nerve gas, Tabun gas, Sarin, Soman, and the marker chemicals zinc sulfide, cadmium sulfide, and QNB. Biologics include *Bacillus globigii, Coxiella burnetti* (which causes Q fever), and Francisella tularensis (which causes tularemia or 'rabbit fever').

Declassification

Revelations concerning Project SHAD were first exposed by Independent Producer and Investigative Journalist Eric Longabardi of TeleMedia News Productions, now based in Los Angeles, CA. Longabardi's 6-year investigation into the still secret program began in early 1994. It ultimately resulted in a series of investigative reports produced by him, which were broadcast on the CBS Evening News in May 2000. After the broadcast of these exclusive reports, the Pentagon and Veteran's Administration opened their own on going investigations into the long classified program. In 2002, Congressional hearings on Project SHAD, in both the Senate and House, further shed media attention on the still classified program. In 2002, a class action federal lawsuit was filed on behalf of the US Navy sailors exposed in the testing. Additional actions, including a multi-year medical study was conducted by National Academy of Sciences/ Institute of Medicine to assess the potential medical harm caused to the thousands of unwitting US Navy sailors, civilians, and others who were exposed in the secret testing. The results of that study were finally released in May 2007.

28 fact sheets have been released, focusing on the Deseret Test Centre in Dugway, Utah, which was built entirely for Project SHAD and was closed after the Project was finished in 1973.

The US Department of Defence (DoD) has come under great scrutiny because those that were involved with Project 112 and SHAD were unaware of any tests being done. No effort was made to ensure the informed consent of the military personnel. Until 1998, the Department of Defence stated officially that Project SHAD did not exist. Because the DoD refused to acknowledge the program, surviving test subjects have been unable to obtain disability payments for health issues related to the project. US Representative Mike Thompson said of the program and the DoD's effort to conceal it, "They told me – they said, but don't worry about it, we only used simulants. And my first thought was, well, you've lied to these guys for 40 years, you've lied to me for a couple of years. It would be a real leap of faith for me to believe that now you're telling me the truth."

The Department of Veterans Affairs has commenced a three-year study comparing known SHAD-affected veterans to veterans of similar ages who were not involved in any way with SHAD or Project 112. The study cost approximately US$3 million, and results are being compiled for future release.

United Kingdom

The United Kingdom has possessed weapons of mass destruction, including nuclear, biological, and chemical weapons. The United Kingdom is one of the five official nuclear weapon states under the Nuclear Non-Proliferation Treaty and has an independent nuclear deterrent. The U.K. has been estimated to have a stockpile of approximately 160 active nuclear warheads and 225 nuclear warheads in total. The United Kingdom renounced the use of chemical and biological weapons in 1956 and subsequently destroyed its general stocks.

Biological Weapons

During the Second World War, British scientists studied the use of biological weapons, including a test using anthrax on the Scottish island of Gruinard which left it contaminated and fenced off for nearly fifty years, until an intensive four-year programme to eradicate the spores was completed in 1990. They also manufactured five million linseed-oil cattle cakes with a hole bored into them for addition of anthrax spores between 1942 and mid-1943.

These were to be dropped on Germany using specially designed containers each holding 400 cakes, in a project known as Operation Vegetarian. It was intended that the disease would destroy the German beef and dairy herds and possibly spread to the human population. Preparations were not complete until early 1944. Operation Vegetarian was only to be used in the event of a German anthrax attack on Britain.

Offensive weapons development continued after the war into the 1950s with tests of plague, brucellosis, tularemia and later equine encephalomyelitis and vaccinia viruses (the latter as a relatively safe simulant for smallpox).

In particular five sets of trials took place at sea using aerosol clouds and animals.

- Operation Harness off Antigua in 1948-1949.
- Operation Cauldron off Stornoway in 1952. The trawler *Carella* accidentally sailed through a cloud of pneumonic plague bacilli (*yersinia pestis*) during this

trial. It was kept under covert observation until the incubation period had elapsed but none of the crew fell ill.

- Operation Hesperus off Stornoway in 1953.
- Operation Ozone off Nassau in 1954.
- Operation Negation off Nassau in 1954-5.

The programme was cancelled in 1956 when the British government renounced the use of biological and chemical weapons. It ratified the Biological and Toxin Weapons Convention in March 1975.

- Porton Down
- Gruinard Island
- Nancekuke.

Former Soviet Union and Russia

The Soviet Union began a biological weapons program in the 1920s at the Leningrad Military Academy in Moscow under the control of the state security apparatus, known as the GPU. This occurred despite the fact that the USSR was a signatory to the 1925 Geneva Convention, which banned both chemical and biological weapons.

During World War II Stalin was forced to move his biological warfare operations out of the path of advancing German forces At the conclusion of the war, Soviet troops invading Manchuria captured many Unit 731 Japanese scientists and learned of their extensive human experimentation through captured documents and prisoner interrogations. Emboldened by these discoveries, Stalin put KGB chief Lavrenty Beria in charge of a new biowarfare program.

By 1960, numerous biological warfare research facilities existed throughout the Soviet Union. Although the USSR also signed the 1972 Biological Weapons Convention (BWC), the Soviets subsequently augmented their biowarfare programs. They doubted the United States' claimed compliance with the BWC, which further motivated their program. The Soviet biological weapons effort became a huge program, comprising various institutions under different ministries along with commercial facilities and collectively known as Biopreparat after 1973. Biopreparat pursued offensive research, development, and production of biological agents under cover of legitimate civil biotechnology research. It conducted its clandestine activities at 52 sites and employed over 50,000 people. Annualized production capacity for weaponized smallpox, for example, was 90 to 100 tons.

In the 1990s, President of the Russian Federation Boris Yeltsin admitted to an offensive bio-weapons program as well as to the true nature of the Sverdlovsk biological weapons accident of 1979, which had resulted in the deaths of at least 64 people Soviet defectors, including Colonel Kanatjan Alibekov, first deputy chief of Biopreparat from 1988 to 1992, confirmed that the program had been massive and still existed. In September 1992, Russia signed an agreement with the United States and Great

Britain promising to end its bio-weapons program and to convert its facilities for benevolent scientific and medical purposes. Compliance with the agreement as well as the fate of the former Soviet bio-agents and facilities, is still mostly undocumented.

History

Timeline

1928 : Revolutionary Military Council signed a decree about weaponization of typhus. Leningrad Military academy began cultivation of typhus in chicken embryos. Human experimentation with typhus, glanders and melioidosis in Solovetsky camp. A laboratory on vaccine and serum research was also established near Moscow in 1928, within Military Chemical Agency. This laboratory was transformed to Red Army's Scientific Research Institute of Microbiology in 1933.

1941: Soviet bioweapons facilities are evacuated to the city of Kirov.

1942: Alleged use of tularemia against German troops.

1945: Japanese documentation from Unit 731 was captured.

1946: A biological weapons facility was established in Sverdlovsk.

1953: Fifteenth directorate of Red Army takes responsibility for the program.

1973: A "civilian" main directorate Biopreparat was founded. Other organizations involved in design and production of biological weapons were Soviet Ministry of Defence, Ministry of Agriculture., Ministry of Health., USSR Academy of Sciences, and KGB.

1990s: specimens of deadly bacteria and viruses were stolen from Western laboratories and delivered by Aeroflot planes to support Russian program of biological weapons. At least one of the pilots was a Russian Foreign Intelligence Service officer". At least two agents died, presumably from the transported pathogens. Beginning of 2000s: Academician "A.S." proposed new biological warfare program "Biological Shield of Russia" to president Vladimir Putin. The program reportedly includes institutes of Russian Academy of Sciences from Pushchino.

Military use During World War II

Tularemia was allegedly used against German troops in 1942 near Stalingrad. Around 10,000 cases of tularemia had been reported in the Soviet Union during the years of 1941 and 1943. However, the number of cases jumped to more than 100,000 in the year of Stalingrad outbreak. German panzer troops fell ill in such significant numbers during the late summer of 1942 that German military campaign came to a temporary halt. German soldiers became ill with the rare pulmonary form of tularemia, which indicate the use of an aerosol biological weapon (the ordinary transmission pathway is through ticks and rodents). According to Kenneth Alibek the used tularemia weapon had been developed in the Kirov military facility. It was suggested by some, however, that the outbreak might be of natural origin, since a pulmonary form of tularemia has also been noted in natural outbreaks in Martha's Vineyard in 2000.

In the Soviet Union the outbreak at Stalingrad was described as a natural outbreak. Crops were left in the field during the German offensive and the rodent population swelled putting many inhabitants into contact with infected rodents. In some parts of the Stalingrad Oblast as many as 75% of the inhabitants became infected. It was noted that before the war there was a so-called threshing tularemia caused by people inhaling infected dusts soiled by rodents while threshing grain.

Post-BWC Developments

Soviet Union continued development and mass production of offensive biological weapons, despite having signed the 1972 Biological Weapons Convention (BWC). The development and production was conducted by main directorate "Biopreparat", Soviet Ministry of Defence, Ministry of Agriculture, Ministry of Health, USSR Academy of Sciences, the KGB, and other state organizations.

In 1980s Soviet Ministry of Agriculture had successfully developed variants of foot-and-mouth disease and rinderpest against cows, African swine fever for pigs, and psittacosis to kill chicken. These agents were prepared to spray them down from tanks attached to airplanes over hundreds of miles. The secret program was code-named "Ecology".

Notable Outbreaks and Accidents

Marburg Virus

The Soviet Union reportedly had a large biological weapons program enhancing the usefulness of the Marburg virus. The development was conducted in Vector Institute under leadership of Dr. Ustinov who accidentally died from the virus. The samples of Marburg taken from Ustinov's organs were more powerful than the original strain. New strain called "Variant U" had been successfully weaponized and approved by Soviet Ministry of Defence in 1990.

Smallpox

The first smallpox weapons factory in the Soviet Union was established in 1947 in the city of Zagorsk, close to Moscow. It was produced by injecting small amounts of the virus into chicken eggs. An especially virulent strain was brought from India in 1967 by a special Soviet medical team that was sent to India to help to eradicate the virus. The pathogen was manufactured and stockpiled in large quantities throughout the 1970s and 1980s.

An outbreak of weaponized smallpox occurred during its testing in the 1970s. General Prof. Peter Burgasov, former Chief Sanitary Physician of the Soviet Army, and a senior researcher within the program of biological weapons described this incident:

> *"On Vozrozhdeniya Island in the Aral Sea, the strongest recipes of smallpox were tested. Suddenly I was informed that there were mysterious cases of mortalities in Aralsk. A research ship of the Aral fleet came 15 km away from the island (it was forbidden to come any closer than*

40 km). The lab technician of this ship took samples of plankton twice a day from the top deck. The smallpox formulation— 400 gr. of which was exploded on the island—"got her" and she became infected. After returning home to Aralsk, she infected several people including children. All of them died. I suspected the reason for this and called the Chief of General Staff of Ministry of Defence and requested to forbid the stop of the Alma-Ata to Moscow train in Aralsk. As a result, the epidemic around the country was prevented. I called Andropov, who at that time was Chief of KGB, and informed him of the exclusive recipe of smallpox obtained on Vozrozhdeniya Island."

A production line to manufacture smallpox on an industrial scale was launched in the Vector Institute in 1990. The development of genetically altered strains of smallpox was presumably conducted in the Institute under leadership of Dr. Sergei Netyosov in the middle of the 1990s, according to Kenneth Alibek.

It was reported that Russia made smallpox available to Iraq in the beginning of 1990s.

Anthrax

Spores of weaponized anthrax were accidentally released from a military facility near the city of Sverdlovsk in 1979. The death toll was at least 105, but no one knows the exact number, because all hospital records and other evidence were destroyed by the KGB, according to former Biopreparat deputy director Kenneth Alibek.

Sverdlovsk Anthrax Leak

The Sverdlovsk anthrax leak is an incident when spores of anthrax were accidentally released from a military facility in the city of Sverdlovsk (formerly, and now again, Yekaterinburg) 1450 km east of Moscow on April 2, 1979. This accident is sometimes called "biological Chernobyl". The ensuing outbreak of the disease resulted in approximately 100 deaths, although the exact number of victims remains unknown. The cause of the outbreak had for years been denied by the Soviet Union, which blamed the deaths on intestinal exposure due to the consumption of tainted meat from the area, and subcutaneous exposure due to butchers handling the tainted meat. All medical records of the victims had been removed in order to avoid revelations of serious violations of the Biological Weapons Convention.

Background

The closed city of Sverdlovsk had been a major production centre of the Soviet military-industrial complex since World War II. It produced tanks, nuclear rockets and other armaments. A major nuclear accident happened in this region in 1958, when a military reactor was damaged, resulting in the spread of radioactive dust over a thousand square kilometres. The biological weapons facility in Sverdlovsk was built after World War II, using documentation captured in Manchuria from the Japanese germ warfare program.

The strain of anthrax produced in *Military Compound 19* near Sverdlovsk was the most powerful in the Soviet arsenal ("Anthrax 836"). It had been isolated as a result of another anthrax leak accident that happened in 1953 in the city of Kirov. A leak from a bacteriological facility contaminated the city sewer system. In 1956, biologist Vladimir Sizov found a more virulent strain in rodents captured in this area. This strain was planned to be used to arm warheads for the SS-18 ICBM, which would target American cities, among other targets.

The Accident

The produced anthrax culture had to be dried to produce a fine powder for use as an aerosol. Large filters over the exhaust pipes were the only barriers between the anthrax dust and the outside environment. On the last Friday of March 1979, a technician removed a clogged filter while drying machines were temporarily turned off. He left a written notice, but did not write this down in the logbook as he was supposed to do. The supervisor of the next shift did not find anything unusual in the logbook, and turned the machines on. In a few hours, someone found that the filter was missing and reinstalled it. The incident was reported to military command, but local and city officials were not immediately informed. Boris Yeltsin, a local Communist Party boss at this time, was helping to cover up the accident. All workers of a ceramic plant across the street fell ill during the next few days. Almost all of them died in a week. The death toll was at least 105, but the exact number is unknown as all hospital records and other evidence were destroyed by the KGB, according to former Biopreparat deputy director Ken Alibek.

The Investigation

In the 1980s, there was vigorous international debate and speculation as to whether the outbreak was natural or an accidental exposure. If accidental, there was discussion of whether it represented violation of the 1972 Biological Weapons Convention. A number of small investigations launched by Russian scientists in the years immediately following the dissolution of the Soviet Union re-opened the case in a number of newspaper articles. A team of Western inspectors lead by Professor Matthew Meselson of Harvard finally gained access to the region in 1992, and determined that all of the victims had been living directly downwind at the time of the release of the spores via aerosol. Livestock in the area were also affected. It was revealed around this time that the accident was caused by the non-replacement of a filter on an exhaust at the facility, and though the problem was quickly rectified it was too late to prevent a release. Had the winds been blowing in the direction of the city at that time, it could have resulted in the pathogen being spread to hundreds of thousands of people. The military facility remains closed to inspection. Professor Meselson's original contention for many years had been that the outbreak was a natural one and that the Soviet authorities were not lying when they disclaimed having an active offensive bio-warfare program, but the information uncovered in the investigation left no room for doubt. Meselson's wife, Jeanne Guillemin (who had participated in the investigation), detailed the events in a 1999 book.

Aftermath

Russian Prime Minister Egor Gaidar issued a decree to begin demilitarization of Compound 19 in 1992. However, the facility continued its work. Not a single journalist has been allowed onto the premises since 1992. About 200 soldiers with Rottweiler dogs still patrol the complex. Classified activities were moved underground, and several new laboratories have been constructed and equipped to work with highly dangerous pathogens. One of their current subjects is reportedly *Bacillus anthracis* strain H-4. Its virulence and antibiotic resistance have been dramatically increased using genetic engineering.

- Biopreparat (18 labs and production centres)
  - Stepnagorsk Scientific and Technical Institute for Microbiology, Stepnogorsk, northern Kazakhstan
  - Institute of Ultra Pure Biochemical Preparations, Leningrad, a weaponized plague centre
  - Vector State Research Centre of Virology and Biotechnology (VECTOR), a weaponized smallpox centre
  - Institute of Applied Biochemistry, Omutninsk
  - Kirov bioweapons production facility, Kirov, Kirov Oblast
  - Zagorsk smallpox production facility, Zagorsk
  - Berdsk bioweapons production facility, Berdsk
  - Sverdlovsk bioweapons production facility (Military Compound 19), Sverdlovsk, a weaponized anthrax centre.

Poison Laboratory of the Soviet Secret Services

Poison laboratory of the Soviet secret services, alternatively known as Laboratory 1, Laboratory 12, and *Kamera* which means "The Chamber" in Russian, was a covert poison research and development facility of the Soviet secret police agencies.

Chronology

- 1921: First poison laboratory within the Soviet secret services was established under the name "Special Office". It was headed by professor of medicine Ignatii Kazakov, according to Pavel Sudoplatov.
- 1926: The laboratory was under the supervision of Genrikh Yagoda, a deputy of OGPU chairman Vyacheslav Menzhinsky, who became NKVD chief in 1934 after Menzhinsky's death.
- February 20, 1939: It becomes *Laboratory 1* headed by Grigory Mairanovsky. The laboratory was under the direct supervision of NKVD director Lavrenty Beria and his deputy Vsevolod Merkulov from 1939 to March 1953.
- December 21, 1951: Grigory Mairanovsky arrested in connection with Viktor Abakumov's arrest, which was presumably a part of Joseph Stalin's campaign to remove NKVD chief, Lavrenty Beria.

- March 14, 1953: It was renamed to *Laboratory 12*. V. Naumov is the newly appointed head. Lavrenty Beria and Vsevolod Merkulov were executed after Stalin's death. Immediate NKVD supervisor of the laboratory, Pavel Sudoplatov, received long prison sentences.
- 1978: Expanded into the *Central Investigation Institute for Special Technology* within the First Chief Directorate of the KGB
- Currently: Several laboratories of the SVR, (headquartered in Yasenevo near Moscow), are responsible for the "creation of biological and toxin weapons for clandestine operations in the West".

Human Experimentation

Mairanovsky and his colleagues tested a number of deadly poisons on prisoners from the Gulag ("enemies of the people"), including mustard gas, ricin, digitoxin and many others. The goal of the experiments was to find a tasteless, odourless chemical that could not be detected *post mortem*. Candidate poisons were given to the victims, with a meal or drink, as "medication"

Finally, a preparation with the desired properties called C-2 was developed According to witness testimonies, the victim changed physically, became shorter, weakened quickly, became calm and silent and died within fifteen minutes. Mairanovsky brought to the laboratory people of varied physical condition and ages in order to have a more complete picture about the action of each poison.

Pavel Sudoplatov and Nahum Eitingon approved special equipment [poisons] only if it had been tested on humans", according to testimony of Mikhail Filimonov. Vsevolod Merkulov said that these experiments were approved by NKVD chief Lavrenty Beria.. Beria himself testified on August 28, 1953, after his arrest that "I gave orders to Mairanovsky to conduct experiments on people sentenced to the highest measure of punishment, but it was not my idea".

In addition to human experimentation, Mairanovsky personally executed people with poisons, under the supervision of Sudoplatov

Prominent Victims

- The leader of the Russian All-Military Union general Alexander Kutepov was drugged and kidnapped in Paris in 1930. He died from a heart attack due to an overdose of the administered drug.
- One of leaders of the White movement, Russian general Evgenii Miller, was drugged and kidnapped in Paris in 1937. He was executed later in Russia.
- Abram Slutsky, head the Soviet foreign intelligence service (GUGB) was poisoned with hydrocyanic acid added in tea in 1938
- Archbishop Theodore Romzha of Ukrainian Catholic Church was killed in 1947 by injection of curare provided by Mairanovsky and administered by a medical nurse who was an Ministry for State Security (USSR) agent.
- In 1978, dissident Bulgarian writer Georgi Markov was assassinated in London

using a tiny pellet poisoned with ricin; the necessary equipment was prepared in this laboratory. In a Discovery Channel television program about his illustrated book of espionage equipment called "The Ultimate Spy," espionage historian H. Keith Melton indicated that once the Bulgarian secret police had decided to kill Markhov, KGB specialists from the Laboratory gave the Bulgarians a choice between two KGB tools that could be provided for the task —either a poisonous topical gelatin to be smeared on Markhov, or an instrument to administer a poison pellet, as was eventually done.

- Attempted poisoning of the second President of Afghanistan Hafizullah Amin on December 13, 1979. Department 8 of KGB succeeded in infiltrating the illegal agent Mitalin Talybov (codenamed SABIR) as a chef of Amin's presidential palace. However, Amin switched his food and drink as if he expected to be poisoned, so his son-in-law became seriously ill, and ironically, was flown to a hospital in Moscow.

Alleged Victims

- Russian writer Maksim Gorky and his son. During the Trial of the Twenty One in 1938, NKVD chief Genrikh Yagoda admitted that he poisoned to death Maksim Gorky and his son and unsuccessfully tried to poison future NKVD boss Nikolay Yezhov. The attempted poisoning of Yezhov was later officially dismissed as falsification, but Vyacheslav Molotov believed that the poisoning accusations were true. Yagoda was never officially rehabilitated (recognized as an innocent victim of political repressions) by Soviet authorities.
- Soviet leader Joseph Stalin. Russian historians Anton Antonov-Ovseenko and Edvard Radzinsky believe that Stalin was poisoned by associates of NKVD chief Lavrentiy Beria, based on the interviews of a former Stalin body guard and numerous circumstantial evidence. Stalin planned to dismiss and execute Beria and other senior members of the Soviet government in 1953. According to Radzinsky, Stalin was poisoned by Khrustalev, a senior bodyguard briefly mentioned in memoirs of Svetlana Alliluyeva, Stalin's daughter.
- Journalist Anna Politkovskaya. During the Beslan school hostage crisis in September 2004 and while on her way to Beslan to help in negotiations with the hostage-takers, Politkovskaya fell violently ill and lost consciousness after drinking tea. She survived. According to a report by The Sunday Times, the drug was prepared in the FSB poison facility.

Planned Victims

- President of Socialist Federal Republic of Yugoslavia Josip Broz Tito. In the late 1940s, the laboratory manufactured a powdered plague for use in a small container and where the assassin was vaccinated against plague. The device was to be used against Tito, but MGB agent Iosif Grigulevich, who previously organized the assault on villa of Leon Trotsky and now have received the assignment to kill Tito, was recalled after the death of Stalin.

- The first democratically elected President of the Republic of Georgia, Zviad Gamsakhurdia. According to former Deputy Director of Biopreparat Ken Alibek, this laboratory was possibly involved in design of undetectable chemical or biological agent to assassinate Gamsakhurdia. BBC News reported that some Gamsakhurdia friends believed he committed suicide, "although his widow insists that he was murdered."

Vozrozhdeniya

Vozrozhdeniya, also known as Rebirth Island, was a former island of the Aral Sea or South Aral Sea. Due to the ongoing shrinkage of the Aral, it became first a peninsula in Mid 2001 and finally part of the mainland. Since the disappearance of the Southeast Aral in 2008, Vozrozhdeniya effectively no longer exists as a distinct geographical feature. The area is now shared by Kazakhstan and Uzbekistan.

Located in the central Aral Sea, Vozrozhdeniya Island was one of the main laboratories and testing sites for the Soviet Union's Microbiological Warfare Group. In 1948, a top-secret Soviet bioweapons laboratory was established here, which tested a variety of agents, including anthrax, smallpox, plague, brucellosis, and tularemia.

In the 1990s, word of the island's danger was spread by Soviet defectors, including Ken Alibek, the former head of the Soviet Union's bioweapons program. It was here, according to just released documents, that anthrax spores and bubonic plague bacilli were made into weapons and stored. The main town on the island was Kantubek, which lies in ruins today, but once had approximately 1,500 inhabitants.

The laboratory staff members abandoned the small island in 1992. Many of the containers holding the spores were not properly stored or destroyed, and over the last decade many of these containers have developed leaks.

In 2002, through a project organized by the United States and with Uzbekistan assistance, 10 anthrax burial sites were decontaminated. According to the Kazakh Scientific Centre for Quarantine and Zoonotic Infections, all burial sites of anthrax were decontaminated.

Part of the campaign in the 2010 video game Call of Duty: Black Ops takes place on Vozrozhdeniya Island.

Japan

Special Japanese military units conducted experiments on civilians and POWs in China. One of the most infamous was Unit 731 under Shirô Ishii. Victims were subjected to vivisection without anesthesia, amputations, and were used to test biological weapons, among other experiments. Anesthesia was not used because it was believed to affect results.

To determine the treatment of frostbite, prisoners were taken outside in freezing weather and left with exposed arms, periodically drenched with water until frozen solid. The arm was later amputated; the doctor would repeat the process on the victim's upper arm to the shoulder. After both arms were gone, the doctors moved on to the legs until only a head and torso remained. The victim was then used for plague and pathogens experiments.

According to GlobalSecurity.org, the experiments carried out by Unit 731 alone caused 3,000 deaths. Furthermore, according to the 2002 *International Symposium on the Crimes of Bacteriological Warfare,* the number of people killed by the Imperial Japanese Army germ warfare and human experiments is around 580,000. According to other sources, "tens of thousands, and perhaps as many as 400,000, Chinese died of bubonic plague, cholera, anthrax and other diseases...", resulting from the use of biological warfare.

One case of human experimentation occurred in Japan itself. At least nine out of 12 crew members survived the crash of a U.S. Army Air Forces B-29 bomber on Kyushu, on May 5, 1945. (This plane was Lt. Marvin Watkins' crew of the 29th Bomb Group of the 6th Bomb Squadron.) The bomber's commander was sent to Tokyo for interrogation, while the other survivors were taken to the anatomy department of Kyushu University, at Fukuoka, where they were subjected to vivisection or killed. On March 11, 1948, 30 people including several doctors were brought to trial by the Allied war crimes tribunal. Charges of cannibalism were dropped, but 23 people were found guilty of vivisection or wrongful removal of body parts. Five were sentenced to death, four to life imprisonment, and the rest to shorter terms. In 1950, the military governor of Japan, General Douglas MacArthur, commuted all of the death sentences and significantly reduced most of the prison terms. All of those convicted in relation to the university vivisection were free by 1958. In addition, many participants who were responsible for these vivisections were never charged by the Americans or their allies in exchange for the information on the experiments.

In 2006, former IJN medical officer Akira Makino stated that he was ordered—as part of his training—to carry out vivisection on about 30 civilian prisoners in the Philippines between December 1944 and February 1945. The surgery included amputations. Ken Yuasa, a former military doctor in China, has also admitted to similar incidents in which he was compelled to participate.

- Unit 731
- Zhongma Fortress
- Unit 100
- Unit 2646
- Unit 8604
- Unit Ei 1644.

Iraq

Saddam Hussein initiated an extensive biological weapons (BW) program in Iraq in the early 1980s, in violation of the Biological Weapons Convention (BWC) of 1972. Details of the BW program — along with a chemical weapons program — surfaced only in the wake of the Gulf War (1990–91) following investigations conducted by the United Nations Special Commission (UNSCOM) which had been charged with the post-war disarmament of Saddam's Iraq. Because of this UN disarmament program, more is known today about the once-secret bioweapons program in Iraq than that of any other nation.

The Program

Startup and Foreign Suppliers

In the early 1980s, five German firms supplied equipment to manufacture botulin toxin and mycotoxin to Iraq. Iraq's State Establishment for Pesticide Production (SEPP) also ordered culture media and incubators from Germany's Water Engineering Trading. Strains of dual-use biological material from France also helped advance Iraq's biological warfare program. From the United States, the non-profit American Type Culture Collection and the U.S. Centres for Disease Control sold or sent biological samples to Iraq up until 1989, which Iraq claimed to need for medical research. These materials included anthrax, West Nile virus and botulism, as well as *Brucella melitensis*, and *Clostridium perfringens*.

Some of these materials were used for Iraq's biological weapons research program, while others were used for vaccine development. In delievering these materials "The CDC was abiding by World Health Organization guidelines that encouraged the free exchange of biological samples among medical researchers..." according to Thomas Monath, CDC lab director. It was a request "which we were obligated to fulfill," as described in WHO and UN treaties.

Facilities, Agents and Production

Iraq's BW facilities included its main biowarfare research centre at Salman Pak (just south of Baghdad), the main bioweapons production facility at Al Hakum (the "Single-Cell Protein Production Plant") and the viral biowarfare research site at Al Manal (the "Foot and Mouth Disease Centre").

The Al Hakum facility began mass production of weapons-grade anthrax in 1989, eventually producing 8,000 litres or more (the 8,000 liter figure is based on declared amounts). Iraq officially acknowledged that it had worked with several species of bacterial pathogen, including *Bacillus anthracis*, *Clostridium botulinum* and *Clostridium perfringens* (gas gangrene) and several viruses (including enterovirus 17 (human conjunctivitis), rotavirus and camelpox).

The program also purified biological toxins, such as botulinum toxin, ricin and aflatoxin. After 1995, it was learned that, in all, Iraq had produced 19,000 litres of concentrated botulinum toxin (nearly 10,000 litres filled into munitions), 8,500 litres of concentrated anthrax (6,500 litres filled into munitions) and 2,200 litres of aflatoxin (1,580 litres filled into munitions). In total, the program grew a half million litres of biological agents.

- Al Hakum
- Salman Pak facility
- Al Manal facility.

Treaties Banning or Restricting BW

- Geneva Protocol
- Biological Weapons Convention.

List of People Associated with Biological Weapons

Bioweaponeers:

- Anton Dilger
- Ira Baldwin
- Paul Fildes
- Rihab Rashid Taha
- William C. Patrick III
- Kanatjan Alibekov, known as Ken Alibek
- Vladimir Pasechnik
- Kurt Blome
- Eugen von Haagen
- Kurt Gutzeit
- Erich Traub
- Shiro Ishii.

AIDS Origins Conspiracy Theories

AIDS conspiracy theories are claims or hypotheses about the origins and/or nature of human immunodeficiency virus (HIV) and AIDS that differ radically from the scientific consensus.

These alternative ideas range from suggestions that AIDS was the inadvertent result of experiments in the development of vaccines, to claims that HIV was developed by scientists working for the U.S. government. While a few reputable mainstream scientists once investigated some of these theories as reasonable hypotheses, this is no longer the case, as continuing research has invalidated the alternative ideas. Recent evidence indicates that AIDS originated in Africa in the mid 1930s from the closely related Simian immunodeficiency virus, today found in monkeys and chimpanzees.

Man-made or Iatrogenic Origins of AIDS

- In an interview by Time magazine with Nobel Peace Prize laureate and environmental activist Wangari Maathai, it was alleged that Maathai had said that "AIDS is a biological weapon manufactured by the developed world to wipe out the black race". Maathai subsequently rejected that in a written statement issued in December 2004: "I neither say nor believe that the virus was developed by white people or white powers in order to destroy the African people. Such views are wicked and destructive."

- Jakob Segal, a former biology professor at Humboldt University in then-East Germany, proposed that HIV was engineered at a U.S. military laboratory at Fort Detrick, by splicing together two other viruses, Visna and HTLV-1. According to his theory, the new virus, created between 1977 and 1978, was tested on prison inmates who had volunteered for the experiment in exchange for early release. He further suggested that it was through these prisoners that the virus was spread to the population at large. He has been accused, however, by KGB defector Vasili Mitrokhin as having been disseminating disinformation on behalf of the Soviet Union, and the disease is known today to have existed in humans since at least 1959.
- Alan Cantwell, in self-published books entitled *AIDS and the Doctors of Death: An Inquiry into the Origin of the AIDS Epidemic* and *Queer Blood: The Secret AIDS Genocide Plot,* says that HIV is a genetically modified organism developed by U.S. Government scientists and that it was introduced into the population through Hepatitis B experiments performed on gay and bisexual men between 1978–1981 in major U.S. cities. Cantwell claims that these experiments were directed by Wolf Szmuness, and that there was an ongoing government cover-up of the origins of the AIDS epidemic. Similar theories have been advanced by Robert B. Strecker, Matilde Krim and Milton William Cooper.
- Leonard G. Horowitz, author of the self-published works *Emerging Viruses: AIDS & Ebola. Nature, Accident or Intentional?* and *Death in the Air: Globalism, Terrorism and Toxic Warfare,* advances the theory that the AIDS virus was engineered by such U.S. Government defence contractors as Litton Bionetics for the purposes of bio-warfare and "population control."
- Smallpox vaccine theory, In 1987 there was some consideration given to the possibility that the "Aids epidemic may have been triggered by the mass vaccination campaign which eradicated smallpox". An article in the Times suggested this, quoting an unnamed "adviser to WHO" with "*I believe the smallpox vaccine theory is the explanation to the explosion of Aids*". It is now thought that the smallpox vaccine causes serious complications for people who already have impaired immune systems, and the Times article described the case of a military recruit with "dormant HIV" who died within months of receiving it. But no citation was provided regarding people who did not previously have HIV. (HIV is now considered to be a contraindication for the smallpox vaccine- both for an infected person and their sexual partners and household members.) Some conspiracy theorists propose an expanded hypothesis in which the smallpox vaccine was deliberately 'laced' with HIV.

In contrast, a research article was published in 2010 suggesting that it might have been the actual eradication of smallpox and the subsequent *ending* of the mass vaccination campaign that contributed to the sudden emergence of HIV, due to the possibility that immunization against smallpox "might play a role in providing an individual with some degree of protection to subsequent HIV infection and/or disease progression". Regardless of the effects of the smallpox vaccine itself, its use in practice in Africa is one of the categories of un-sterile injection that may have contributed to the spread and mutation of the immunodeficiency viruses.

Alternative Ideas Regarding Causation, Origin or Treatment

- The Duesberg hypothesis promoted by biologist Peter Duesberg argues that AIDS is not caused by HIV, but rather that HIV is a harmless passenger virus, and that AIDS is caused by non-infectious agents like illegal drug usage. The scientific consensus is that the Duesberg hypothesis has been discredited.
- Thabo Mbeki, former President of South Africa, along with other prominent members of the ruling African National Congress party, has argued that AIDS is the result of poverty, chronic disease, malnutrition and other environmental factors. Mbeki based his views on the discredited beliefs of AIDS denialists, especially Peter Duesberg. It has been suggested that the ANC leadership adopted this position as a political expedient, intended to deflect criticism that the ANC had not done enough to fight AIDS in South Africa. In 2000, two statements by government spokespeople, (one later retracted), placed the financial cost of treating pregnant HIV positive women and the subsequent cost to the state of raising the child as central in the decision of whether to provide anti-retroviral drug treatment. Also in 2000, the Johannesburg *Mail & Guardian* reported that in a leaked text for a speech Mbeki was to give to an ANC caucus, Mbeki claimed that the CIA and Western drug companies were secretly promoting the view that HIV causes AIDS in order to increase sales of anti-AIDS drugs.

Antibiotic Resistance

Antibiotic resistance is a type of drug resistance where a microorganism is able to survive exposure to an antibiotic. Genes can be transferred between bacteria in a horizontal fashion by conjugation, transduction, or transformation. Thus a gene for antibiotic resistance which had evolved via natural selection may be shared. Evolutionary stress such as exposure to antibiotics then selects for the antibiotic resistant trait. Many antibiotic resistance genes reside on plasmids, facilitating their transfer. If a bacterium carries several resistance genes, it is called multiresistant or, informally, a superbug or super bacterium. The primary cause of antibiotic resistance is genetic mutation in bacteria. The prevalence of antibiotic resistant bacteria is a result of antibiotic use both within medicine and veterinary medicine. The greater the duration of exposure the greater the risk of the development of resistance irrespective of the severity of the need for antibiotics. As resistance becomes more common there becomes a greater need for alternative treatments. However despite a push for new antibiotic therapies there has been a continued decline in the number of newly approved drugs. Antibiotic resistance therefore poses a significant problem.

Causes

The widespread use of antibiotics both inside and outside of medicine is playing a significant role in the emergence of resistant bacteria. Antibiotics are often used in rearing animals for food and this use among others leads to the creation of resistant strains of bacteria. In some countries antibiotics are sold over the counter without a prescription which also leads to the creation of resistant strains. In supposedly well-regulated human medicine the major problem of the emergence of resistant bacteria is due to misuse and overuse of antibiotics by doctors as well as patients. Other practices contributing towards resistance include the addition of antibiotics to the feed of livestock. Household use of antibacterials in soaps and other products, although not clearly contributing to resistance, is also discouraged (as not being effective at infection control). Also unsound practices in the pharmaceutical manufacturing industry can contribute towards the likelihood of creating antibiotic resistant strains.

Certain antibiotic classes are highly associated with colonisation with superbugs compared to other antibiotic classes. The risk for colonisation increases if there is a lack of sensitivity (resistance) of the superbugs to the antibiotic used and high tissue penetration as well as broad spectrum activity against "good bacteria". In the case of MRSA, increased rates of MRSA infections are seen with glycopeptides, cephalosporins and especially quinolones. In the case of colonisation with C difficile the high risk antibiotics include cephalosporins and in particular quinolones and clindamycin.

In medicine

The volume of antibiotic prescribed is the major factor in increasing rates of bacterial resistance rather than compliance with antibiotics. A single dose of antibiotics leads to a greater risk of resistant organisms to that antibiotic in the person for up to a year.

Inappropriate prescribing of antibiotics has been attributed to a number of causes including: people who insist on antibiotics, physicians simply prescribe them as they feel they do not have time to explain why they are not necessary, physicians who do not know when to prescribe antibiotics or else are overly cautious for medical legal reasons. A third of people for example believe that antibiotics are effective for the common cold and 22% of people do not finish a course of antibiotics primarily due to that fact that they feel better (varying from 10% to 44% depending on the country). Compliance with once daily antibiotics is better than with twice daily antibiotics. Sub optimum antibiotic concentrations in critically ill people increase the frequency of antibiotic resistance organisms. While taking antibiotics doses less than those recommended may increase rates of resistance, shortening the course of antibiotics may actually decrease rates of resistance. Poor hand hygiene by hospital staff has been associated with the spread of resistant organisms and an increase in hand washing compliance results in decreased rates of these organisms.

Role of other Animals

Drugs are used in animals that are used as human food, such as cows, pigs, chickens, fish, etc., and these drugs can affect the safety of the meat, milk, and eggs produced from those animals and can be the source of superbugs. For example, farm animals, particularly pigs, are believed to be able to infect people with MRSA. The resistant bacteria in animals due to antibiotic exposure can be transmitted to humans via three pathways, those being through the consumption of meat, from close or direct contact with animals, or through the environment.

The World Health Organization concluded that antibiotics as growth promoters in animal feeds should be prohibited (in the absence of risk assessments). In 1998, European Union health ministers voted to ban four antibiotics widely used to promote animal growth (despite their scientific panel's recommendations). Regulation banning the use of antibiotics in European feed, with the exception of two antibiotics in poultry feeds, became effective in 2006. In Scandinavia, there is evidence that the ban has led to a lower prevalence of antimicrobial resistance in (non-hazardous) animal bacterial populations. In the USA federal agencies do not collect data on antibiotic use in animals but animal to human spread of drug resistant organisms has been demonstrated in research studies. Antibiotics are still used in U.S. animal feed—along with other ingredients which have safety concerns.

Growing U.S. consumer concern about using antibiotics in animal feed has led to a niche market of "antibiotic-free" animal products, but this small market is unlikely to change entrenched industry-wide practices.

In 2001, the Union of Concerned Scientists estimated that greater than 70% of the antibiotics used in the US are given to food animals (e.g. chickens, pigs and cattle) in the absence of disease. In 2000 the US Food and Drug Administration (FDA) announced their intention to revoke approval of fluoroquinolone use in poultry production because of substantial evidence linking it to the emergence of fluoroquinolone resistant campylobacter infections in humans. The final decision to ban fluoroquinolones from use in poultry production was not made until five years later because of challenges from the food animal and pharmaceutical industries. Today, there are two federal bills (S. 549 and H.R. 962) aimed at phasing out "non-therapeutic" antibiotics in US food animal production.

Mechanisms

Schematic representation of how antibiotic resistance evolves via natural selection. The top section represents a population of bacteria before exposure to an antibiotic. The middle section shows the population directly after exposure, the phase in which selection took place. The last section shows the distribution of resistance in a new generation of bacteria. The legend indicates the resistance levels of individuals.

Antibiotic resistance can be a result of horizontal gene transfer, and also of unlinked point mutations in the pathogen genome and a rate of about 1 in 10^8 per chromosomal replication. The antibiotic action against the pathogen can be seen as an environmental pressure; those bacteria which have a mutation allowing them to survive will live on to reproduce. They will then pass this trait to their offspring, which will result in a fully resistant colony.

The four main mechanisms by which microorganisms exhibit resistance to antimicrobials are:

1. Drug inactivation or modification: e.g. enzymatic deactivation of *Penicillin* G in some penicillin-resistant bacteria through the production of β-lactamases.
2. Alteration of target site: e.g. alteration of PBP—the binding target site of penicillins—in MRSA and other penicillin-resistant bacteria.
3. Alteration of metabolic pathway: e.g. some sulfonamide-resistant bacteria do not require para-aminobenzoic acid (PABA), an important precursor for the synthesis of folic acid and nucleic acids in bacteria inhibited by sulfonamides. Instead, like mammalian cells, they turn to utilizing preformed folic acid.
4. Reduced drug accumulation: by decreasing drug permeability and/or increasing active efflux (pumping out) of the drugs across the cell surface.

There are three known mechanisms of fluoroquinolone resistance. Some types of efflux pumps can act to decrease intracellular quinolone concentration. In gram-negative bacteria, plasmid-mediated resistance genes produce proteins that can bind to DNA gyrase, protecting it from the action of quinolones. Finally, mutations at key sites in DNA gyrase or Topoisomerase IV can decrease their binding affinity to quinolones, decreasing the drug's effectiveness. Research has shown that the bacterial protein LexA may play a key role in the acquisition of bacterial mutations giving resistance to quinolones and rifampicin.

Antibiotic resistance can also be introduced artificially into a microorganism through laboratory protocols, sometimes used as a selectable marker to examine the mechanisms of gene transfer or to identify individuals that absorbed a piece of DNA that included the resistance gene and another gene of interest.

Resistant Pathogens

Staphylococcus Aureus

Staphylococcus aureus (colloquially known as "Staph aureus" or a *Staph infection*) is one of the major resistant pathogens. Found on the mucous membranes and the human skin of around a third of the population, it is extremely adaptable to antibiotic pressure. It was one of the earlier bacteria in which penicillin resistance was found—in 1947, just four years after the drug started being mass-produced. Methicillin was then the antibiotic of choice, but has since been replaced by oxacillin due to significant kidney toxicity. MRSA (methicillin-resistant *Staphylococcus aureus*) was first detected in Britain in 1961 and is now "quite common" in hospitals. MRSA was responsible for 37% of fatal cases of sepsis in the UK in 1999, up from 4% in 1991. Half of all *S. aureus* infections in the US are resistant to penicillin, methicillin, tetracycline and erythromycin.

This left vancomycin as the only effective agent available at the time. However, strains with intermediate (4-8 μg/ml) levels of resistance, termed GISA (glycopeptide intermediate *Staphylococcus aureus*) or VISA (vancomycin intermediate *Staphylococcus aureus*), began appearing in the late 1990s. The first identified case was in Japan in 1996, and strains have since been found in hospitals in England, France and the

US. The first documented strain with complete (>16 μg/ml) resistance to vancomycin, termed VRSA (Vancomycin-resistant *Staphylococcus aureus*) appeared in the United States in 2002.

A *new* class of antibiotics, oxazolidinones, became available in the 1990s, and the first commercially available oxazolidinone, linezolid, is comparable to vancomycin in effectiveness against MRSA. Linezolid-resistance in *Staphylococcus aureus* was reported in 2003.

CA-MRSA (Community-acquired MRSA) has now emerged as an epidemic that is responsible for rapidly progressive, fatal diseases including necrotizing pneumonia, severe sepsis and necrotizing fasciitis. Methicillin-resistant *Staphylococcus aureus* (MRSA) is the most frequently identified antimicrobial drug-resistant pathogen in US hospitals. The epidemiology of infections caused by MRSA is rapidly changing. In the past 10 years, infections caused by this organism have emerged in the community. The 2 MRSA clones in the United States most closely associated with community outbreaks, USA400 (MW2 strain, ST1 lineage) and USA300, often contain Panton-Valentine leukocidin (PVL) genes and, more frequently, have been associated with skin and soft tissue infections. Outbreaks of community-associated (CA)-MRSA infections have been reported in correctional facilities, among athletic teams, among military recruits, in newborn nurseries, and among men who have sex with men. CA-MRSA infections now appear to be endemic in many urban regions and cause most CA-S. aureus infections.

Streptococcus and Enterococcus

Streptococcus pyogenes (Group A Streptococcus: GAS) infections can usually be treated with many different antibiotics. Early treatment may reduce the risk of death from invasive group A streptococcal disease. However, even the best medical care does not prevent death in every case. For those with very severe illness, supportive care in an intensive care unit may be needed. For persons with necrotizing fasciitis, surgery often is needed to remove damaged tissue. Strains of *S. pyogenes* resistant to macrolide antibiotics have emerged, however all strains remain uniformly sensitive to penicillin.

Resistance of *Streptococcus pneumoniae* to penicillin and other beta-lactams is increasing worldwide. The major mechanism of resistance involves the introduction of mutations in genes encoding penicillin-binding proteins. Selective pressure is thought to play an important role, and use of beta-lactam antibiotics has been implicated as a risk factor for infection and colonization. Streptococcus pneumoniae is responsible for pneumonia, bacteremia, otitis media, meningitis, sinusitis, peritonitis and arthritis.

Penicillin-resistant pneumonia caused by *Streptococcus pneumoniae* (commonly known as *pneumococcus*), was first detected in 1967, as was penicillin-resistant gonorrhea. Resistance to penicillin substitutes is also known as beyond *S. aureus*. By 1993 *Escherichia coli* was resistant to five fluoroquinolone variants. *Mycobacterium tuberculosis* is commonly resistant to isoniazid and rifampin and sometimes universally resistant to the common treatments. Other pathogens showing some resistance include *Salmonella, Campylobacter,* and *Streptococci.*

Enterococcus faecium is another superbug found in hospitals. Penicillin-Resistant Enterococcus was seen in 1983, vancomycin-resistant enterococcus (VRE) in 1987, and Linezolid-Resistant Enterococcus (LRE) in the late 1990s.

Pseudomonas Aeruginosa

Pseudomonas aeruginosa is a highly prevalent opportunistic pathogen. One of the most worrisome characteristics of *P. aeruginosa* consists in its low antibiotic susceptibility. This low susceptibility is attributable to a concerted action of multidrug efflux pumps with chromosomally-encoded antibiotic resistance genes and the low permeability of the bacterial cellular envelopes. Besides intrinsic resistance, *P. aeruginosa* easily develop acquired resistance either by mutation in chromosomally-encoded genes, or by the horizontal gene transfer of antibiotic resistance determinants. Development of multidrug resistance by *P. aeruginosa* isolates requires several different genetic events that include acquisition of different mutations and/or horizontal transfer of antibiotic resistance genes. Hypermutation favours the selection of mutation-driven antibiotic resistance in *P. aeruginosa* strains producing chronic infections, whereas the clustering of several different antibiotic resistance genes in integrons favours the concerted acquisition of antibiotic resistance determinants. Some recent studies have shown that phenotypic resistance associated to biofilm formation or to the emergence of small-colony-variants may be important in the response of *P. aeruginosa* populations to antibiotics treatment.

Clostridium Difficile

Clostridium difficile is a nosocomial pathogen that causes diarrheal disease in hospitals world wide. Clindamycin-resistant *C. difficile* was reported as the causative agent of large outbreaks of diarrheal disease in hospitals in New York, Arizona, Florida and Massachusetts between 1989 and 1992. Geographically dispersed outbreaks of *C. difficile* strains resistant to fluoroquinolone antibiotics, such as Cipro (ciprofloxacin) and Levaquin (levofloxacin), were also reported in North America in 2005.

Salmonella and E. Coli

Escherichia coli and *Salmonella* come directly from contaminated food. Of the meat that is contaminated with *E. coli*, eighty percent of the bacteria are resistant to one or more drugs made; it causes bladder infections that are resistant to antibiotics ("HSUS Fact Sheet"). *Salmonella* was first found in humans in the 1970s and in some cases is resistant to as many as nine different antibiotics ("HSUS Fact Sheet"). When both bacterium are spread, serious health conditions arise. Many people are hospitalized each year after becoming infected, and some die as a result.

Acinetobacter Baumannii

On November 5, 2004, the Centres for Disease Control and Prevention (CDC) reported an increasing number of *Acinetobacter baumannii* bloodstream infections in patients at military medical facilities in which service members injured in the Iraq/Kuwait region during Operation Iraqi Freedom and in Afghanistan during Operation Enduring Freedom were treated. Most of these showed multidrug resistance (MRAB), with a few isolates resistant to all drugs tested.

Alternatives

Prevention

Rational use of antibiotics may reduce the chances of development of opportunistic infection by antibiotic-resistant bacteria due to dysbacteriosis. In one study the use of fluoroquinolones are clearly associated with *Clostridium difficile* infection, which is a leading cause of nosocomial diarrhea in the United States, and a major cause of death, worldwide.

There is clinical evidence that topical dermatological preparations such as those containing tea tree oil and thyme oil may be effective in preventing transmittal of CA-MRSA. In addition, other phytotherapeutic medicines too can reduce the use of antibiotics or eliminate their use entirely.

Vaccines do not suffer the problem of resistance because a vaccine enhances the body's natural defences, while an antibiotic operates separately from the body's normal defences. Nevertheless, new strains may evolve that escape immunity induced by vaccines; for example an update Influenza vaccine is needed each year.

While theoretically promising, anti-staphylococcal vaccines have shown limited efficacy, because of immunological variation between *Staphylococcus* species, and the limited duration of effectiveness of the antibodies produced. Development and testing of more effective vaccines is under way.

The Australian Commonwealth Scientific and Industrial Research Organization (CSIRO), realizing the need for the reduction of antibiotic use, has been working on two alternatives. One alternative is to prevent diseases by adding cytokines instead of antibiotics to animal feed. These proteins are made in the animal body "naturally" after a disease and are not antibiotics so they do not contribute to the antibiotic resistance problem. Furthermore, studies on using cytokines have shown that they also enhance the growth of animals like the antibiotics now used, but without the drawbacks of non-therapeutic antibiotic use. Cytokines have the potential to achieve the animal growth rates traditionally sought by the use of antibiotics without the contribution of antibiotic resistance associated with the widespread non-therapeutic uses of antibiotics currently utilized in the food animal production industries. Additionally, CSIRO is working on vaccines for diseases.

Phage Therapy

Phage therapy, an approach that has been extensively researched and utilized as a therapeutic agent for over 60 years, especially in the Soviet Union, is an alternative that might help with the problem of resistance. Phage therapy was widely used in the United States until the discovery of antibiotics, in the early 1940s. Bacteriophages or "phages" are viruses that invade bacterial cells and, in the case of lytic phages, disrupt bacterial metabolism and cause the bacterium to lyse. Phage therapy is the therapeutic use of lytic bacteriophages to treat pathogenic bacterial infections.

Bacteriophage therapy is an important alternative to antibiotics in the current era of multidrug resistant pathogens. A review of studies that dealt with the therapeutic use of phages from 1966–1996 and few latest ongoing phage therapy projects via internet showed: phages were used topically, orally or systemically in Polish and Soviet studies. The success rate found in these studies was 80–95% with few gastrointestinal or allergic side effects. British studies also demonstrated significant efficacy of phages against *Escherichia coli, Acinetobacter* spp., *Pseudomonas* spp and *Staphylococcus aureus*. US studies dealt with improving the bioavailability of phage. Phage therapy may prove as an important alternative to antibiotics for treating multidrug resistant pathogens.

Research

New Medications

Until recently, research and development (R&D) efforts have provided new drugs in time to treat bacteria that became resistant to older antibiotics. That is no longer the case. The potential crisis at hand is the result of a marked decrease in industry R&D, and the increasing prevalence of resistant bacteria. Infectious disease physicians are alarmed by the prospect that effective antibiotics may not be available to treat seriously ill patients in the near future.

The pipeline of new antibiotics is drying up.Major pharmaceutical companies are losing interest in the antibiotics market because these drugs may not be as profitable as drugs that treat chronic (long-term) conditions and lifestyle issues.

The resistance problem demands that a renewed effort be made to seek antibacterial agents effective against pathogenic bacteria resistant to current antibiotics. One of the possible strategies towards this objective is the rational localization of bioactive phytochemicals. Plants have an almost limitless ability to synthesize aromatic substances, most of which are phenols or their oxygen-substituted derivatives such as tannins. Most are secondary metabolites, of which at least 12,000 have been isolated, a number estimated to be less than 10% of the total. In many cases, these substances serve as plant defence mechanisms against predation by microorganisms, insects, and herbivores. Many of the herbs and spices used by humans to season food yield useful medicinal compounds including those having antibacterial activity.

Traditional healers have long used plants to prevent or cure infectious conditions. Many of these plants have been investigated scientifically for antimicrobial activity and a large number of plant products have been shown to inhibit growth of pathogenic bacteria.A number of these agents appear to have structures and modes of action that are distinct from those of the antibiotics in current use, suggesting that cross-resistance with agents already in use may be minimal. For example the combination of 5'-methoxyhydnocarpine and berberine in herbs like Hydrastis canadensis and Berberis vulgaris can block the MDR-pumps that cause multidrug resistance. This has been shown for Staphylococcus aureus. Archaeocins is the name given to a new class of potentially useful antibiotics that are derived from the Archaea group of organisms. Eight archaeocins have been partially or fully characterized, but hundreds of archaeocins are believed to exist, especially within the haloarchaea. The prevalence of archaeocins is unknown simply because no one has looked for them. The discovery of new archaeocins hinges on recovery and cultivation of archaeal organisms from the environment. For example, samples from a novel hypersaline field site, Wilson Hot Springs, recovered 350 halophilic organisms; preliminary analysis of 75 isolates showed that 48 were archaeal and 27 were bacterial.

In research published on October 17, 2008 in *Cell*, a team of scientists pinpointed the place on bacteria where the antibiotic myxopyronin launches its attack, and why that attack is successful. The myxopyronin binds to and inhibits the crucial bacterial enzyme, RNA polymerase. The myxopyronin changes the structure of the switch-2 segment of the enzyme, inhibiting its function of reading and transmitting DNA code. This prevents RNA polymerase from delivering genetic information to the ribosomes, causing the bacteria to die. One of the major causes of antibiotic resistance is the decrease of effective drug concentrations at their target place, due to the increased action of ABC transporters. Since ABC transporter blockers can be used in combination with current drugs to increase their effective intracellular concentration, the possible impact of ABC transporter inhibitors is of great clinical interest. ABC transporter blockers that may be useful to increase the efficacy of current drugs have entered clinical trials and are available to be used in therapeutic regimes.

Applications

Antibiotic resistance is an important tool for genetic engineering. By constructing a plasmid which contains an antibiotic resistance gene as well as the gene being engineered or expressed, a researcher can ensure that when bacteria replicate, only the copies which carry along the plasmid survive. This ensures that the gene being manipulated passes along when the bacteria replicates. The most commonly used antibiotics in genetic engineering are generally "older" antibiotics which have largely fallen out of use in clinical practice. These include:

- ampicillin
- kanamycin
- tetracycline
- chloramphenicol.

Industrially the use of antibiotic resistance is disfavored since maintaining bacterial cultures would require feeding them large quantities of antibiotics. Instead, the use of auxotrophic bacterial strains (and function-replacement plasmids) is preferred.

Biological Hazard

Biological hazards, also known as biohazards, refer to biological substances that pose a threat to the health of living organisms, primarily that of humans. This can include medical waste or samples of a microorganism, virus or toxin (from a biological source) that can impact human health. It can also include substances harmful to animals. The term and its associated symbol is generally used as a warning, so that those potentially exposed to the substances will know to take precautions.

In Unicode, the biohazard sign is U+2623.

Biohazardous agents are classified for transportation by UN number:

- UN 2814 (Infectious substance to Humans)
- UN 2900 (Infectious substance to Animals)
- UN 3291 (Medical Waste)
- Category A, UN 2814-Infectious substances affecting humans and animals: An infectious substance in a form capable of causing permanent disability or life-threatening or fatal disease in otherwise healthy humans or animals when exposure to it occurs.
- Category B, UN 2900-Infectious substances affecting animals only: An infectious substance that is not in a form generally capable of causing permanent disability of life-threatening or fatal disease in otherwise healthy humans and animals when exposure to themselves occurs.
- Category B, UN 3373-Biological substance transported for diagnostic or investigative purposes.
- Regulated Medical Waste, UN 3291-Waste or reusable material derived from medical treatment of an animal or human, or from biomedical research, which includes the production and testing of biological products.

Levels of Biohazard

The United States' Centres for Disease Control and Prevention (CDC) categorizes various diseases in levels of biohazard, Level 1 being minimum risk and Level 4 being extreme risk. Laboratories and other facilities are categorized as BSL (Biosafety Level) 1-4 or as *P1* through *P4* for short (Pathogen or Protection Level).

- Biohazard Level 1: Bacteria and viruses including *Bacillus subtilis*, canine hepatitis, *Escherichia coli*, varicella (chicken pox), as well as some cell cultures and non-infectious bacteria. At this level precautions against the biohazardous materials in question are minimal, most likely involving gloves and some sort of facial protection. Usually, contaminated materials are left in open (but separately indicated) waste receptacles. Decontamination procedures for this level are similar in most respects to modern precautions against everyday viruses (i.e.: washing one's hands with anti-bacterial soap, washing all exposed surfaces of the lab with disinfectants, etc). In a lab environment, all materials used for cell and/or bacteria cultures are decontaminated via autoclave.

- Biohazard Level 2: Bacteria and viruses that cause only mild disease to humans, or are difficult to contract via aerosol in a lab setting, such as hepatitis A, B, and C, influenza A,Lyme disease, salmonella, mumps, measles, scrapie, dengue fever, and HIV. "Routine diagnostic work with clinical specimens can be done safely at Biosafety Level 2, using Biosafety Level 2 practices and procedures. Research work (including co-cultivation, virus replication studies, or manipulations involving concentrated virus) can be done in a BSL-2 (P2) facility, using BSL-3 practices and procedures. Virus production activities, including virus concentrations, require a BSL-3 (P3) facility and use of BSL-3 practices and procedures".
- Biohazard Level 3: Bacteria and viruses that can cause severe to fatal disease in humans, but for which vaccines or other treatments exist, such as anthrax, West Nile virus, Venezuelan equine encephalitis, SARS virus, variola virus (smallpox), tuberculosis, typhus, Rift Valley fever, Rocky Mountain spotted fever, yellow fever, and malaria. Among parasites *Plasmodium falciparum*, which causes Malaria, and *Trypanosoma cruzi*, which causes trypanosomiasis, also come under this level.
- Biohazard Level 4: Viruses and bacteria that cause severe to fatal disease in humans, and for which vaccines or other treatments are *not* available, such as Bolivian and Argentine hemorrhagic fevers, (Argentine Hemorragic does have a vaccine "The Candid #1 vaccine for AHF was created in 1985 by Argentine virologist Dr. Julio Barrera Oro. The vaccine was manufactured by the Salk Institute in the United States, and became available in Argentina since 1990.

Candid #1 has been applied to adult high-risk population and is 95.5% effective. On 29 August 2006, the Maiztegui Institute obtained certification for the production of the vaccine in Argentina. A vaccination plan is yet to be outlined, but the budget for 2007 allows for 390,000 doses, at AR$8 each (about US$2.6 or €2). The Institute has the capacity to manufacture, in one year, the 5 million doses required to vaccinate the entire population of the endemic area." Argentine_hemorrhagic_fever#Vaccine H5N1(bird flu), Dengue hemorrhagic fever, Marburg virus, Ebola virus, hantaviruses, Lassa fever, Crimean-Congo hemorrhagic fever, and other hemorrhagic diseases. When dealing with biological hazards at this level the use of a Hazmat suit and a self-contained oxygen supply is mandatory. The entrance and exit of a Level Four biolab will contain multiple showers, a vacuum room, an ultraviolet light room, autonomous detection system, and other safety precautions designed to destroy all traces of the biohazard. Multiple airlocks are employed and are electronically secured to prevent both doors opening at the same time. All air and water service going to and coming from a Biosafety Level 4 (P4) lab will undergo similar decontamination procedures to eliminate the possibility of an accidental release.

Biosecurity

Biosecurity is a set of preventive measures designed to reduce the risk of transmission of infectious diseases, quarantined pests, invasive alien species, living modified organisms. While biosecurity does encompass the prevention of the intentional removal (theft) of biological materials from research laboratories, this definition is narrower in scope than the definition used by many experts, including the United Nations Food and Agriculture Organization. These preventative measures are a combination of systems and practices put into its place at legitimate bioscience laboratories to prevent the use of dangerous pathogens and toxins for malicious use, as well as by customs agents and agricultural and natural resource managers to prevent the spread of these biological agents in natural and managed. Reference no. 123 ecosystems. Although security is usually thought of in terms of "Guards, Gates, and Guns", biosecurity encompasses much more than that and requires the cooperation of scientists, technicians, policy makers, security engineers, and law enforcement officials.

Components of a laboratory biosecurity program include:

- Physical security
- Personnel security
- Material control & accountability
- Transport security
- Information security
- Program management.

Animal Biosecurity

Animal biosecurity is the product of all actions undertaken by an entity to prevent introduction of disease agents into a specific area. Animal biosecurity differs from biosecurity which are measures taken to reduce the risk of infectious agent theft and dispersal by means of bioterrorism. Animal biosecurity is a comprehensive approach, encompassing different means of prevention and containment. A critical element in animal biosecurity, biocontainment, is the control of disease agents already present in a particular area, and works to prevent novel transmissions. Animal biosecurity may protect organisms from infectious agents or noninfectious agents such as toxins or pollutants, and can be executed in areas as large as a nation or as small as a local farm.

Animal biosecurity takes into account the epidemiological triad for disease occurrence: the individual host, the disease, and the environment in contributing to disease susceptibility. It aims to improve nonspecific immunity of the host to resist the introduction of an agent, or limit the risk that an agent will be sustained in an environment at adequate levels. Biocontainment, an element of animal biosecurity, works to improve specific immunity towards already present pathogens.

Biosecurity means the prevention of the illicit use of pathogenic bioorganisms by laboratory staff or others. Biosafety means the protection of laboratory staff from being infected by pathogenic bioorganisms.

Challenges

The destruction of the World Trade Centre in Manhattan on September 11, 2001 by terrorists, and subsequent wave of anthrax attacks on U.S. media and government outlets (both real and hoax), led to increased attention on the risk of bioterror attacks in the United States. Proposals for serious structural reforms, national and/or regional border controls, and a single co-ordinated system of biohazard response abounded.

One of the major challenges in biosecurity is the increasing availability and accessibility of potentially harmful technology. Biomedical advances and the globalization of scientific and technical expertise have made it possible to greatly improve public health. However, there is also the risk that advances can lead to make biological weapons. The proliferation of high biosafety level laboratories around the world has many experts worried about availability of targets for those that might interested in stealing dangerous pathogens. Emerging and Re-emerging disease is also a serious biosecurity concern. The recent growth in containment laboratories is often in response to emerging diseases, many new containment lab's main focus is to find ways to control these diseases. By strengthening national disease surveillance, prevention, control and response systems, these labs are raising international public health to new heights.

UNU/IAS Research into Biosecurity & Biosafety emphasizes "long-term consequences of the development and use of biotechnology" and need for "an honest broker to create avenues and forums to unlock the impasses."

Biosecurity Incidents

- 1984 Rajneeshee religious cult attacks, The Dalles, Oregon
- Objective: Gain control of the Wasco County Court by affecting the election
- Organism: Salmonella typhimurium, purchased from commercial supplier
- Dissemination: Restaurant salad bars
- 751 illnesses, Early investigation by CDC suggested the event was a naturally occurring outbreak. Cult member arrested on unrelated charge confessed involvement with the event.
- 1990s Aum Shinrikyo attempts in Tokyo, Japan:
- Objective: Fulfill apocalyptic prophecy
- Organisms: Bacillus anthracis (Vaccine strain), Clostridium botulinum (Environmental isolate, Avirulent strain) Aum Shinrikyo ordered Clostridium botulinum from a pharmaceutical company, and Ebola virus (Attempted to acquire from Zaire outbreak under guise of an "Humanitarian mission")
- Dissemination: Aerosolization in Tokyo (B. anthracis and Botulinum toxin)
- Leader Asahara was convicted of criminal activity.
    - o 2001 Anthrax attacks in the US
    - o 1995—Larry Wayne Harris, a white supremacist, ordered 3 vials of Yersinia pestis from the ATCC

- o 1995—Laboratory technician Diane Thompson removed Shigella dysenteriae Type 2 from hospital's collection and infected co-workers
- o Professor Thomas Butler, United States, 2003.
  - * 30 vials of Yersinia pestis missing from lab (never recovered); Butler served 19 months in jail.
  - * Dr. Mario Jascalevich, New Jersey doctor, accused of poisoning 5 patients with this plant-derived toxin-Tubocurarine: 1966
  - * Arnfinn Nesset, nursing home operator in Norway, killed 27 residents at a nursing home with curacit: May 1977 – November 1980
  - * Dr. David Acer, Florida dentist, infects 6 patients with HIV, unclear if this was a deliberate act: 1987–1990
  - * Dr. Ray W. Mettetal, Jr., a neurologist in Virginia, was found in possession of ricin after arrest on another issue: 1995
  - * Debora Green, a physician, convicted of trying to murder her estranged husband with ricin
  - * Richard Schmidt, a gastroenterologist in Louisiana, convicted of attempted second degree murder for infecting nurse Janice Allen with HIV by injecting her with blood from an AIDS patient: 1998
  - * Brian T. Stewart, a phlebotomist, sentenced to life in prison for deliberately infecting his 11-month-old baby with HIV-infected blood to avoid child support payments: 1999
  - * Physician reports theft of a vial of Mycobacterium tuberculosis: June 1999
  - * Japan 1964-1966, Dr. Mitsuru Suzuki was a physician with training in bacteriology
  - * Objective: Revenge due to deep antagonism to what he perceived as a prevailing seniority system
  - * Organisms: Shigella dysenteriae and Salmonella typhi
  - * Dissemination: Sponge cake, other food sources
  - * He was later implicated in 200 – 400 illnesses and 4 deaths
  - * Official investigation started after anonymous tip to Ministry of Health and Welfare. He was charged, but was not convicted of any deaths
- Hospital in Dallas, TX, 1996, Diane Thompson was a clinical laboratory technician
  - o Objective: Unclear, possibly revenge against former boyfriend and cover-up by infecting co-workers
  - o Organism: Shigella dysenteriae Type 2, acquired from clinical laboratory

- o Dissemination: Contaminated pastries in the office breakroom
- o Infected 12 of her coworkers, She was arrested, convicted, 20 year sentence.

The Role of Education in Biosecurity

The advance of the life sciences and biotechnology has the potential to bring great benefits to humankind through responding to societal challenges. However, it is also possible that such advances could be exploited for hostile purposes, something evidenced in a small number of incidents of bioterrorism, but more particularly by the series of large-scale offensive biological warfare programmes carried out by major states in the last century.

Dealing with this challenge, which has been labelled the 'dual-use' dilemma requires a number of different activities such as those identified above as being require for biosecurity. However, one of the essential ingredients in ensuring that the life sciences continue to generate great benefits and do not become subject to misuse for hostile purposes is a process of engagement between scientists and the security community and the development of strong ethical and normative frameworks to compliment legal and regulatory measures that are being developed by states.

Biosecurity Regulations

- US Select Agent Regulations:
  - o Facility registration if it possesses one of 81 Select Agents
  - o Facility must designate a Responsible Official
  - o Background checks for individuals with access to Select Agents
  - o Access controls for areas and containers that contain Select Agents
  - o Detailed inventory requirements for Select Agents
  - o Security, safety, and emergency response plans
  - o Safety and security training
  - o Regulation of transfers of Select Agents
  - o Extensive documentation and recordkeeping
  - o Safety and security inspections.
- Biological Weapons Convention addresses three relevant issues:
  - o National Implementing Legislation
  - o National Pathogen Security (biosecurity)
  - o International Cooperation
  - o States Parties agree to pursue national implementation of laboratory and transportation biosecurity (2003)
- UN 1540:
  - o Urges States to take preventative measures to mitigate the threat of WMD proliferation by non-state actors

  - o "Take and enforce effective measures to establish domestic controls to prevent the proliferation of... biological weapons...; including by establishing appropriate controls over related materials"
  - o European Commission Green Paper on Bio-Preparedness (November 2007)
  - o recommends developing European standards on laboratory biosecurity including Physical protection, access controls, accountability of pathogens, and registration of researchers.
- Organization for Economic Cooperation and Development:
  - o published "Best Practice Guidelines for Biological Resource Centres" including a section on biosecurity in February 2007
- Kampala Compact (October 2005) and the Nairobi Announcement (July 2007)
  - o Stress importance of implementing laboratory biosafety and biosecurity in Africa.

Agents of Concern

The following agents are deemed a biosecurity concern by the US Government through the US Select Agent List. The list is divided between agents that can infect only humans, zoonotic agents (which can infect both humans and animals), agents that can only infect animals, and agents that can infect only plants.

Human Agents

- Abrin
- Cercopithecine herpesvirus 1 (Herpes B Virus)
- Coccidioides posadasii
- Conotoxins
- Crimean-Congo haemorrhagic fever virus
- Diacetoxyscirpenol
- Ebola virus
- Lassa fever virus
- Marburg virus
- Monkeypox virus
- Reconstructed replication competent forms of the 1918 pandemic influenza virus containing any portion of the coding regions of all eight gene segments (Reconstructed 1918 Influenza virus)
- Ricin
- Rickettsia prowazekii
- Rickettsia rickettsii
- Saxitoxin
- Shiga-like ribosome inactivating proteins

- South American Haemorrhagic Fever viruses;
  - o Flexal Virus
  - o Guanarito virus
  - o Junin virus
  - o Machupo
  - o Sabia
- Tetrodotoxin
- Tick-borne encephalitis complex (flavi) viruses
  - o Central European Tick-borne encephalitis
  - o Far Eastern Tick-borne encephalitis
  - o Kyasanur Forest disease
  - o Omsk hemorrhagic fever
  - o Russian spring-summer encephalitis
  - o Variola major virus (Smallpox virus) and
  - o Variola minor virus (Alastrim)
  - o Yersinia pestis.

Zoonotic Agents

(Overlap Select Agents And Toxins):

- Bacillus anthracis
- Botulinum neurotoxin
- Botulinum neurotoxin producing species of Clostridium
- Brucella abortus
- Brucella melitensis
- Brucella suis
- Burkholderia mallei (formerly Pseudomonas mallei)
- Burkholderia pseudomallei (formerly Pseudomonas pseudomallei)
- Clostridium perfringens epsilon toxin
- Coccidioides immitis
- Coxiella burnetii
- Eastern equine encephalitis virus
- Francisella tularensis
- Hendra virus
- Nipah virus
- Rift Valley fever virus
- Shiga toxin

- Staphylococcal enterotoxins
- T-2 toxin
- Venezuelan equine encephalitis virus.

Animal agents

(Usda select agents and Toxins):

- African horse sickness virus
- African swine fever virus
- Akabane virus
- Avian influenza virus (highly pathogenic)
- Bluetongue virus (Exotic)
- Bovine spongiform encephalopathy
- Camel pox virus
- Classical swine fever virus
- Cowdria ruminantium (Heartwater)
- Foot-and-mouth disease virus
- Goat pox virus
- Japanese encephalitis virus
- Lumpy skin disease virus
- Malignant catarrhal fever virus (Alcelaphine herpesvirus type 1)
- Menangle virus
- Mycoplasma capricolum/M.F38/M. mycoides Capri (contagious caprine pleuropneumonia)
- Mycoplasma mycoides mycoides (contagious bovine pleuropneumonia)
- Newcastle disease virus (velogenic)
- Peste des petits ruminants virus
- Rinderpest virus
- Sheep pox virus
- Swine vesicular disease virus
- Vesicular stomatitis virus (Exotic).

Plant Agents

(Usda plant protection and quarantine (PPQ) select agents and toxins):

- Candidatus *Liberobacter africanus*
- Candidatus *Liberobacter asiaticus*
- *Peronosclerospora philippinensis*
- *Ralstonia solanacearum* race 3, biovar. 2

- *Sclerophthora rayssiae* var. *zeae*
- *Synchytrium endobioticum*
- *Xanthomonas oryzae* pv. *oryzicola*
- *Xylella fastidiosa* (citrus variegated chlorosis strain).

Entomological Warfare

Entomological warfare (EW) is a type of biological warfare that uses insects to attack the enemy. The concept has existed for centuries and research and development have continued into the modern era. EW has been used in battle by Japan and several other nations have developed and been accused of using an entomological warfare program.

Description

Entomological warfare (EW) is a specific type of biological warfare (BW) that uses insects in a direct attack or as vectors to deliver a biological agent, such as plague or cholera. Essentially, EW exists in three varieties. One type of EW involves infecting insects with a pathogen and then dispersing the insects over target areas.

The insects then act as a vector, infecting any person or animal they might bite. Another type of EW is a direct insect attack against crops; the insect may not be infected with any pathogen but instead represents a threat to agriculture. The final method of entomological warfare is to use uninfected insects, such as bees, to directly attack the enemy.

History

Entomological warfare is not a new concept; historians and writers have studied EW in connection to multiple historic events. A 14th century plague epidemic in Asia Minor that eventually became known as the Black Death is one such event that has drawn attention from historians as a possible early incident of entomological warfare. That plague's spread over Europe may have been the result of a biological attack on the Crimean city of Kaffa. According to Jeffrey Lockwood, author of *Six-Legged Soldiers* (a book about EW), the earliest incident of entomological warfare was probably the use of bees by early humans.

The bees or their nests were thrown into caves to force the enemy out and into the open. Lockwood theorizes that the Ark of Covenant may have been deadly when opened because it contained deadly fleas.

During the American Civil War the Confederacy accused the Union of purposely introducing the harlequin bug in the South. These accusations were never proven, and modern research has shown it more likely that the insect arrived by other means. The world did not experience large-scale entomological warfare until World War II; Japanese attacks in China were the only verified instance of BW or EW during the war. During, and following, the war other nations began their own EW programs.

World War II

Canada

Among the Allied Powers, Canada led the pioneering effort in vector-borne warfare. After Japan became intent on developing the plague flea as a weapon, Canada, and then the United States followed suit. Cooperating closely with the United States, Dr. G.B. Reed, chief of Kingston's Queen's University's Defence Research Laboratory, focused his research efforts on mosquito vectors, biting flies, and plague infected fleas during World War II. Much of this research was shared with or conducted in concert with the United States.

Canada's entire bio-weapons program was ahead of the British and the Americans during the war. The Canadians tended to work in areas their allies ignored, entomological warfare was one of these areas. As the U.S. and British programs evolved, the Canadians worked closely with both nations. The Canadian BW work would continue well after the war, including entomological research.

France

France is known to have pursued entomological warfare programs during World War II. Like Germany, the nation suggested that the Colorado potato beetle, aimed at the enemy's food sources, would be an asset during the war. As early as 1939 biological warfare experts in France suggested that the beetle be used against German crops.

Germany

Germany is known to have pursued entomological warfare programs during World War II. The nation pursued the mass-production, and dispersion, of the Colorado potato beetle (*Lepinotarsa decemlineata*), aimed at the enemy's food sources. The beetle was first found in Germany in 1914, as an invasive species from North America. There are no records that indicate the beetle was ever employed as a weapon by Germany, or any other nation during the war. Regardless, the Germans had developed plans to drop the beetles on English crops.

Germany carried out testing of its Colorado potato beetle weaponization program south of Frankfurt, where they released 54,000 of the beetles. In 1944, an infestation of Colorado potato beetles was reported in Germany. The source of the infestation is unknown, speculation has offered three alternative theories as to the origin of the infestation. One option is Allied action, an entomological attack, another is that it was the result of the German testing, and still another more likely explanation is that it was merely a natural occurrence.

Japan

Japan used entomological warfare on a large-scale during World War II in China. Unit 731, Japan's infamous biological warfare unit, used plague-infected fleas and flies covered with cholera to infect the population in China. The Japanese military dispersed the insects by spraying them from low-flying airplanes and dropping bombs filled with a mixture of insects and disease. Localized and deadly epidemics resulted and nearly 500,000 Chinese died of disease. An "international symposium" of historians declared in 2002 that Japanese entomological warfare in China was responsible for the deaths of 440,000.

United Kingdom

A British scientist, J.B.S. Haldane, suggested that Britain and Germany were both vulnerable to entomological attack via the Colorado potato beetle. In 1942 the United States shipped 15,000 Colorado potato beetles to Britain for study as a weapon.

Cold War

Soviet Union

The Soviet Union researched, developed and tested an entomological warfare program as a major part of an anti-crop and anti-animal BW program. The Soviets developed techniques for using insects to transmit animal pathogens, such as foot and mouth disease — which they used ticks to transmit. The nation also used avian ticks to transmit *Chlamydophila psittaci* to chickens. In addition, the Soviet Union claimed to have developed an automated mass insect breeding facility, capable of outputting millions of parasitic insects per day.

United States

The United States seriously researched the potential of entomological warfare during the Cold War. The United States military developed plans for an entomological warfare facility, designed to produce 100 million yellow fever-infected mosquitoes per month. A U.S. Army report titled "Entomological Warfare Target Analysis" listed vulnerable sites within the Soviet Union that the U.S. could attack using entomological vectors. The military also tested the mosquito biting capacity by dropping uninfected mosquitoes over U.S. cities.

North Korean and Chinese officials levelled accusations that during the Korean War the United States engaged in biological warfare, including EW, in North Korea. The claim is dated to the period of the war, and has been thoroughly denied by the U.S. In 1998, Stephen Endicott and Edward Hagermann claimed that the accusations were true in their book, *The United States and Biological Warfare: Secrets from the Early Cold War and Korea* The book received mixed reviews, some called it "bad history" and "appalling", while other praised the case the authors made. Other historians have revived the claim in recent decades as well. The same year Endicotts' book was published Kathryn Weathersby and Milton Leitenberg of the Cold War International History Project at the Woodrow Wilson Centre in Washington released a cache of Soviet and Chinese documents which revealed the North Korean claim was an elaborate disinformation campaign.

During the 1950s the United States conducted a series of field tests using entomological weapons. Operation Big Itch, in 1954, was designed to test munitions loaded with uninfected fleas (*Xenopsylla cheopis*). Big Itch went awry when some of the fleas escaped into the plane and bit all three members of the air crew. In May 1955 over 300,000 yellow fever mosquitoes (*Aedes aegypti*) were dropped over parts of the U.S. state of Georgia to determine if the air-dropped mosquitoes could survive to take meals from humans. The mosquito tests were known as Operation Big Buzz. The U.S. engaged in at least two other EW testing programs, Operation Drop Kick and Operation May Day. A 1981 Army report outlined these tests as well as multiple cost-associated issues that occurred with EW. The report is partially declassified — some information is blacked out, including everything concerning "Drop Kick" — and included "cost per death" calculations. The cost per death, according to the report, for a vector-borne biological agent achieving a 50% mortality rate in an attack on a city was $0.29 in 1976 dollars. Such an attack was estimated to result in 625,000 deaths.

The United States has also applied entomological warfare research and tactics in non-combat situations. In 1990 the U.S. funded a $6.5 million program designed to research, breed and drop caterpillars. The caterpillars were to be dropped in Peru on coca fields as part of the American War on Drugs. As recently as 2002 U.S. entomological anti-drug efforts at Fort Detrick were focused on finding an insect vector for a virus that affects the opium poppy.

Bioterrorism

Clemson University's Regulatory and Public Service Program listed "diseases vectored by insects" among bioterrorism scenarios considered "most likely". Because invasive species are already a problem worldwide one University of Nebraska entomologist considered it likely that the source of any sudden appearance of a new agricultural pest would be difficult, if not impossible, to determine. Lockwood considers insects a more effective means of transmitting biological agents for acts of bioterrorism than the actual agents. Insect vectors are easily gathered and their eggs easily transportable without detection. Isolating and delivering biological agents, on the other hand, is extremely challenging and hazardous.

In one of the few suspected acts of entomological bioterrorism an eco-terror group known as The Breeders claimed to have released Mediterranean fruit flies (medflies) amidst an ongoing California infestation. Lockwood asserts that there is some evidence the group played a role in the event. The pest attacks a variety of crops and the state of California responded with a large-scale pesticide spraying program. At least one source asserted that there is no doubt that an outside hand played a role in the dense 1989 infestation. The group stated in a letter to then Los Angeles Mayor Tom Bradley that their goals were twofold. They sought to cause the medfly infestation to grow out of control which, in turn, would render the ongoing Malathion spraying program financially infeasible.

Legal Status

The Biological and Toxic Weapons Convention (BWC) of 1972 does not specifically mention insect vectors in its text. The language of the treaty, however, does cover vectors. Article I bans, "Weapons, equipment or means of delivery designed to use such agents or toxins for hostile purposes or in armed conflict." It would appear, due to the text of the BWC, that insect vectors as an aspect of entomological warfare are covered and outlawed by the Convention. The issue is less clear when warfare with uninfected insects against crops is considered.

Ethnic Bioweapon

An ethnic bioweapon (*biogenetic weapon*) aims to harm only or primarily persons of specific ethnicities or genotypes.

History of Ethnic Bioweapons

One of the first fictional discussions of ethnic weapons is in Robert A. Heinlein's 1942 novel *Sixth Column* (republished as *The Day After Tomorrow*) in which a race-specific radiation weapon is used against a so-called "Pan-Asian" invader.

Genetic Weapon

In 1997, U.S. Secretary of Defence William Cohen referred to the concept as a possible risk. In 1998 some biological weapon experts considered such a "genetic weapon" a plausible possibility, and believed the former Soviet Union had undertaken some research on the influence of various substances on human genes.

The possibility of a "genetic bomb" is presented in Vincent Sarich's and Frank Miele's book, *Race: The Reality of Human Differences*, published in 2004. The authors believe that information from the Human Genome Project will be used in just such a manner.

In 2005 the official view of the International Committee of the Red Cross was "The potential to target a particular ethnic group with a biological agent is probably not far off. These scenarios are not the product of the ICRC's imagination but have either occurred or been identified by countless independent and governmental experts."

Israeli "Ethno-bomb" Controversy

In November 1998, *The Sunday Times* reported that Israel was attempting to build an "ethno-bomb" containing a biological agent that could specifically target genetic traits present amongst Arab populations. *Wired News* also reported the story, as did *Foreign Report*. The article was quickly denounced as a hoax. Microbiologists and geneticists were skeptical towards the scientific plausibility of such a biological agent. The *New York Post*, describing the claims as "blood libel", reported that the likely source for the story was a work of science fiction by Israeli academic Doron Stanitsky. Stanitsky had sent his completely fictional work about such a weapon to Israeli newspapers two years before. The article also noted the views of genetic researchers who claimed the idea as "wholly fantastical", still others admit that the weapon was theoretically possible. A planned second installment of the article never appeared, and no sources were ever identified. Neither of the authors of the Sunday Times story, Uzi Mahnaimi and Marie Colvin, have spoken publicly on the matter.

Russian Ban on Export of Biological Samples

In May 2007, Russian newspaper *Kommersant* reported that the Russian government banned all exports of human biosamples. The report claims that the reason for the ban was a secret FSB report about on-going development of "genetic bioweapons" targeting Russian population by Western institutions. The report mentions the Harvard School of Public Health, American International Health Alliance, United States Department of Justice Environment and Natural Resources Division, Karolinska Institutet and United States Agency for International Development.

Deinococcus Radiodurans: A Marvelous Berry that Withstands Radiations

Deinococcus radiodurans is an extremophilic bacterium, one of the most radioresistant organisms known. It can survive cold, dehydration, vacuum, and acid, and is therefore known as a polyextremophile and has been listed as the world's toughest bacterium in *The Guinness Book Of World Records.*

Name and Classification

The name *Deinococcus radiodurans* derives from the Greek deinos and kokkos meaning "terrible grain" and the Latin radius and durare meaning "radiation surviving". The species was formerly called *Micrococcus radiodurans.* As a consequence of its hardiness, it has been nicknamed "Conan the Bacterium", a play on "Conan the Barbarian".

Initially it was placed in the genus *Micrococcus.* After evaluation of ribosomal RNA sequences and other evidence, it was placed in its own genus *Deinococcus,* which is closely related to the genus *Thermus* of heat-resistant bacteria; the group consisting of the two is accordingly known as Deinococcus-Thermus.

Deinococcus is the only genus in the order *Deinococcales. D. radiodurans* is the type species of this genus, and the best studied member. All known members of the genus are radioresistant: *D. proteolyticus, D. radiopugnans, D. radiophilus, D. grandis, D. indicus, D. frigens, D. saxicola, D. marmoris, D. deserti, D. geothermalis* and *D. murrayi*; the latter two are also thermophilic.

History

D. radiodurans was discovered in 1956 by Arthur W. Anderson at the Oregon Agricultural Experiment Station in Corvallis, Oregon. Experiments were being performed to determine if canned food could be sterilized using high doses of gamma radiation. A tin of meat was exposed to a dose of radiation that was thought to kill all known forms of life, but the meat subsequently spoiled, and *D. radiodurans* was isolated.

The complete DNA sequence of *D. radiodurans* was published in 1999 by TIGR. A detailed annotation and analysis of the genome appeared in 2001. The sequenced strain was ATCC BAA-816. *Deinococcus radiodurans* has a unique quality in which it can repair DNA both single and double stranded. When a mutation is apparent to the cell it brings it into a compartmental ring like structure where the DNA is repaired and then is able to fuse the nucleoids from the outside of the compartment with the damaged DNA.

Description

D. radiodurans is a rather large spherical bacterium, with a diameter of 1.5 to 3.5 µm. Four cells normally stick together, forming a tetrad. The bacteria are easily cultured and do not appear to cause disease. Colonies are smooth, convex, and pink to red in colour. The cells stain gram positive, although its cell envelope is unusual and is reminiscent of the cell walls of gram negative bacteria.

D. radiodurans does not form endospores and is nonmotile. It is an obligate aerobic chemoorganoheterotroph, i.e. it uses oxygen to derive energy from organic compounds in its environment. It is often found in habitats rich in organic materials, such as soil, feces, meat, or sewage, but has also been isolated from dried foods, room dust, medical instruments and textiles.

It is extremely resistant to ionizing radiation, ultraviolet light, desiccation, oxidizing and electrophilic agents.

Its genome consists of two circular chromosomes, one 2.65 million base pairs long and the other 412,000 base pairs long, as well as a megaplasmid of 177,000 base pairs and a plasmid of 46,000 base pairs. It has about 3,195 genes. In its stationary phase each bacterial cell contains four copies of this genome; when rapidly multiplying, each bacterium contains 8-10 copies of the genome.

Ionizing Radiation Resistance

D. radiodurans is capable of withstanding an acute dose of 5,000 Gy of ionizing radiation with almost no loss of viability, and an acute dose of 15,000 Gy with 37% viability. A dose of 5,000 Gy is estimated to introduce several hundred double strand breaks (DSBs) into the organism's DNA (~0.005 DSB/Gy/Mbp (haploid genome)). For comparison, a chest X-ray or Apollo mission involves about 1 milligray, 5 Gy can kill a human, 200-800 Gy will kill *E. coli*, and over 4,000 Gy will kill the radiation-resistant tardigrade.

Several bacteria of comparable radioresistance are now known, including some species of the genus *Chroococcidiopsis* (phylum cyanobacteria) and some species of *Rubrobacter* (phylum actinobacteria); among the archaea, the species *Thermococcus gammatolerans* shows comparable radioresistance. *Deinocuccus radiodurans* also has a unique ability to repair damaged DNA. It isolates the damaged segments in a controlled area and repairs it. This bacteria can also repair many small fragments from an entire chromosome.

Ionizing Radiation Resistance Mechanisms

Deinococcus accomplishes its resistance to radiation by having multiple copies of its genome and rapid DNA repair mechanisms. It usually repairs breaks in its chromosomes within 12–24 hours through a 2-step process. First, *D. radiodurans* reconnects some chromosome fragments through a process called single-strand annealing. In the second step, a protein mends double-strand breaks through homologous recombination. This process does not introduce any more mutations than a normal round of replication would.

A persistent question regarding *D. radiodurans* is how such a high degree of radioresistance could evolve. Natural background radiation levels are very low—in most places, on the order of 0.4 mGy per year, and the highest known background radiation, near Ramsar, Iran is only 260 mGy per year. With naturally-occurring background radiation levels so low, organisms evolving mechanisms specifically to ward off the effects of high radiation are unlikely.

Valerie Mattimore and John R. Battista of Louisiana State University have suggested that the radioresistance of *D. radiodurans* is simply a side-effect of a mechanism for dealing with prolonged cellular desiccation (dryness). To support this hypothesis, they performed an experiment in which they demonstrated that mutant strains of *D. radiodurans* which are highly susceptible to damage from ionizing radiation are also highly susceptible to damage from prolonged desiccation, while the wild type strain is resistant to both. In addition to DNA repair, *D. radiodurans* use LEA proteins (Late Embryogenesis Abundant proteins) expression to protect against desiccation.

Scanning electron microscopy analysis has shown that DNA in *D. radiodurans* is organized into tightly packed toroids, which may facilitate DNA repair.

A team of Croatian and French researchers led by Miroslav Radman have bombarded *D. radiodurans* to study the mechanism of DNA repair. At least two copies of the genome, with random DNA breaks, can form DNA fragments through annealing. Partially overlapping fragments are then used for synthesis of homologous regions through a moving D-loop that can continue extension until they find complementary partner strands. In the final step there is crossover by means of RecA-dependent homologous recombination.

Michael Daly has suggested that the bacterium uses manganese complexes as antioxidants to protect itself against radiation damage. In 2007 his team showed that high intracellular levels of manganese(II) in *D. radiodurans* protect proteins from being oxidized by radiation, and proposed the idea that "protein, rather than DNA, is the principal target of the biological action of [ionizing radiation] in sensitive bacteria, and extreme resistance in Mn-accumulating bacteria is based on protein protection".

A team of Russian and American scientists proposed that the radioresistance of *D. radiodurans* had a Martian origin. Evolution of the microorganism could have taken place on the Martian surface until it was delivered to Earth on a meteorite. However, apart from its resistance to radiation, Deinococcus is genetically and biochemically very similar to other terrestrial life forms, arguing against an extraterrestrial origin.

In 2009 it was reported that nitric oxide plays an important role in the bacteria's recovery from radiation exposure: the gas is required for division and proliferation after DNA damage has been repaired. A gene was described that increases nitric oxide production after UV radiation, and in the absence of this gene the bacteria were still able to repair DNA damage but would not grow.

Applications

Deinococcus has been genetically engineered for use in bioremediation to consume and digest solvents and heavy metals, even in a highly radioactive site. For example, the bacterial mercuric reductase gene has been cloned from *Escherichia coli* into *Deinococcus* to detoxify the ionic mercury residue frequently found in radioactive waste generated from nuclear weapons manufacture. Those researchers developed a strain of *Deinococcus* that could detoxify both mercury and toluene in mixed radioactive wastes. The Craig Venter Institute has used a system derived from the rapid DNA repair mechanisms of *D. radiodurans* to assemble synthetic DNA fragments into chromosomes, with the ultimate goal of producing a synthetic organism they call *Mycoplasma laboratorium*.

In 2003, U.S. scientists demonstrated that *D. radiodurans* could be used as a means of information storage that might survive a nuclear catastrophe. They translated the song *It's a Small World* into a series of DNA segments 150 base pairs long, inserted these into the bacteria, and were able to retrieve them without errors 100 bacterial generations later.Thermococcus gammatolerans.

Thermococcus gammatolerans is an archaea extremophile and the most radiation resistant known organism. Recently discovered in a submarine hydrothermal vent in the Guaymas Basin about 2,000 meters deep off the coast of California. Thermococcus gammatolerans sp. nov thrives in temperatures between 55-95 °C with an optimum development at approximately 88 °C. The optimal growth pH is 6, favoring the presence of sulfur (S), which is reduced to hydrogen sulfide (H2S). It is the organism with the strongest resistance to radiation, supporting a radiation of gamma rays from 30 KGy. It belongs to the family of Thermococci they are a group of microorganisms in the Archaea domain and phylum Euryarchaeota. The Thermococci live in extremely hot environments such as hydrothermal vents with a growth optimum temperature above 80 °C. Thermococcus and Pyrococcus (literally "ball of fire") are both chemoorganotrophic anaerobic required. Thermococcus prefer 70-95 °C, whereas the range is better Pyrococcus 70-100 °C. The resistance to ionizing radiation of T. Gammatolerans is enormous, while a dose of 10 Gy is sufficient to kill a human being, and a dose of 60 Gy is able to kill all cells in a colony of E. coli, the Thermococcus gammatolerans sp. nov can withstand an instantaneous dose of up to 5,000 Gy with no loss of viability and doses of up to 30 000 Gy

History

The gammatolerans Thermococcus sp. nov, was discovered in 2003 on samples collected from a hydrothermal chimney collected at the Guaymas Basin about 2,000 meters deep off the coast of California. (27 degrees 01 'N, 111 ° 24' W).

Mechanisms of Resistance to Radiation

Unlike other organisms Thermococcus gammatolerans cell survival is not altered by the changing conditions in its growth phase, but the lack of ideal conditions and nutrients decreases its radioresistance. The system of chromosomal DNA repair shows that cells in stationary phase of growth reconstitute the DNA more rapidly than cells in exponential growth phase. T. Gammatolerans can slowly or quickly rebuild damaged chromosomes without loss of viability.

Applications

It has been studied its application to the development of new enzymatic markers are resistant to high temperatures. Their application in the study of carcinogenesis and the study of the development of mitochondrial diseases. Has been speculated that DNA repair mechanisms of T. gammatolerans could be incorporated into the genome of higher species in order to improve DNA repair and reduce cellular aging.

Radiotrophic Fungus

Radiotrophic fungi are fungi which appear to use the pigment melanin to convert gamma radiation into chemical energy for growth. This proposed mechanism may be similar to anabolic pathways for the synthesis of reduced organic carbon (e.g., carbohydrates) in phototrophic organisms, which capture photons from visible light with pigments such as chlorophyll whose energy is then used in photolysis of water to generate usable chemical energy (as ATP) in photophosphorylation of photosynthesis. However, whether melanin-containing fungi employ a similar multi-step pathway as photosynthesis, or some chemosynthesis pathways, is unknown. These were first discovered in 2007 as black molds growing inside and around the Chernobyl Nuclear Power Plant. Research at the Albert Einstein College of Medicine showed that three melanin-containing fungi, *Cladosporium sphaerospermum*, *Wangiella dermatitidis*, and *Cryptococcus neoformans*, increased in biomass and accumulated acetate faster in an environment in which the radiation level was 500 times higher than in the normal environment. Exposure of *C. neoformans* cells to these radiation levels rapidly (within 20–40 minutes of exposure) altered the chemical properties of its melanin and increased melanin-mediated rates of electron transfer (measured as reduction of ferricyanide by NADH) 3 to 4-fold compared with unexposed cells. Similar effects on melanin electron-transport capability were observed by the authors after exposure to non-ionizing radiation, suggesting that melanotic fungi might also be able to use light or heat radiation for growth. However, melanization may come at some metabolic cost to the fungal cells: in the absence of radiation, some non-melanized fungi (that had been mutated in the melanin pathway) grew faster than their melanized counterparts. Limited uptake of nutrients due to the melanin molecules in the fungal cell wall or toxic intermediates formed in melanin biosynthesis have been suggested to contribute to this phenomenon.

6

Biosensors: Concepts and Applications

A biosensor is an analytical device for the detection of an analyte that combines a biological component with a physicochemical detector component.

It consists of 3 parts:

- the *sensitive biological element* (biological material (eg. tissue, microorganisms, organelles, cell receptors, enzymes, antibodies, nucleic acids, etc), a biologically derived material or biomimic) The sensitive elements can be created by biological engineering.
- the *transducer* or the *detector element* (works in a physicochemical way; optical, piezoelectric, electrochemical, etc.) that transforms the signal resulting from the interaction of the analyte with the biological element into another signal (i.e., transducers) that can be more easily measured and quantified;
- associated electronics or signal processors that are primarily responsible for the display of the results in a user-friendly way.. This sometimes accounts for the most expensive part of the sensor device, however it is possible to generate a user friendly display that includes transducer and sensitive element.

A common example of a commercial biosensor is the blood glucose biosensor, which uses the enzyme glucose oxidase to break blood glucose down. In doing so it first oxidizes glucose and uses two electrons to reduce the FAD (a component of the enzyme) to FADH2. This in turn is oxidized by the electrode (accepting two electrons from the electrode) in a number of steps. The resulting current is a measure of the concentration of glucose. In this case, the electrode is the transducer and the enzyme is the biologically active component.

Recently, arrays of many different detector molecules have been applied in so called electronic nose devices, where the pattern of response from the detectors is used to fingerprint a substance.. Current commercial electronic noses, however, do not use biological elements.

A canary in a cage, as used by miners to warn of gas, could be considered a biosensor. Many of today's biosensor applications are similar, in that they use organisms which respond to toxic substances at a much lower concentrations than humans can detect to warn of the presence of the toxin. Such devices can be used in environmental monitoring, trace gas detection and in water treatment facilities.

Principles of Detection

Photometric

Many optical biosensors based on the phenomenon of surface plasmon resonance are evanescent wave techniques. This utilises a property of gold and other materials; specifically that a thin layer of gold on a high refractive index glass surface can absorb laser light, producing electron waves (surface plasmons) on the gold surface. This occurs only at a specific angle and wavelength of incident light and is highly dependent on the surface of the gold, such that binding of a target analyte to a receptor on the gold surface produces a measurable signal.

Surface plasmon resonance sensors operate using a sensor chip consisting of a plastic cassette supporting a glass plate, one side of which is coated with a microscopic layer of gold. This side contacts the optical detection apparatus of the instrument. The opposite side is then contacted with a microfluidic flow system. The contact with the flow system creates channels across which reagents can be passed in solution. This side of the glass sensor chip can be modified in a number of ways, to allow easy attachment of molecules of interest. Normally it is coated in carboxymethyl dextran or similar compound.

Light of a fixed wavelength is reflected off the gold side of the chip at the angle of total internal reflection, and detected inside the instrument. This induces the evanescent wave to penetrate through the glass plate and some distance into the liquid flowing over the surface.

The refractive index at the flow side of the chip surface has a direct influence on the behaviour of the light reflected off the gold side. Binding to the flow side of the chip has an effect on the refractive index and in this way biological interactions can be measured to a high degree of sensitivity with some sort of energy.

Other evanescent wave biosensors have been commercialised using waveguides where the propagation constant through the waveguide is changed by the absorption of molecules to the waveguide surface. One such example, Dual Polarisation Interferometry uses a buried waveguide as a reference against which the change in propagation constant is measured. Other configurations such as the Mach-Zehnder have reference arms lithographically defined on a substrate. Higher levels of integration can be achieved using resonator geometries where the resonant frequency of a ring resonator changes when molecules are absorbed.

Other optical biosensors are mainly based on changes in absorbance or fluorescence of an appropriate indicator compound and do not need a total internal reflection geometry. For example, a fully operational prototype device detecting casein in milk has been fabricated. The device is based on detecting changes in absorption of a gold layer. A widely used research tool, the micro-array, can also be considered a biosensor.

Biological biosensors often incorporate a genetically modified form of a native protein or enzyme. The protein is configured to detect a specific analyte and the ensuing signal is read by a detection instrument such as a fluorometer or luminometer. An example of a recently developed biosensor is one for detecting cytosolic concentration of the analyte cAMP (cyclic adenosine monophosphate), a second messenger involved in cellular signalling triggered by ligands interacting with receptors on the cell membrane. Similar systems have been created to study cellular responses to native ligands or xenobiotics (toxins or small molecule inhibitors). Such "assays" are commonly used in drug discovery development by pharmaceutical and biotechnology companies. Most cAMP assays in current use require lysis of the cells prior to measurement of cAMP. A live-cell biosensor for cAMP can be used in non-lysed cells with the additional advantage of multiple reads to study the kinetics of receptor response.

Electrochemical

Electrochemical biosensors are normally based on enzymatic catalysis of a reaction that produces or consumes electrons (such enzymes are rightly called redox enzymes). The sensor substrate usually contains three electrodes; a reference electrode, a working electrode and a sink electrode. An auxiliary electrode (also known as a counter electrode) may also be present as an ion source. The target analyte is involved in the reaction that takes place on the active electrode surface, and the ions produced create a potential which is subtracted from that of the reference electrode to give a signal. We can either measure the current (rate of flow of electrons is now proportional to the analyte concentration) at a fixed potential or the potential can be measured at zero current (this gives a logarithmic response). Note that potential of the working or active electrode is space charge sensitive and this is often used. Further, the label-free and direct electrical detection of small peptides and proteins is possible by their intrinsic charges using biofunctionalized ion-sensitive field-effect transistors.

Another example, the potentiometric biosensor, works contrary to the current understanding of its ability. Such biosensors are screenprinted, conducting polymer coated, open circuit potential biosensors based on conjugated polymers immunoassays. They have only two electrodes and are extremely sensitive and robust. They enable the detection of analytes at levels previously only achievable by HPLC and LC/MS and without rigorous sample preparation. The signal is produced by electrochemical and physical changes in the conducting polymer layer due to changes occurring at the surface of the sensor. Such changes can be attributed to ionic strength, pH, hydration and redox reactions, the latter due to the enzyme label turning over a substrate.

Ion Channel Switch

The use of ion channels has been shown to offer highly sensitive detection of target biological molecules.. By imbedding the ion channels in supported or tethered bilayer membranes (t-BLM) attached to a gold electrode, an electrical circuit is created. Capture molecules such as antibodies can be bound to the ion channel so that the binding of the target molecule controls the ion flow through the channel.

This results in a measurable change in the electrical conduction which is proportional to the concentration of the target.

An Ion Channel Switch (ICS) biosensor can be created using gramicidin, a dimeric peptide channel, in a tethered bilayer membrane.. One peptide of gramicidin, with attached antibody, is mobile and one is fixed. Breaking the dimer stops the ionic current through the membrane. The magnitude of the change in electrical signal is greatly increased by separating the membrane from the metal surface using a hydrophilic spacer. Quantitative detection of an extensive class of target species, including proteins, bacteria, drug and toxins has been demonstrated using different membrane and capture configurations.

Others

Piezoelectric sensors utilise crystals which undergo an elastic deformation when an electrical potential is applied to them. An alternating potential (A.C.) produces a standing wave in the crystal at a characteristic frequency. This frequency is highly dependent on the elastic properties of the crystal, such that if a crystal is coated with a biological recognition element the binding of a (large) target analyte to a receptor will produce a change in the resonance frequency, which gives a binding signal. In a mode that uses surface acoustic waves (SAW), the sensitivity is greatly increased. This is a specialised application of the Quartz crystal microbalance as a biosensor.

Applications

There are many potential applications of biosensors of various types. The main requirements for a biosensor approach to be valuable in terms of research and commercial applications are the identification of a target molecule, availability of a suitable biological recognition element, and the potential for disposable portable detection systems to be preferred to sensitive laboratory-based techniques in some situations. Some examples are given below:

- Glucose monitoring in diabetes patients □!historical market driver
- Other medical health related targets
- Environmental applications e.g. the detection of pesticides and river water contaminants
- Remote sensing of airborne bacteria e.g. in counter-bioterrorist activities
- Detection of pathogens
- Determining levels of toxic substances before and after bioremediation
- Detection and determining of organophosphate
- Routine analytical measurement of folic acid, biotin, vitamin B12 and pantothenic acid as an alternative to microbiological assay
- Determination of drug residues in food, such as antibiotics and growth promoters, particularly meat and honey.
- Drug discovery and evaluation of biological activity of new compounds.

Glucose Monitoring

Commercially available gluocose monitors rely on amperometric sensing of glucose by means of glucose oxidase, which oxidises glucose producing hydrogen peroxide which is detected by the electrode. To overcome the limitation of ameperometric sensors, a flurry of research is present into novel sensing methods, such as fluorescent glucose biosensors.

Biosensors in Food Analysis

There are several applications of biosensors in food analysis. In food industry optic coated with antibodies are commonly used to detect pathogens and food toxins. The light system in these biosensors has been fluorescence, since this type of optical measurement can greatly amplify the signal.

A range of immuno-and ligand-binding assays for the detection and measurement of small molecules such as water-soluble vitamins and chemical contaminants (drug residues) such as sulfonamides and Beta-agonists have been developed for use on SPR based sensor systems, often adapted from existing ELISA or other immunological assay. These are in widespread use across the food industry.

Surface Attachment of the Biological Elements

An important part in a biosensor is to attach the biological elements (small molecules/protein/cells) to the surface of the sensor (be it metal, polymer or glass). The simplest way is to functionalize the surface in order to coat it with the biological elements. This can be done by polylysine, aminosilane, epoxysilane or nitrocellulose in the case of silicon chips/silica glass. Subsequently the bound biological agent may be for example fixed by Layer by layer deposition of alternatively charged polymer coatings

Alternatively three dimensional lattices (hydrogel/xerogel) can be used to chemically or physically entrap these (where by chemically entraped it is meant that the biological element is kept in place by a strong bond, while physically they are kept in place being unable to pass through the pores of the gel matrix). The most commonly used hydrogel is sol-gel, a glassy silica generated by polymerization of silicate monomers (added as tetra alkyl orthosilicates, such as TMOS or TEOS) in the presence of the biological elements (along with other stabilizing polymers, such as PEG) in the case of physical entrapment.

Another group of hydrogels, which set under conditions suitable for cells or protein, are acrylate hydrogel, which polymerize upon radical initiation. One type of radical initiator is a peroxide radical, typically generated by combining a persulfate with TEMED (Polyacrylamide gel are also commonly commonly used for protein electrophoresis), alternatively light can be used in combination with a photoinitiator, such as DMPA (2,2-dimethoxy-2-phenylacetophenone). Smart materials that mimic the biological components of a sensor can also be classified as biosensors using only the active or catalytic site or analogous configurations of a biomolecule.

Intellectual Property Rights and Microbes

Intellectual property (IP) is a term referring to a number of distinct types of creations of the mind for which property rights are recognized—and the corresponding fields of law. Under intellectual property law, owners are granted certain exclusive rights to a variety of intangible assets, such as musical, literary, and artistic works; discoveries and inventions; and words, phrases, symbols, and designs. Common types of intellectual property include copyrights, trademarks, patents, industrial design rights and trade secrets in some jurisdictions.

Although many of the legal principles governing intellectual property have evolved over centuries, it was not until the 19th century that the term *intellectual property* began to be used, and not until the late 20th century that it became commonplace in the United States. The British Statute of Anne 1710 and the Statute of Monopolies 1623 are now seen as the origin of copyright and patent law respectively.

History

Modern usage of the term *intellectual property* goes back at least as far as 1867 with the founding of the North German Confederation whose constitution granted legislative power over the protection of intellectual property (*Schutz des geistigen Eigentums*) to the confederation. When the administrative secretariats established by the Paris Convention (1883) and the Berne Convention (1886) merged in 1893, they located in Berne, and also adopted the term intellectual property in their new combined title, the United International Bureaux for the Protection of Intellectual Property. The organisation subsequently relocated to Geneva in 1960, and was succeeded in 1967 with the establishment of the World Intellectual Property Organization (WIPO) by treaty as an agency of the United Nations. According to Lemley, it was only at this point that the term really began to be used in the United States (which had not been a party to the Berne Convention), and it did not enter popular usage until passage of the Bayh-Dole Act in 1980.

"The history of patents does not begin with inventions, but rather with royal grants by Queen Elizabeth I (1558-1603) for monopoly privileges... Approximately 200 years after the end of Elizabeth's reign, however, a patent represents a legal [right] obtained by an inventor providing for exclusive control over the production and sale of his mechanical or scientific invention... [demonstrating] the evolution of patents from royal prerogative to common-law doctrine."

In an 1818 collection of his writings, the French liberal theorist, Benjamin Constant, argued against the recently introduced idea of "property which has been called intellectual." The term *intellectual property* can be found used in an October 1845 Massachusetts Circuit Court ruling in the patent case *Davoll et al. v. Brown.*, in which Justice Charles L. Woodbury wrote that "only in this way can we protect intellectual property, the labours of the mind, productions and interests are as much a man's own...as the wheat he cultivates, or the flocks he rears." The statement that "discoveries are...property" goes back earlier. Section 1 of the French law of 1791 stated, "All new discoveries are the property of the author; to assure the inventor the property and temporary enjoyment of his discovery, there shall be delivered to him a patent for five, ten or fifteen years." In Europe, French author A. Nion mentioned *propriete intellectuelle* in his *Droits civils des auteurs, artistes et inventeurs,* published in 1846.

The concept's origins can potentially be traced back further. Jewish law includes several considerations whose effects are similar to those of modern intellectual property laws, though the notion of intellectual creations as property does not seem to exist – notably the principle of Hasagat Ge'vul (unfair encroachment) was used to justify limited-term publisher (but not author) copyright in the 16th century. The Talmud contains the prohibitions against certain mental crimes (further elaborated in the Shulchan Aruch), notably Geneivat da'at, which some have interpreted as prohibiting theft of ideas, though the doctrine is principally concerned with fraud and deception, not property.

Objectives

Financial Incentive

These exclusive rights allow owners of intellectual property to benefit from the property they have created, providing a financial incentive for the creation of and investment in intellectual property, and, in case of patents, pay associated research and development costs. Some commentators, such as David Levine and Michele Boldrin, dispute this justification.

Economic Growth

The existence of IP laws is credited with significant contributions toward economic growth. Economists estimate that two-thirds of the value of large businesses in the U.S. can be traced to intangible assets. "IP-intensive industries" are estimated to generate 72 percent more value added (price minus material cost) per employee than "non-IP-intensive industries". A joint research project of the WIPO and the United Nations University measuring the impact of IP systems on six Asian countries found "a positive correlation between the strengthening of the IP system and subsequent economic growth." Other models, such as the Nash equilibrium, would not expect that this correlation necessarily means causation: The Nash equilibrium model predicts that patent holders will prefer to operate in countries with stronger IP laws. In some of the cases, as was shown for Taiwan after the 1986 reform, the economic growth that comes with a stronger IP system might be due to an increase in stock capital from direct foreign investment.

In 2005 the RSA launched the Adelphi Charter, aimed at creating an international policy statement to frame how governments should make balanced intellectual property law.

Biological Patent

A biological patent is a patent relating to an invention or discovery in biology. It can be a composition of matter, a method for obtaining or using one or more thereof, or a product combining such things. Even when a natural biological substance itself is patented (apart from any associated process or usage), this has been permitted in the United States as long as they are sufficiently "isolated" from their naturally-occurring states. Prominent historical examples of such patents on isolated products of nature include adrenaline, insulin, vitamin B_{12}, and unmodified genes (the latter becoming its own notable controversy as of late).

History

The 1970s marked the first time when scientists patented methods on their biotechnological inventions with recombinant DNA. It wasn't until 1980 that patents for whole-scale living organisms were permitted. In Diamond v. Chakrabarty, the U.S. Supreme Court established the patentability of living matter, provided it was truly "man-made." The subject for this particular case was a genetically-engineered bacterium that was specifically modified to help clean up and degrade oil spills.

Since that 1980 court case, there has been much patenting of biological matter. Companies and organizations, like the University of California, have patented entire genomes. In 1998, the U.S. Patent and Trademark Office (PTO) issued a broad patent claiming primate (including human) embryonic stem cells, entitled "Primate Embryonic Stem Cells". On 13 March 2001, a second patent (6,200,806) was issued with the same title but focused on human embryonic stem cells.

Recently, there has been a slowdown and backlash against patenting biological material worldwide.

Controversy

Some believe that natural occurrences are not invented and thus should not be patentable. This is especially true if the biological matter being patented can be found in humans, such as sequences of DNA, which invokes a range of political and ethical arguments. The legal reasons for the controversy can vary by different countries' standards for patentability. In the US, an invention must have some "utility" but not necessarily an identified industrial application. According to Nielson and Whittaker (2002), the "notion of 'utility' is less specific, it means it is useful". Beyond utility/usefulness, the US also requires that inventions be new/novel and non-obvious.

Many outside the United States feel that the patenting of stem cells was rash and are seeking to reverse patent rulings. In December 2006, Germany made an important precedent when it annulled the validity of a stem cell patent. The German Federal Patent Court declared that anything made from human tissue cannot be patented. The European Patent Office has ruled that certain stem cell lines, derived from destruction of human embryos, were not to be granted a European patent. Worldwide, people are questioning the validity of stem cell patents. Wisconsin Alumni Research Foundation has sought to gain approval for its US patent in Europe without success.

The backlash against stem cell patents is also occurring in the United States, but to a lesser degree. Currently, an appeal against the patents is being processed by two non-profit organizations: The Foundation for Taxpayer & Consumer Rights and Public Patent Foundation along with molecular biologist Jeanne Loring of the Burnham Institute. They are fighting the validity of the patents by arguing that two of the patents cover a technique published in 1992, already patented by an Australian scientist. Another claim is that the techniques tied up with the patents are rendered obvious under a 1990 paper and two textbooks.

Advocates who speak against biological patents suggest that the techniques and processes associated with the discovery could be patentable but not the actual biological matter itself. For example, an advocate against biological patents would suggest that a gene associated with cancer should not be patented, but the test used to detect the gene could be.

Effects on Research

Some believe that the increase in patenting biological information leads to inefficiency in research. Many scientists are coming up against patent thickets, which are masses of information that they must obtain permission (and often pay large fees to utilize) before they can ever work with the information. Michael Heller and Rebecca Eisenberg (2005) explain that there is a recent trend of patenting more and more steps along the research path. This creates a "tragedy of the anticommons," whereby "each upstream patent allows its owner to set up another tollbooth on the road to product development, adding to the cost and slowing the pace of downstream... innovation".

A report shows that notwithstanding escalating funding, in the past half-decade biomedical innovation has slowed markedly. The number of drugs approved by the Food and Drug Administration has fallen below previous eras. The technologies approved, it continues, are less influential than previous innovations approved. The current trend of patenting what previously were thought of as basic science insights have raised the financial bar for other scientists wanting to use such insight. The overall trend of more patents may be slowing innovation.

American Type Culture Collection (ATCC)

The American Type Culture Collection (ATCC) is a private, not-for-profit biological resource centre whose mission focuses on the acquisition, authentication, production, preservation, development and distribution of standard reference microorganisms, cell lines and other materials for research in the life sciences.

Established in 1914 and originally incorporated by scientists in 1925 to serve as a worldwide repository and distribution centre for cultures of microorganisms, the ATCC has developed into the global leader in research and development expertise for identifying, characterizing, preserving and distributing a wide range of cell lines and microbes. Aside from maintaining the biorepository, an R&D program and a product development team, the ATCC also competes for federal grants and contracts and engages in partnerships and collaborations with academic institutions and private companies.

Biological Materials

ATCC serves U.S. and international researchers by characterizing cell lines, bacteria, viruses, fungi and protozoa, as well as developing and evaluating assays and techniques for validating research resources and preserving and distributing biological materials to the public and private sector research communities. Their management philosophy emphasizes customer satisfaction, value addition, cost-effective operations and competitive benchmarking for all areas of the enterprise.

ATCC's collections include a wide range of biological materials for research, including cell lines, molecular genomics tools, microorganisms and bioproducts. The organization holds a collection of more than 4,000 human, animal and plant cell lines and an additional 1,200 hybridomas. The molecular genomics collection at ATCC contains 8 million cloned genes from a host of species, including human, mouse, soybean, rat, monkey, zebrafish and several disease vectors. ATCC's microorganism collection includes a collection of more than 18,000 strains of bacteria from 900 genera, as well as 2,000 different types of animal viruses and 1,000 plant viruses. In addition, ATCC maintains collections of protozoans, yeasts and fungi with over 49,000 yeast and fungi strains from 1,500 genera and 2,000 strains of protists.

In addition to serving as a biorepository and distributor, ATCC provides specialized services related to its overall mission as a biological resource centre. Individuals and groups can employ a safe deposit service for their own cell cultures, providing a secure back-up for valuable biomaterials on a cGMP basis if required. ATCC also is able to retain secure samples of patented materials and distribute them according to instructions and approval of the patent holder. ATCC provides expert biological repository management services to institutions, agencies and companies wishing to outsource the handling of their own culture collections.

Biological Standards

ATCC biological standards are vital to assuring reliability of research results, reproducibility of experimentation and consistency in the scientific method. Standards from ATCC also help scientists in a wide range of industries ensure safety and quality in their products. ATCC reagents are cited as standards by such federal agencies as the U.S. Food and Drug Administration and the U.S. Department of Agriculture, as well as organizations such as AOAC International, the Clinical Laboratory Standards Institute, the U.S. Pharmacopeia, and the World Health Organization. ATCC-produced standards are used by in a host of applications relevant to the lives of citizens, including in developing therapeutic and diagnostic products, testing the quality of food, water and environmental samples, making accurate medical diagnoses and obtaining sound forensic information.

Facilities

ATCC activities are housed in a custom-built facility in Manassas, Virginia. The 126,000 sq ft (11,700 m^2). building is monitored constantly by on-site security staff, and electric power is backed up by on-site generators. The repository portion of the facility occupies 18,000 sq ft (1,700 m^2). and contains 200 freezers to store biomaterials, including 65 vapour-phase liquid nitrogen freezers and 50 mechanical freezers, as well as cold rooms for storage at 4°-8 °C. The repository space is complemented by 35,000 sq ft (3,300 m^2). of laboratory space. ATCC also occupies 45,000 sq ft (4,200 m^2) of research and administrative space at the Prince William County campus of George Mason University, which is located across the street from ATCC's Manassas building.

The community of users of ATCC products and services is international, and includes researchers in academia and government, as well as private industry. Over 80% of ATCC's 17,000-strong customer base represents academia and industry – 42% from universities and colleges and 41% from private industry. Government customers comprise 6% of the organization's total. Three-quarters of ATCC customers hail from the U.S., while the remaining 25% are international customers. ATCC maintains authorized distributors in Europe, Japan, Australia, New Zealand, Hong Kong, Singapore, South Korea, Israel, and Taiwan, and makes other international shipments directly from its Virginia facilities. Among the industries represented ATCC's customer base are the pharmaceutical, biotechnology, agricultural and diagnostics industries, as well as food, beverage and cosmetics makers and reference and testing laboratories.

The ATCC also has close working links with several other international culture collections, such as the UK's National Collection of Plant Pathogenic Bacteria (NCPPB), Landcare Research in New Zealand, the Collection Française de Bacteries Phytopathogènes (CFBP, or Collection of Plant Pathogenic Bacteria) in France, Belgian Co-ordinated Collections of Micro-organisms (BCCM), IBSF in Brazil, the Deutsche Sammlung von Mikroorganismen und Zellkulturen (DSMZ, or German Collection of Microorganisms and Cell Cultures), the Japanese Collection of Research Bioresources (JCRB), and others.

Biopiracy

Biopiracy is a situation where indigenous knowledge of nature is exploited for commercial gain with no compensation to the indigenous people themselves. Detractors of utilization of natural knowledge such as Greenpeace claim these practices contribute to inequality between developing countries rich in biodiversity, and developed countries hosting companies which engage in *biopiracy*.

Bioprospecting is an umbrella term describing the discovery of new and useful biological mechanisms, either with or without the help of indigenous knowledge, and with or without compensation. In this way, bioprospecting includes biopiracy and also includes the search for previously unknown compounds in organisms that have never been used in traditional medicine.

Famous Cases

The Rosy Periwinkle

The rosy periwinkle case dates from the 1950s. The rosy periwinkle, while native to Madagascar, had been widely introduced into other tropical countries around the world well before the discovery of vincristine. This meant that researchers could obtain local knowledge from one country and plant samples from another. The locally known medical properties of the plant were not the same as the medical properties discovered and commercially used by Eli Lilly. The use of the plant as a cure for diabetes was the original stimulus for research, but cures for cancer were the most important results. Different countries are reported as having acquired different beliefs about the medical properties of the plant. The Hodgkin's Lymphoma chemotherapeutic drug vinblastine is also derivable from the rosy periwinkle.

The Neem Tree

In 1995 the U.S. Department of Agriculture and a pharmaceutical research firm received a patent on a technique to extract an anti-fungal agent from the neem tree (Azadirachta indica), which grows throughout [[India][Nepal]]; Indian villagers have long understood the tree's medicinal value. Although the patent had been granted on an extraction technique, the Indian press described it as a patent on the neem tree itself; the result was widespread public outcry, which was echoed throughout the developing world. Legal action by the Indian government followed, with the patent eventually being overturned in 2005.

Importantly, the pharmaceutical company involved in the neem case argued that as traditional Indian knowledge of the properties of the neem tree had never been published in an academic journal, such knowledge did not amount to "prior art" (*prior art* is the term used when previously existing knowledge bars a patent). Public knowledge and public disclosure (verbal or written) is prior art.

The Enola Bean

The Enola bean is a variety of Mexican yellow bean, so called after the wife of the man who patented it in 1999.. The allegedly distinguishing feature of the variety is seeds of a specific shade of yellow. The patent-holder subsequently sued a large number of importers of Mexican yellow beans with the following result: "...export sales immediately dropped over 90% among importers that had been selling these beans for years, causing economic damage to more than 22,000 farmers in northern Mexico who depended on sales of this bean." A lawsuit was filed on behalf of the farmers, and on April 14, 2005 the US-PTO ruled in favour of the farmers. An appeal was heard on 16 January 2008, and the patent was revoked in May 2008. An appeal to the court against the revocation was unsuccessful.

Basmati Rice

In September 1997, a Texas company called RiceTec won a patent (U.S. Patent No. 5,663,484) on "basmati rice lines and grains." The patent secures lines of basmati and basmati-like rice and ways of selecting that rice for breeding. RiceTec, owned by Prince Hans-Adam of Liechtenstein, faced international outrage over allegations of biopiracy. It had also caused a brief diplomatic crisis between India and United States with India threatening to take the matter to WTO as a violation of TRIPS which could have resulted in a major embarrassment for the United States. Both voluntarily and due to review decisions by the United States Patent Office, RiceTec has lost most of the claims of the patent.

Hoodia

Hoodia, a succulent plant, originates from the Kalahari Desert of South Africa. For generations it has been known to the traditionally-living San people as an appetite suppressant. In recent years (2004 onwards) there has been sensationalist media coverage of the plant. Derived products may be introduced into developed countries to help obesity. The long-term benefits are controversial.

Further Cases

The following is a selection of some of the further cases in recent biopiracy studies.

- A large selection of African biopiracy cases are discussed at GhanaWeb's *Public Agenda* (March 31, 2006)
- The case of the Maya people's *pozol* drink is discussed in *The New England Journal of International and Comparative Law* and at Global Exchange
- The case of the Maya and other people's use of *Mimosa tenuifolia*, including many other such cases in general are discussed at GRAIN
- The case of the Andean *maca* radish is discussed in the American University's Trade Environment Database
- The *United Kingdom Select Committee on Environmental Audit 1999; Appendices to the Minutes of Evidence, Appendix 7: Trade Related Intellectual Property Rights (TRIPs) and Farmers' Rights* interestingly lists and describes the cases of turmeric (India), karela (India), quinoa (Bolivia), brazzein berries (Gabon), and others.
- Captopril.

Legal and Political Aspects

Patent Law

One common misunderstanding is that pharmaceutical companies patent the plants they collect. While obtaining a patent on a naturally occurring organism as previously known or used is not possible, patents may be taken out on specific chemicals isolated or developed from plants. Often these patents are obtained with a stated and researched use of those chemicals. Generally the existence, structure and synthesis of those compounds is not a part of the indigenous medical knowledge that led researchers to analyse the plant in the first place. As a result, even if the indigenous medical knowledge is taken as prior art, that knowledge does not by itself make the active chemical compound "obvious," which is the standard applied under patent law.

In the United States, patent law can be used to protect "isolated and purified" compounds. In 1873, Louis Pasteur patented a "yeast" which was "free from disease" (patent #141072). Patents covering biological inventions have been treated similarly. In the 1980 case of *Diamond v. Chakrabarty*, the Supreme Court upheld a patent on a bacterium that had been genetically modified to consume petroleum, reasoning that U.S. law permits patents on "anything under the sun that is made by man." The United States Patent and Trademark Office (USPTO) has observed that "a patent on a gene covers the isolated and purified gene but does not cover the gene as it occurs in nature". Also possible under US law is patenting a cultivar, a new variety of an existing organism. The patent on the enola bean (now revoked) was an example of this sort of patent. The intellectual property laws of the US also recognize plant breeders' rights under the Plant Variety Protection Act, 7 U.S.C. §§ 2321-2582.

Convention on Biological Diversity (CBD)

The CBD came into force in 1993. It secured rights to control access to genetic resources for the countries in which those resources are located. One objective of the CBD is to enable lesser-developed countries to better benefit from their resources and traditional knowledge. Under the rules of the CBD, bioprospectors are required to obtain informed consent to access such resources, and must share any benefits with the biodiversity-rich country. However, some critics believe that the CBD has failed to establish appropriate regulations to prevent biopiracy. Others claim that the main problem is the failure of national governments to pass appropriate laws implementing the provisions of the CBD. The CBD has been ratified all countries in the world except for Andorra, Holy See and United States. The 1994 Agreement on Trade-Related Aspects of Intellectual Property Rights (TRIPs) and the 2001 International Treaty on Plant Genetic Resources for Food and Agriculture are further relevant international agreements.

Bioprospecting Contracts

The ethical debate has sparked a new branch of international patent and trade law. Bioprospecting contracts lay down the rules, between researchers and countries, of benefit sharing and can bring royalties to lesser-developed countries. However, the fairness of these contracts has been a subject of debate. Unethical bioprospecting contracts (as distinct from ethical ones) can be viewed as a new form of biopiracy.

An extensively discussed example of a bioprospecting contract is the INBio agreement between Merck and Costa Rica.

Ownership

The rights at issue in the biopiracy debates are primarily ownership rights. Who (if anyone) owns the Earth's biodiversity? Under what circumstances is it appropriate to talk of 'ownership' of biodiversity, or of particular examples of it? Various and sometimes inconsistent regional and international laws and regulations assign ownership rights to many entities. Moreover, regardless of the legal situation, the case for any party having an ethical right to ownership of biological resources is a separate issue.

Ownership rights of National Governments

Under current international law, national governments exercise a degree of physical control over the biological resources within their country, just as they control mineral rights. However, it is less clear that governments have the right (or even the power) to control knowledge of the application of biological resources, nor are governments able in practice to control the extraction of biological resources in small amounts for research purposes.

An advantage of national government ownership is that some national governments may be strong enough to defend property rights (e.g. against pharmaceutical corporations or other nations). Among objections to national government ownership is the fact that there may be conflicts of interest in developing countries between national governments and local communities. High biodiversity tends to occur in the least developed regions. National governments tend to represent the more developed and urbanised populations of a country. Ethnic, historical, and cultural gaps between governmentally well-represented groups and the populations of the least developed regions are not infrequent. The knowledge at issue in the biopiracy debates is often the knowledge of local communities, not the knowledge of their governments.

Ownership rights of Local Communities

Many groups argue that it is the local communities who possess the traditional biomedical knowledge should benefit from the commercial use of such knowledge. Ownership rights should be attributed to these communities in order to safeguard their interests.

One argument against this is that patent and copyright law are widely implemented as temporary legal mechanisms. Under most intellectual property laws, patents and copyrights expire. In many cases where local communities have been using traditional medicines for generations, these kinds of intellectual property rights would have expired. Many such indigenous groups maintain that their knowledge of the medicines should be protected on an international scale according to each group's internal intellectual property laws.

All Humankind as the Owner of Biodiversity

A more humanistic view of this debate is the claim that biodiversity is something that should be held in common by people in general. Under this view, any and all people who have a need for an advantage reaped by scientific exploration should be granted access to it.

Consequentialist Arguments

Consequentialist arguments typically consider the maximisation of utility or other benefits.

In the case of the Rosy Periwinkle, the most noticeable consequence of the plant's widespread medicinal use is the numbers of cancer patients whose lives it saved. Most ownership rights entail the right to refuse certain uses, and a Consequentialist might conclude that the benefit of saving people's lives should outweigh and supersede property rights. In the case of the Enola bean, however, Consequentialist arguments favour the interests of Mexican bean farmers.

An alternative approach to the utilitarian Consequentialist argument assesses claims of biopiracy by examining the economic incentives underlying each step of the inventive transformation of genetic information into commercially valuable applications. In this reasoning, the benefit of awarding a temporary patent as an incentive for parties that use this information to develop a medicine that can be widely distributed outweighs claims to compensation for alleged biopiracy, especially given the premise that most ethnobiological information already lies in the public domain.

Virtue-based Arguments

Many virtue-based arguments hold that there is something fundamentally and intuitively offensive or wrong about outsiders using traditional knowledge from indigenous communities to make a profit without compensating the communities in question. Virtue-based debates underlie many of this controversy's ethical issues. Many virtue-based approaches recognize the importance of human knowledge of biodiversity while encouraging some kind of exchange between bioprospectors and indigenous peoples.

Traditional Knowledge Database

In response to biopiracy threats faced in cases of turmeric, neem and basmati rice, Government of India has been translating and publishing ancient manuscripts containing old remedies in electronic form, and in 2001 the Traditional Knowledge Digital Library was set up as repository of 1200 formulations of various systems of Indian medicine, such as Ayurveda, Unani and Siddha. The texts are being recorded from Sanskrit, Urdu, Persian and Arabic; made available to patent offices in English, German, French, Japanese and Spanish. The aim is to protect India's heritage from being exploited by foreign companies. Hundreds of Yoga poses are also kept in the collection. The project has been criticized by a spokesman for the pharmaceutical industry as "a solution in search of a problem". The library has also signed agreements with leading international patent offices such as European Patent Office (EPO), United Kingdom Trademark & Patent Office (UKPTO) and the United States Patent and Trademark Office to protect traditional knowledge from biopiracy as it allows patent examiners at International Patent Offices to access TKDL databases for patent search and examinations purposes.

- Diamond v. Chakrabarty was a United States Supreme Court case dealing with whether genetically modified micro-organisms can be patented.

Background

Genetic engineer Ananda Mohan Chakrabarty, working for General Electric, had developed a bacterium (derived from the *Pseudomonas* genus) capable of breaking down crude oil, which he proposed to use in treating oil spills. He requested a patent for the bacterium in the United States but was turned down by a patent examiner, because the law dictated that living things were not patentable.

The Board of Patent Appeals and Interferences agreed with the original decision; however, the United States Court of Customs and Patent Appeals overturned the case in Chakrabarty's favour, writing that "the fact that micro-organisms are alive is without legal significance for purposes of the patent law." Sidney A. Diamond, Commissioner of Patents and Trademarks, appealed to the Supreme Court.

The Supreme Court case was argued on March 17, 1980 and decided on June 16, 1980.

Decision

In a 5–4 ruling, the court ruled in favour of Chakrabarty, and upheld the patent, holding that:

A live, human-made micro-organism is patentable subject matter under [Title 35 U.S.C.] 101. Respondent's micro-organism constitutes a "manufacture" or "composition of matter" within that statute.

Ruling

Chief Justice Warren E. Burger wrote the decision, and was joined by Potter Stewart, Harry Blackmun, William Rehnquist, and John Paul Stevens.

Burger wrote that the question before the court was a narrow one—the interpretation of 35 U.S.C. 101, which says:

> *"Whoever invents or discovers any new and useful process, machine, manufacture, or composition of matter, or any new and useful improvement thereof, may obtain a patent therefor, subject to the conditions and requirements of this title."*

He wrote that:

> *"We have [in earlier ruling 289 U.S. 178] cautioned that courts "should not read into the patent laws limitations and conditions which the legislature has not expressed."*

Regarding the scope of the original legislation, he wrote:

In choosing such expansive terms as "manufacture" and "composition of matter," modified by the comprehensive "any," Congress plainly contemplated that the patent laws would be given wide scope.

Finding that Congress had intended patentable subject matter to "include anything under the sun that is made by man," he concluded that:

Judged in this light, respondent's micro-organism plainly qualifies as patentable subject matter. His claim is... to a nonnaturally occurring manufacture or composition of matter—a product of human ingenuity.

Dissent

The dissenting opinion was written by William J. Brennan, who was joined by Byron White, Thurgood Marshall, and Lewis Franklin Powell.

Brennan's dissent focused on the argument that there is evidence in the legislative record that the United States Congress did not intend living organisms to be patented.

We must be careful to extend patent protection no further than Congress has provided.

Brennan noted that "we do not confront a complete legislative vacuum", and commented on the 1930 Plant Patent Act and 1970 Plant Variety Protection Act, which explicitly allow patents for plants in certain cases:

The Acts evidence Congress' understanding, at least since 1930, that 101 does not include living organisms. If newly developed living organisms not naturally occurring had been patentable under 101, the plants included in the scope of the 1930 and 1970 Acts could have been patented without new legislation.

Therefore:

Because Congress thought it had to legislate in order to make agricultural "human-made inventions" patentable and because the legislation Congress enacted is limited, it follows that Congress never meant to make items outside the scope of the legislation patentable.

And with regard to the specifics of the 1970 act:

Congress specifically excluded bacteria from the coverage of the 1970 Act... The fact is that Congress, assuming that animate objects as to which it had not specifically legislated could not be patented, excluded bacteria from the set of patentable organisms.

Gene Patent

A gene patent is a patent on a specific isolated gene sequence discovery, its chemical composition, processes for obtaining or using it, or a combination of such claims. Gene patents are a prominent subset of the broader category of biological patents. Patents on genes have only been granted on isolated gene sequences with known functions.

History

The United States has been patenting chemical compositions based upon human extractions for over 100 years. The first patent for such a chemical was granted on March 20, 1906 for adrenaline. Judge Learned Hand justified permitting patents on purified natural substances on the rationale that they can become substantially more useful than their non-extracted or less-pure states. Insulin followed adrenaline as the second patent in 1923. Vitamin B12 was also patented (by Merck).

But prior to the 1970s it was almost unheard of to submit an application to the United States Patent and Trademark Office (USPTO) for an organism. The United States Supreme Court in *Diamond v. Chakrabarty* clarified that as long as the organism is truly "man-made," such as through genetic engineering, then it is indeed patentable. As a rule, raw natural material is generally rejected for patent approval by the USPTO. It is only after a DNA product is isolated and purified (or modified) that a patent is considered.

As of 2010, approximately 40,000 United States patents exist that relate to an estimated 2,000 human genes, or about 20 percent of the human genome. As noted above, these patents do not cover the genes in any human body as the genes in the human body are natural products and never have been patentable. The issued patents cover things like (i) isolated genes; (ii) methods of using the isolated genes (for instance, to manufacture protein drugs); (iii) methods to diagnose a disease based on an association between a gene and a disease.

Controversy

There is some political and ethical controversy over whether these patents advance technology by providing scientists at companies and universities with an incentive to create, or if they hinder basic bench research via litigation or threat of litigation. The case that there are negative effects on basic research appears to be mostly driven by academic law professors, and not by practicing university scientists. There is data showing that gene patents have little to no effect on basic university research provided at the Science Progress Blog.

There is also controversy over whether and to what extent companies controlling gene patents clinical research. One of the points raised in the 2009 Myriad case, was that Myriad made efforts to prevent clinical trials to explore just how good their test was and that patents denied patients from receiving second opinions in regards their test results. Finally, there is controversy over the price of medical treatments.

While there is some controversy concerning the patenting of isolated genes and the way those patents are used, and there is controversy concerning patents on the diagnostic uses of genes (the real source of dispute in the Myriad court case), it is difficult to find controversy surrounding patents on genes that are used to manufacture therapeutic proteins (for an example of patents on therapeutic proteins, the drug candidate that is the subject of the early part of the movie 'Extraordinary Measures' was covered in part by a classic gene patent, US Patent 6,770,468) There is also little controversy concerning the role of gene patents in the chemical industry—for instance in the manufacture of enzymes used in consumer products or industrial processes. In descriptions of the controversy over "gene patents", it is rare to find references to these everyday uses of genes and gene patents.

In September 2010 in Australia, Melissa Parke, a member of Parliament, announced that she intends to make a motion calling for amendments to the Patents Act so gene patents would no longer be recognized for reasons similar to those in the Myriad case, discussed below. A news article on this is here.

Gene Patents Ruled Invalid by a U.S. Federal Court; Ruling being Appealed

A complaint against Myriad Genetics and the U.S. Patent and Trademark Office filed in 2009 by professional medical organizations, doctors, and patients, all represented by the American Civil Liberties Union, sought to invalidate and discontinue all patents for naturally-occurring genes, which have thus far been issued on the grounds that such genes are "isolated and purified" to a non-naturally-occurring state. Myriad owns some of its patents; some were licensed from University of Utah. Specific claims, but not all claims, in seven of Myriad's 23 patents on BRCA1 and BRCA2 were challenged in the complaint. The challenged claims covered the isolated genes as well as diagnostic methods.

Many people working in the patent field had predicted that the courts would throw out this case.

However, the case was accepted, and all the challenged claims were ruled invalid on March 29, 2010 by United States District Court Judge Robert W. Sweet, a judge in the U.S. District Court for the Southern District of New York.

The claims that had not been challenged still stand.

Judge Sweet's 152–page decision ruled that the challenged claims to the isolated gene sequences had been "improperly granted" because they claimed unpatentable subject matter; the claims to the diagnostic methods were found invalid under the recent In re Bilsky decision. Because the case could be decided with patent law, Judge Sweet did not look at the challenge on First Amendment grounds and dismissed them without prejudice. In his decision handed down March 29, 2010, the judge rejected the legal equivalency between 'chemical compositions' like purified adrenaline and DNA. In his opinion Judge Sweet said: *"The information encoded in DNA is not information about its own molecular structure incidental to its biological function, as is the case with adrenaline or other chemicals found in the body...this informational quality (of DNA) is unique among the chemical compounds found in our bodies, and it would be erroneous to view DNA as 'no different' than other chemicals previously the subject of patents....DNA, in particular the ordering of its nucleotides, therefore serves as the physical embodiment of laws of nature-those that define the construction of the human body...the preservation of this defining characteristic of DNA in its native and isolated forms mandates the conclusion that the challenged composition claims are to unpatentable products of nature."*

The decision could prove far-reaching for the biotechnology field.

However, on March 30, 2010, Myriad announced that it will appeal the decision. On June 16, 2010, Myriad filed its Notice of Appeal.

On October 29, 2010 the United States Department of Justice filed an amicus brief in this case.

The U.S. amicus brief supports Judge Sweet's decision on gene patents in some respects, and argues against it in others. The amicus brief makes a distinction, on the one hand, between DNA that is isolated and altered, and on the other hand, DNA that has simply been isolated. It argues that isolated and altered DNA should be patentable, whereas DNA that is simply isolated should not be patentable. It reads:

"...the district court erroneously cast doubt on the patent-eligibility of a broad range of manmade compositions of matter whose value derives from the information encoding capacity of DNA. Such compositions—e.g., cDNAs, vectors, recombinant plasmids, and chimeric proteins, as well as countless industrial products, such as vaccines and genetically modified crops, created with the aid of such molecules—are in every meaningful sense the fruits of human ingenuity and thus qualify as '"human-made inventions"' eligible for patent protection.... The district court therefore erred in invalidating the challenged composition claims, such as claim 2 of the '282 patent, that are directed solely to cDNAs. The district court correctly held, however, that genomic DNA that has merely been isolated from the human body, without further alteration or manipulation, is not patent-eligible."

Fuel from Microbes – another LS9 Production

Many plants, insects and microbes naturally produce small quantities of alkanes and alkenes, the long-chain carbon and hydrogen molecules that are major components of diesel, petrol (gasoline) and jet fuel.

The biotechnology company LS9, based in South San Francisco, California, has now pinpointed the biochemical pathway bacteria use to do this – after receiving patents for the genes responsible, of course.

In a paper published today in *Science* (*Science*, 329, 559; 2010 doi:10.1126/science.1187936), they report that they expressed these genes in the bacterium *Escherichia coli*, fed it glucose, and showed that it directly secreted diesel-like fuel – a mixture of alkanes and alkenes with chain lengths from 13 to 17 carbon atoms. This is about what you need for diesel, though the company would have to do some more tweaking to generate the hydrocarbons used in petrol and jet fuel.

The pathway involves two cyanobacteria genes which code for enzymes that convert metabolites of fatty acids (which would normally be used to build cell membranes) into alkanes and alkenes.

The only other known pathway that allows engineered microbes to make pure hydrocarbons, works via a different biochemical route: the conversion of isoprenoid compounds, which are used to fine-tune cell membranes, into hydrocarbons with multiple double bonds. This pathway – favoured by other synthetic biofuels companies, such as Amyris – requires further chemical conversion outside the fermenter, whereas LS9's process involves direct, one-step conversion of feedstock to fuel.

It's LS9's second biofuel tour de force this year: in January, the team showed (*Nature*, 468, 559-562; 2010 and *Nature* news article) that they could engineer *E. coli* to convert sugars or hemicellulose into biodiesel, which is clean and road-ready. The biodiesel consists of a non-hydrocarbon molecule: methyl esters. That project is further advanced, industrially, than the alkane/alkenes pathway that LS9 announce today.

Have LS9 found the only direct *E.coli*-to-alkane pathway that exists? "There are most likely other pathways in other organisms," says the company's Andreas Schirmer.

Insects, for example, can use P450 enzymes to make alkanes, though no-one has yet published the genes involved in this pathway. Moreover, it will be hard to find pathways that can be transferred to an industrial organism like *E. coli*, Schirmer notes. So LS9 may have sown up the intellectual property rights on this one.

The big question for all of these microbial fuel efforts (whose competition is summarised by GreentechMedia) is whether the processes can be scaled up to produce large amounts of fuel, cheaply enough to compete with oil. So far, nobody knows.

"From a purely business perspective, LS9 is doing reasonably well," says Mark Bunger, research director at business consultancy Lux Research in San Francisco. "They claim their costs [for biodiesel] are under $100 per barrel of oil equivalent today (still a bit high – crude sells for about $79); but they are hoping to produce at $50 per barrel by 2012," he says. The company has won $25 million in new funding from various sources, including a strategic partnership with oil giant Chevron last September. They also purchased a bankrupt plant in Florida for $2 million in 2009, which they are now retrofitting in order to produce up to 100,000 gallons of biodiesel capacity in 2010.

Mashelkar Report says Modified Organisms are Intellectual Property

THE expert group on patent law headed by R A Mashelkar has recommended that microbes like bacteria, fungi and virus should be patented if they have been modified. The committee has also said India cannot limit its patents to new chemical substances in pharmaceutical products. The recommendations were submitted to the Union Ministry of Commerce and Industry in early April.

A similar report submitted by the committee in December 2008 had to be withdrawn following charges of plagiarism. The committee had then sought time from the government to resubmit the report.

The expert committee was set up in April 2005 to look into two contentious issues following a parliamentary debate on Patents (Amendment) Bill, 2005.One, whether it is possible to exclude microbes from patent protection and if patents should be restricted to new chemical substances in the pharma sector; two, whether these would be in line with Trade Related Intellectual Property Rights (trips) backed by wto.

The expert group in its report has said new products, processes, uses and compositions using biological materials could be patented. The rationale, the report said, is that India is rich in biodiversity and protecting biotechnological inventions would help biotechnology research compete globally. "Not patenting microbes would be against the international trade regime. What's more, the amendment to the Patents Act clearly identify micro-organisms as patentable," Mashelkar said.

Experts are sceptical about the report and said strict guidelines would be needed to examine patent applications. Inventions should be considered for patent only if they fulfil the criteria of novelty, inventive steps and commercial utility, said the Association of Biotechnology Led Enterprises, a forum of leading Indian biotech companies. Suman Sahai, the convener of Gene Campaign, a non-profit working for farmers' rights and bio-resources, said patents should be issued conditional to the patent holder sharing the economic benefits with the community.

Soil and Microbial Biodiversity

Though Seldom Acknowledged in discussions of agricultural genetic resources, soils are "the critical life-support surface on which all terrestrial biodiversity depends". Soils are providers, storers and generators of biodiversity-but they are also one of the most undervalued and poorly researched habitats on earth. At the very time soil ecologists are beginning to uncover the magnitude and importance of life in the soil, the resource itself is literally disappearing off the face of the earth. Human activities are the greatest threat to soil biodiversity:

- Human-induced soil degradation by wind, water and pollution affects about 24% of the inhabited land area of the globe.
- A recent study by the UK's Royal Commission on Environmental Pollution concluded that some 10% of the world's soil has already been lost this century through deforestation, erosion, urban development and other abuses of the land.
- Approximately 30% of the world's arable crop land has been abandoned because of severe soil erosion in the last 40 years.
- Worldwide, soil is being lost at a rate 13 to 80 times faster than it is being formed. It takes about 500 years to form 25 mm of soil under agricultural conditions, and about 1,000 years to form the same amount in forest habitats.

The staggering diversity of soil biota may be orders of magnitude higher than above ground diversity of plants and animals, but no one has yet made an exhaustive census of even one natural habitat. According to the Global Biodiversity Assessment, "a single gram of temperate forest soil could contain 10,000 million individual cells comprising 4,000-5,000 bacterial types, of which less than 10% have been isolated and are known to science;" more than 500 species of soil invertebrates (e.g. snails, earthworms, termites, mites, nematodes, etc.) have been recorded from a beech forest; over 2,500 species of fungi have been identified from a few hectares of land in southwest England. Even moss tussock communities in the Antarctic Peninsula are home to over a hundred species of soil microorganisms and invertebrates. Tropical soil biota, though perhaps richer than in temperate regions, is still relatively unknown and undocumented.

Microbial diversity encompasses a spectrum of microscopic organisms including bacteria, fungi, algae and protozoa. An estimated 50 percent of all living protoplasm on Earth is microbial. There may be 1.5 million species of fungi yet only 5% are described; as many as one million species of bacteria may exist, but only about 5,000 have been described in the last century. According to new estimates by the Centre for Microbial Ecology at Michigan State University (USA) a gram of typical soil contains about 1 billion bacteria, but only 1 percent can be successfully grown (cultured) in the laboratory. Fewer than 5% of all microbial species have been discovered and named- and even less is known about the diversity within those species. So little is known about most of the microbial world that no one has ever documented the extinction of a bacterium.

Life in the Soil and Life on Earth

Soil biodiversity influences a huge range of ecosystem processes that contribute to the sustainability of life on earth. For example, soil organisms maintain critical processes such as carbon storage, nutrient cycling and plant species diversity. Soil biodiversity plays a role in soil fertility, soil erosion, nutrient uptake by plants, formation of soil organic matter, nitrogen fixation, the biodegradation of dead plant and animal material, reducing hazardous waste, the production of organic acids that weather rocks, and control of plant and insect populations through natural biocontrol.

Through production of food, fibre and renewable forms of energy, soil-based plant productivity supports the livelihood of every person on earth. Soil biota enhance crop productivity because they recycle the basic nutrients required for all ecosystems, including nitrogen, phosphorous, potassium and calcium. Soil organisms enhance the productivity of the soil by increasing water infiltration, thereby reducing surface water runoff and decreasing soil erosion.

Termites, earthworms and other burrow-building soil organisms enhance soil productivity by churning and mixing the upper soil, which redistributes nutrients, aerates the soil and increases surface water infiltration. Earthworms and other invertebrates can bring to the surface from 10 to 500 tonnes/per hectare/per year of soil, and thus play a critical role in the formation of topsoil. Cornell University entomologist David Pimentel estimates that the value of soil biota to soil formation on agricultural land worldwide is US$50,000 million per annum.

Nitrogen from natural and commercial sources is vital to plants and animals. It is the main nutrient required for growth in plants and for building proteins in animals. Biologically fixed nitrogen (primarily nitrogen-fixing microorganisms that live symbiotically on the roots of leguminous plants and trees) makes an enormous contribution to global agricultural productivity. In poor soils, where alternative sources of fertilizer are either unavailable or unaffordable, biological nitrogen-fixation is vital to crop production. Worldwide, an estimated 140 to 170 million tonnes of nitrogen, valued at approximately US$90,000 million are fixed by microorganisms in agricultural and natural systems each year.

Soil biota play a major role in stabilizing and regulating the earth's climate. Global warming is the result of increasing levels of carbon dioxide and other greenhouse gases in the Earth's atmosphere-primarily caused by the burning of fossil fuels by humans. The rate of exchange of carbon between the earth's surface, the oceans and the atmosphere, known as "the carbon cycle", is the primary mediating force with regard to climate change. Through the process of photosynthesis, green plants absorb carbon dioxide from the atmosphere. It is well known that trees and forests store the absorbed carbon in woody biomass. But it is actually soil organic matter that is the major global storage reservoir for carbon. The living microbes, fungi and invertebrates found in the soil are responsible for decomposing carbon and nitrogen and making them available for plant growth, while at the same time contributing to the rate of production and consumption of carbon dioxide, methane and nitrogen.

Despite the importance of soil biodiversity to life-sustaining ecosystem processes, soils are one of the most neglected habitats on earth. In most cases, soil biologists simply don't know which organisms or groups of organisms play the most important roles in ecological processes, they don't know which soil taxa are being lost, or what impact these losses will have in the future.

There is general consensus that we are losing soil biodiversity. Many microbes live symbiotically with higher organisms. Every plant and animal that becomes extinct is likely to take several species of microorganism with it. According to soil ecologist Diana Freckman of Colorado State University, knowledge of soil species remains a "black box" in our understanding of how soil systems function. A study published by the US National Research Council in 1993 noted that "Our lack of knowledge of microorganisms and invertebrates, which are estimated to make up as much as 88% of all species, seriously hampers our ability to understand and manage ecosystems".

Soil ecologists believe that it is essential and urgent to establish the cause and effect relationships between the loss of soil biodiversity and the impact on terrestrial and global ecosystem processes. Only by knowing and understanding life in the soil can we begin to conserve and better utilize its life-sustaining services.

Industrial agriculture has contributed to the neglect of soil biodiversity because conventional soil science has generally relied on the use of purchased farm inputs to overcome constraints and modify the soil environment. (For example, if the soil is dry, irrigate; if soil fertility is low, buy synthetic fertilizer; if pests and weeds invade, spray chemicals). With growing awareness and need for low-input and sustainable agriculture, knowledge of soil biodiversity is increasingly important to future farming systems. A better understanding of soil biota will enable farmers to depend less on modification of the natural environment and place greater emphasis on using biological processes to optimize nutrient cycling, minimize the use of purchased inputs, and maximize the efficiency of their use.

The Value of Microbial Genetic Resources

Microorganisms (or microbes) are tiny living things that are not visible except with a microscope. They include algae, bacteria, fungi (including yeasts), certain protists (one celled animals that are not bacteria) and viruses. Microbial biodiversity is a vast frontier and a potential goldmine for the biotechnology industry because it offers countless new genes and biochemical pathways to probe for enzymes, antibiotics and other useful molecules.

Worldwide, the economic value of microorganisms is estimated to be "at least many tens of billions of US dollars" Pharmaceuticals of microbial origin account for sales of approximately $35-50 billion per annum in the North. It is the invisible world of microbes that has given us more than 3,222 antibiotics, for example, many derived from soil samples. In 1993, five of the pharmaceutical industry's top-selling drugs were derived from microbes; accounting for more than $4,500 million in annual sales. The commercial value of microbials extends beyond pharmaceuticals. The total world market for industrial enzymes, all produced by microorganisms, is $1,300 million. Enzymes are natural catalysts that can speed up a chemical reaction. Because the process is biological, they are biodegradable and can be used instead of synthetic chemicals. For example, industrial enzymes are used to enhance detergents, as biological pesticides, to clean up toxic wastes, to replace chemicals in paper and pulp processing, and for oil extraction.

With the use of modern biotechnnology, the potential applications of microorganisms is vast. Scientists are experimenting with genetically engineered bacteria that are capable of producing products such as biodegradable plastics, artificial skin, and fibres that are as strong as spider silk. Maize, rice, potato and cotton are among the crops that have been genetically engineered to produce insecticidal genes from a common soil bacterium, Bacillus thuringiensis (Bt). The Bt genes enable the crops to produce a toxic protein that kill insects which feed on the plant. Microbial diversity can play an important role in the decomposition of hazardous wastes. Molecular biologists are attempting to harness specific organisms, or groups of organisms, to clean-up toxic wastes in the environment, or reduce hazardous waste production in industrial processes.

Today, transnational microbe hunters are especially interested in exotic and hostile environments-including boiling hot springs, undersea hydrothermal vents, alkali lakes and the frozen tundra of Antarctica-as a source of unexplored microbial diversity. Bioprospecting for microbes goes, quite literally, to the ends of the earth. The following are just a few examples:

- A subsidiary of German agrochemical giant AgrEvo is conducting intensive soil sampling in India. According to German researcher Michael Flitner, the company has already screened over 90,000 Indian soil samples and is building a new, high-efficiency system in Frankfurt for screening plants and soil collected in India.
- When employees of Denmark's Novo Nordisk corporation go on holiday, they take soil-collection kits to gather exotic, enzyme-producing microbes. The father of one Novo Nordisk scientist collected a soil sample from an Indonesian temple which yielded an enzyme that is now widely used by soft-drink suppliers to change starch into sugar. In 1949, Filipino scientist Abelardo Aguilar sent his employer, Eli Lilly Co., samples of an antibiotic isolated from a soil sample that he collected in his home province of Iloilo. Three years later, Eli Lilly sent a congratulatory letter to Aguilar, promising to name the new antibiotic "Ilosone" after the Filipino province where the soil sample was found. The drug, erythromycin, sold under the brand name "Ilosone" has since earned Eli Lilly millions of dollars, but neither Aguilar nor the Philippines received any

royalties, despite Aguilar's 40-year battle to be recognized and rewarded.

- Sponges growing on a coral reef off the coast of Papua New Guinea are the source of a powerful antifungal agent "Papuamine." Because the sponges yield only minute quantities of the antifungal agent, Myco Pharmaceuticals (USA) is now attempting to synthesize papuamine in the laboratory.
- Bacteria found in the whale gut from the last legal Eskimo whale hunt are capable of breaking down toxic petrochemicals. Scientist A. Morrie Craig of Oregon State University has applied for patents on some of the whale gut bacteria, and Pioneer Hi-bred has already secured rights over commercial products that may someday result from the bacteria.
- Bacteriologist Thomas D. Brock discovered a bacterium, Thermus aquaticus in the boiling hot springs of Yellowstone National Park (USA) in 1966. An enzyme isolated from Thermus aquaticus is the catalyst for the polymerase chain reaction, or PCR, a technique that is widely and routinely used for producing copies of any DNA sequence. Although Thomas Brock donated the bacterium he discovered to the scientific world, it was later patented and now brings royalties valued at hundreds of millions of dollars annually to Swiss pharmaceutical corporation Hoffman-LaRoche.

Microbial Genetic Resources in the International Policy Arena

Despite its growing economic importance, microbial genetic diversity has been under-valued and under-recognized in biodiversity debates. There is an obvious policy gap in the international arena, and it is poor farmers who will likely pay the greatest price for this oversight. The vast majority of microbial culture collections are located in the North, and there is a growing trend toward privatization and patenting of this material. Microbial genetic resources can no longer be disregarded as ubiquitous life forms outside of the mainstream of biodiversity policy debates. Today, the genetic resources of microorganisms are very much an issue in the international policy arena.

Microbial Biodiversity-where's the Political Debate?

The Convention on Biological Diversity excludes from its scope all ex situ germplasm collected prior to the Convention coming into force at the end of 1993. This means that all microbial culture collections, the vast majority of which are located in the industrialized world, are the legal property of the depositor and not of the donor country, regardless of where the germplasm was collected. The U.S.-based American Type Culture Collection, the world's largest microbial culture collection, contains thousands of biological specimens from the South, dozens of which are the subject of patent claims by Northern pharmacueticals such as Bristol-Myers, Pfizer and Eli Lilly. Patent culture depositories are regulated internationally by the Budapest Treaty on International Recognition of the Deposit of Microorganisms for the Purposes of Patent Procedure administered by the World Intellectual Property Organization in Geneva. Currently, 32 countries are signatories to the Budapest Treaty. An estimated 86% of global microbial collections is held in industrialized nations.

A network of microbial resource centres for the developing world (MIRCENs) was established in the early 1970s by UNEP and UNESCO. Today, there are no policies in place to protect these microbial genetic resources from privatization or to insure equitable exchange of microbial genetic resources in culture collections worldwide. Normally, MIRCENs have a policy of free exchange of microbial materials within the network, but each MIRCEN may decide on a case-by-case basis.

The Uruguay Round of the General Agreement on Tariff and Trade (GATT) incorporates an element called Trade Related Aspects of Intellectual Property Rights (TRIPs) which specifies that microorganisms may not be excluded from patent protection (Section 5, Article 27.2). All countries that are signatories to the World Trade Agreement are now obligated to adopt and implement patent laws for microorganisms and for biotechnology processes applied to living organisms. What is the definition of a microorganism? When is a microorganism patentable? For the purposes of patent protection, there is considerable uncertainty and controversy regarding the answer to these questions. In many countries, the term microorganism extends to cell lines and plasmids-including human genetic material.

The patenting of human genetic material is no longer a theoretical concern, but a shocking reality. On March 14, 1995 the US Patent and Trademark Office granted a patent to the US National Institutes of Health (NIH) for an unmodified human cell line drawn from a 20-year old Hagahai man from Papua New Guinea. It is the first time that an indigenous person's cells have ever been patented. Not only plants, animals and microorganisms from gene-rich ecosystems of the South, but also the genes and cells of indigenous peoples have become targets of Northern scientists and industrial bioprospectors. Private ownership of human biological materials raises many profound moral, ethical, and political issues. There is no international protocol to protect human subjects from patent claims and unjust commercial exploitation. And there is no mechanism to compensate individuals or communities from whom DNA samples are taken.

Signatories to the World Trade Agreement must determine whether or not human genetic materials are included in its definition of microbial materials. At the Jakarta meeting of the Conference of Parties to the Convention on Biological Diversity held in November, 1995, delegates made it clear that they did not wish to regard human genetic materials as part of the Convention, despite the fact that the legally-binding Convention does not explicitly exclude human biodiversity from its mandate. The World Health Organization has yet to establish internationally-accepted medical ethics protocols covering the commericialization or patenting of human genetic material. There is a serious policy vacuum that some international body must fill.

7

Industrial Aspects of Amylase Enzymes

Amylase is a digestive enzyme classified as a saccharidase (an enzyme that cleaves polysaccharides). It is mainly a constituent of pancreatic juice and saliva, needed for the breakdown of long-chain carbohydrates (such as starch) into smaller units.

Amylase is a digestive enzyme made primarily by the pancreas and salivary glands. Enzymes are substances made and used by the body to trigger specific chemical reactions. The primary function of the enzyme amylase is to break down starches in food so that they can be used by the body. Amylase is also synthesised in the fruit of plants during ripening, causing them to become sweeter.

Amylases are enzymes that catalyse the hydrolysis of alpha-1, 4-glycosidic linkages of polysaccharides to yield dextrins, oligosaccharides, maltose and D-glucose. Amylases are derived from animal, fungal and plant sources. Pancreatin and pancrelipase contain amylase derived from the pancreas of animals, usually porcine pancreas. Amylase is also derived from barley malt and the fungus Aspergillus oryzae. There are a few different amylases. These enzymes are classified according to the manner in which the glysosidic bond is attacked. Alpha-amylases hydrolyze alpha-1, 4-glycosidic linkages, randomly yielding dextrins, oligosaccharides and monosaccharides. Alpha-amylases are endo-amylases. Exoamylases hydrolyze the alpha-1, 4-glycosidic linkage only from the non-reducing outer polysaccharide chain ends. Exoamylases include beta-amylases and glucoamylases (gamma-amylases, amyloglucosidases). Beta-amylases yield beta-limit dextrins and maltose. Gamma-amylases yield glucose. Amylases are used as digestants.

Amylase is a calcium dependent enzyme which hydrolyzes complex carbohydrates at alpha 1,4-linkages to form maltose and glucose. Amylase is filtered by renal tubules and resorbed (inactivated) by tubular epithelium. Active enzyme does not appear in urine. Small amounts of amylase are taken up by Kupffer cells in the liver. In healthy dogs, 14% of amylase is bound to globulins. Because of this polymerization, canine amylase has variable (high) molecular weights and is not normally filtered by the kidney. In dogs with renal disease, this polymerized (macroamylase) amylase is found in higher concentration (from 5-62% of total amylase activity) and contributes to the hyperamylasemia seen in these disorders. There are 4 different isoenzymes of amylase in the dog: isoenzyme 3 is found in the pancreas (>50%), whereas isoenzyme 4 is found in all tissues. a-Amylase catalyses the hydrolysis of internal (endoglycosidase) a-1,4-glucan links in polysaccharides containing 3 or more a-1,4-linked D-glucose units (amylose and amylopectin) yielding a mixture of maltose and glucose. Both of these monosaccharides can then be converted to about 30 ATP units of energy by encountering the glycolysis mechanism and the citric acid cycle. Both amylose and amylopectin are examples of polysaccharides, that is, polymers made of many sugar units. Both are composed entirely of glucose monomers. Amylose is a straight chain polymer of glucose units bonded by a-1,4 glycosidic linkages. This means that neighboring glucose units are joined from the C1 carbon of one to the C4 carbon of the other, the bond having an orientation referred to as alpha. Amylopectin is a branched chain molecule. The glucose units in amylopectin are bonded in the same manner as in amylose, except for the branch points where the chains are joined by a-1,6 glycosidic linkages. Amylose composes from 20 to 30% of starch. Amylopectin comprises the remaining 70 to 80%, and because of its branching, is less soluble and more viscous than amylose.

Alpha-Amylase (1,4-a-D-glucan glucanohydrolase) in animals are monomeric, calcium-binding proteins that are present in both salivary and pancreatic secretions. In the gastrointestinal tract, it is a digestive enzyme that is secreted into the Duodenum where it starts to degrade (hydrolyze) the a-1,4 glucosidic linkages of starch to maltose and maltotriose. a-Amylases are used to hydrolyze glycogen, the reserve carbohydrates in animals, when the blood glucose levels are low. Salivary amylase, a major component of human salivary secretions, possesses multiple functions in the oral cavity. It is the only enzyme in the saliva capable of degrading oligosaccharides, which are used by the oral microflora for nutritional purposes. The structural neighbour, b-amylase, breaks down starch during the germination of seeds (rich in starch) into sugars. These sugars constitute the chief energy source in the early development of the plant. b-Amylase is able to break down the a-1,4 linkages of starch polymers in plants and seeds.

Amylase is also used in industry. It is used in brewing and fermentation industries for the conversion of starch to fermentable sugars, in the textile industry for designing textiles, in the laundry industry in a mixture with protease and lipase to launder clothes, in the paper industry for sizing, and in the food industry for preparation of sweet syrups, to increase diastase content of flour, for modification of food for infants, and for the removal of starch in jelly production.

The pancreas has the highest amylase concentration and largest total amount of amylase of any organ in the body. Unfortunately, other tissues, particularly the parotid salivary glands, also contain large quantities. Pancreatic or salivary amylase is not absorbed by intact gut mucosa, and normal serum amylase activity levels are low. Any amylase that is present presumably comes from enzyme leakage directly into the blood from the ascinar cells or via lymphatics draining the glands. With ductular obstruction and/or inflammation of either the pancreas or parotid glands, enzyme leakage directly into the blood or via the lymphatics increases. In rare situations gut permeability changes, whereby a large quantitiy of amylase leaks into the peritoneal fluid and is eventually delivered via thoracic duct into the circulation causing hyperamylasemia. When amylase levels are too high in the urine and someone has abdominal pain then they are automatically diagnosed with pancreatitis.

The major role of the calcium-containing metalloenzyme amylase in the oral cavity is the hydrolysis of the alpha-1,4 linkages of starch. Amylase can play an important role in the colonization and metabolism of oral bacteria leading to dental plaque formation. Amylase bound to the bacterial surface retains its enzymatic activity. The bacterial-bound amylase can hydrolyze starch to glucose, which is then metabolized to lactic acid by the bacteria, thereby leading to dental plaque formation and progression.

History of Enzymes

The history of modern enzyme technology really began in 1874 when the Danish chemist Christian Hansen produced the first specimen of rennet by extracting dried calves' stomachs with saline solution. Apparently this was the first enzyme preparation of relatively high purity used for industrial purposes.

This significant event had been preceded by a lengthy evolution. Enzymes have been used by man throughout the ages, either in the form of vegetables rich in enzymes, or in the form of microorganisms used for a variety of purposes, for instance in brewing processes, in baking, and in the production of alcohol. It is generally known that enzymes were already used in the production of cheese since old times.

Even though the action of enzymes has been recognised and enzymes have been used throughout history, it was quite recently that their importance were realised. Enzymatic processes, particularly fermentation, were the focus of numerous studies in the 19th century and many valuable discoveries in this field were made. A particularly important experiment was the isolation of the enzyme complex from malt by Payen and Persoz in 1833. This extract, like malt itself, converts gelatinised starch into sugars, primarily into maltose, and was termed 'diastase'.

Development progressed during the following decades, particularly in the field of fermentation where the achievements by Schwann, Liebig, Pasteur and Kuhne were of the greatest importance. The dispute between Liebig and Pasteur concerning the fermentation process caused much heated debate. Liebig claimed that fermentation resulted from chemical process and that yeast was a non-viable substance continuously in the process of breaking down. Pasteur, on the other hand, argued that fermentation did not occur unless viable organisms were present.

The dispute was finally settled in 1897, after the death of both adversaries, when the Buchner brothers demonstrated that cell free yeast extract could convert glucose into ethanol and carbon dioxide just like viable yeast cells. In other words, the conversion was not ascribable to yeast cells as such, but to their non-viable enzymes.

In 1876, William Kuhne proposed that the name 'enzyme' be used as the new term to denote phenomena previously known as 'unorganised ferments', that is, ferments isolated from the viable organisms in which they were formed. The word itself means 'in yeast' and is derived from the Greek 'en' meaning 'in', and 'zyme' meaning 'yeast' or 'leaven'.

Early Developments in Japan

During the early part of this century, enzyme technology was also developing slowly but surely outside Europe. In the Far East, an age-old tradition prevailed where mould fungi, the so-called koji, were (and still are) used in the production of certain foodstuffs and flavour additives based on soya protein (shoyu, miso, tempeh) and fermented beverages (sake, alcohol). Koji is prepared from steamed rice into which a mixture of mould fungi is inoculated, the composition of the mixture being passed down from generation to generation. This formed the basis which the Japanese scientist Takamine developed a fermentation process for the industrial production of fungal amylase; the process included the culture of Aspergillus oryzae on moist rice or wheat bran. The product was called "Takadiastase' and it is still used as a digestive aid. The method of fermentation suggested by Takamine, the 'surface culture' or 'semi-solid culture' is still actively used in the production of various enzymes.

Textile Desizing

At about the same time as Takamine was developing his novel fermentation technique, another field was being opened up for the use of enzymes-the desizing of textiles. Previously, textiles were treated with acid, alkali or oxidising agents, or soaked in water for several days so that naturally occurring microorganisms could break down the starch. However, both of these methods were difficult to control and sometimes damaged or discoloured the material. It represented great progress, therefore, when crude enzyme extracts in the form of malt extract, or later, in the form of pancreas extract, were first used to carry out desizing.

Bacterial amylase derived from Bacillus subtilis was used for desizing, the first time by Boidin and Effront as early as 1917.

Leather Bating

Investigations carried out by the German chemist and industrial magnate Otto Rohm before World War I were of great importance for the further development of the industrial use of enzymes. Among other things, he studied the so called 'bating' process, a step in the preparation of hides and skins prior to tanning.

According to tradition, bating required the excrement of dogs and pigeons, a fact that did not improve the image of tanning which was considered a stinking and unpleasant activity. Rohm's theory was that these excrements exerted their effect because they contained residual amounts of the animals' digestive enzymes. If this was so, it might be possible to use extracts of the pancreas directly for bating. Such extracts were tried and produced the expected positive results. Naturally, Rohm accepted this as confirmation of the correctness of his theory, but later experiments showed that it was not the animals' enzymes that were active, but rather enzymes of bacteria growing in the intestinal tract.

The First Detergent Enzyme

Parallel to his studies of the problems involved in tanning, Rohm investigated other processes where enzymes would prove even more valuable. Nevertheless, his efforts were not to score a success until 50 years later. Rohm actually developed the first method for washing protein stained cloth in detergents containing enzymes and manufactured the first detergent preparation containing enzymes.

The enzyme preparation used was pancreatin (extracted from pancreatic glands), which contains the protein degrading enzyme trypsin.

Breakthrough in detergents was made in 1959, when a Swiss chemist Dr. Jaag, developed a new product called Bio 40 containing a bacterial protease instead of trypsin.

Sugars from Starch

A very important field in which enzymes have proved to be of great value over the last 15-20 years is the starch industry. In 1950s, fungal amylase was used in the manufacture of specific types of syrup, i.e., those containing a range of sugars, which could not be produced by conventional acid hydrolysis. The real turning point was reached early in the 1960s when an enzyme glucoamylase, was launched for the first time, which could completely break down starch into glucose. Within a few years, almost all glucose production was re-organised and enzyme hydrolysis was used instead of acid hydrolysis because of the more benefits such as greater yield, higher degree of purity and easier crystallisation.

The process was further improved by the introduction of a new technique used for the enzymatic pre-treatment (liquefaction) of starch by using a heat-stable alpha amylase.

Years of research in biochemistry and biotechnology have boosted knowledge of enzymes for industries as well as research. Many new techniques have been established to modify enzymes or increase their yields. New techniques for purification of enzymes are constantly developing and so are being discovered new application of enzymes in medicine, research and industries.

Enzymes for Textile

At Maps, we continuously develop our product line in order to have innovative enzymes with unique performance features for existing and new applications within the textile industry. Our R&D aims to provide innovative products for fabric treatment reducing process time, chemical consumption and energy costs in compliance with sustainable development.

We provide a range of enzymes like amylases, cellulases, catalase, pectinase and protease for various textile wet-processing applications like desizing, bio-polishing, denim finishing, bleach clean-up, bio-scouring and de-wooling.

Desizing

For fabrics made from cotton or blends, the warp threads are coated with an adhesive substance know as 'size'; to prevent the threads breaking during weaving. Although many different compounds have been used to size fabrics, starch and its derivatives have been the most common sizing agent. After weaving, the size must be removed again in order to prepare the fabric for dyeing and finishing.

This process (desizing) must be carried out by treating the fabric with chemicals such as acids, alkali or oxidising agents. However starch-breaking enzymes (amylases) are preferred for desizing due to their high efficiency and specific action. Amylases bring about complete removal of the size without any harmful effects on the fabric. Another benefit of enzymes compared to strong chemicals mentioned above is that enzymes are environment friendly.

Maps offers a range of amylases for desizing which work at different temperatures and for different equipments.

Palkozyme	Alpha amylase for low-medium temperature conventional desizing.
Palkozyme Ultra	Alpha amylase for low-medium temperature desizing
Palkozyme Plus	Alpha amylase for high temperature desizing
Palkozyme HT	Heat-stable alpha amylase for high temperature desizing
Palkozyme CLX	Alpha amylase for low temperature desizing

Bio-Polishing

Cotton and other natural fibres based on cellulose can be improved by an enzymatic treatment known as Bio-Polishing. This treatment gives the fabric a smoother and glossier appearance. The treatment is used to remove 'fuzz'-the tiny strands of fibre that protrude from the surface of yarn. A ball of fuzz is called a 'pill' in the textile trade. After Bio-Polishing, the fuzz and pilling are reduced. The other benefits of removing fuzz are a softer and smoother handle, and superior colour brightness.

Maps offers a range of cellulases for bio-polishing which work on depending on fibre, fabric type and equipments.

Palkofeel	Cellulase for bio-polishing cotton and blended fabric and garment
Palkofeel C	Cellulase for bio-polishing cotton fabric and garments
Palkosoft	Cellulase for bio-polishing cotton and blended fabric and garment

Denim Finishing

Many garments are subjected to a wash treatment to give them a slightly worn look; example is the stonewashing of denim jeans. In the traditional stonewashing process, the blue denim was faded by the abrasive action of pumice stones on the garment surface. Nowadays, denim finishers are using a special cellulase.

Cellulase works by loosening the indigo dye on the denim in a process know as 'Bio-Stonewashing'. A small dose of enzyme can replace several kilograms of pumice stones. The use of less pumice stones results in less damage to garment, machine and less pumice dust in the laundry environment.

Bio-Stonewashing has opened up new possibilities in denim finishing by increasing the variety of finishes available. For example, it is now possible to fade denim to a greater degree without running the risk of damaging the garment. Productivity can also be increased because laundry machines contain fewer stones or no stones and more garments.

Maps offers a range of cellulases for denim finishing, each with its own special properties. These can be used either alone or in combination with pumice stones in order to obtain a specific look.

Palkowash	Cellulase for bio-stonewashing denims used in garment wet-processing
Palkostone	Cellulase for bio-stonewashing denims used in garment wet-processing
Palkocel	Cellulase for bio-stonewashing denims used in garment wet-processing

Bleach Clean-up

Natural fabrics such as cotton are normally bleached with hydrogen peroxide before dyeing. Bleaches are highly reactive chemicals and any peroxide left on the fabric can interfere with the dyeing process. A thorough 'Bleach Cleanup' is necessary. The traditional method is to neutralize the bleach with a reducing agent, but the dose has to be controlled precisely. Enzymes present a more convenient alternative because they are easier and quicker to use. A small dose of catalase is capable of breaking down hydrogen peroxide into water and oxygen. Compared with the traditional clean-up methods, the enzymatic process results in cleaner waste water or reduced water consumption.

Maps offer catalase for removing residual hydrogen peroxide after the bleaching of cotton. It reduces the rinsing necessary to remove bleach or it can be used to replace chemical treatments.

Palkoperox	Catalase for bleach clean-up i.e. removal residual hydrogen peroxide after the bleaching of cotton.

Bio-Scouring

Cotton yarn or fabric, prior to dyeing or printing, goes through a number of processes in a textile processing unit. A very important process is scouring. In this process, non-cellulosic components from native cotton are completely or partially removed.

Scouring gives a fabric with a high and even wet ability so that it can be bleached and dyed successfully. Today, highly alkaline chemicals caustic soda are used for scouring. These chemicals not only remove the non-cellulosic impurities from the cotton, but also attack the cellulose leading to heavy strength loss and weight loss in the fabric. Furthermore, using these hazardous chemicals result in high COD (chemical oxygen demand), BOD (biological oxygen demand) and TDS, in the waste water.

Recently a new enzymatic scouring process know as 'Bio-Scouring' is used in textile wet-processing with which all non-cellulosic components from native cotton are completely or partially removed. After this Bio-Scouring process, the cotton has an intact cellulose structure, with lower weight loss and strength loss. The fabric gives better wetting and penetration properties, making subsequent bleach process easy and resultantly giving much better dye uptake.

Enzymes for Leather

One of the oldest applications of industrial enzymes is processing hides and skins for leather. Hides and skins contain proteins and fat in between collagen fibres and before tanning; these substances should be partially and fully removed. The proteins can be removed by proteases and lipases as well as other chemicals can remove the fat. Today, proteases and lipases are mainly used for soaking, bating and enzyme assisted un-hairing. Using lipases to dissolve and remove fat is a recent development and lipases are now extensively used for leather processing in many parts of the world.

Maps is a major supplier of enzyme to the leather industry in India and across the globe. We offer total enzyme solutions for bating, un-haring, degreasing and soaking in the beam-house processes. With the introduction of our new range of products based on Microorganisms, we assure to provide clean and green leather tanneries.

Bating

To make leather pliable, the hides and skins require an enzymatic treatment before tanning know as bating. During bating, scud is loosened and other unwanted proteins are removed. Bating de-swells swollen pelts and prepares leather for tanning. It makes the grain surface of the finished leather clean, smooth and fine. Bating with enzymes is an indispensable operation of leather processing to obtain best quality of leather and cannot be substituted with a chemical process.

Traditional methods for bating employed manure of dog, pigeon or hen. These were very unpleasant, unreliable and slow methods. Bio-technical developments in science have now completely replaced these methods with use of industrial enzymes.

Maps offers a range of proteases for bating which work in different pH conditions.

Palkobate	Protease for bating in alkaline pH conditions
Palkocid	Protease for bating in acidic pH conditions

Soaking

Soaking is first important operation of leather processing. Hides and skins received into a tannery are in the four conditions, as green or fresh, as wet salted, as dry salted or as dried. It is advisable to carry out soaking for all types of skin and hides to obtain best quality leather. Soaking cleans hides and skins by removing dirt, blood, flesh, grease, dung etc. and most importantly, re-hydrates them to bring skins as far as possible back to state of green hides. Soaking agents fall into three categories, like Chemical Agents, Surface-active agents and Enzymatic agents. Enzymatic agents are biocatalyst. Specific protease and lipase enzymes enhance water uptake by dissolving intrafibrillary proteins that cement fibres together and disperse fats and oils together with dirt and other contaminants present on skin.

Maps offers a range of protease and lipase for soaking which work in different pH conditions.

Palkosoak	A mixture of protease and lipase for soaking in alkaline pH conditions
Palkosoak ACP	A mixture of protease and lipase for soaking in acidic pH conditions

Un-hairing

The conventional and most wide spread way to remove hair from bovine hides is to use lime and sodium sulphide in a hair-burning process. They dissolve the hair and open up the fibre structure.

Most importantly, enzyme-assisted un-hairing results in a cleaner grain surface and improved area yield and softness. The use of a specific protease also offers tanneries a number of options. For instance, the sulphide and lime requirements can be reduced by as much as 40% while maintaining the same liming time. Alternatively tanners can shorten the liming time by at least half without any loss of quality. Another possibility is to avoid the use of amines, which can be converted into carcinogenic compounds. The hair-burning process is the most widespread but a better alternative to this, is the hair-saving process, which is environment friendly, where the hair is not dissolved but can be filtered out from the liming float. It is possible to reduce the COD up to 50% and BOD up to 30% in waste discharges.

Maps offers a specific protease for un-hairing which can be used either alone or in combination lime and sodium sulphide:

Palkodehair	Protease for un-hairing in high alkaline pH conditions

Degreasing

Lipases are a type of enzyme that specifically degrades fat and so cannot damage the leather itself. Lipases hydrolyse not just the fat on the outside of the hides and skins, but also the fat inside the skin structure. Once most of the natural fat has been removed, subsequent chemical treatments such as tanning, re-tanning and dyeing have a better effect.

The main advantages of using lipases are a more uniform colour and a cleaner appearance. Lipases also improve the production of hydrophobic (waterproof) leather; makers of leather for car upholstery have commented that 'fogging' is reduced. This is the term for the build-up of a film of chemicals on the inside of car windscreens

Lipases represent a more environmentally sound method of removing fat. For bovine hides, lipases allow tensides to be replaced completely. For sheepskins, which contain up to 40% fat, the use of solvents is very common and these can also be replaced with lipases and surfactants. Solvents tend to dry out the skin and give it a pale colour.

If surfactants are used for sheepskins, they are usually not as effective and may be harmful to the environment. Stronger surfactants such as nonyl phenol ethoxylate have a better effect but they are more detrimental to the environment. When using lipases, the original surfactant dosage can be reduced by at least 50% in the case of both sheepskins and pigskins. In addition, nonyl phenol ethoxylate can be substituted with more biodegradable surfactants.

Enzymes for Baking

As a result of our R&D over the past few years, today Maps has a versatile product range of enzymes for the baking industry. Apart from individual enzymes like amylase, xylanase, protease, cellulase, we also have a range of tailored enzyme cocktails (mixtures of different enzymes) to solve problems in certain baking applications. Our customers for our baking enzymes are the baking improver industry or the milling industry, the Wafer, Biscuit and Cracker industry, etc.

Like all other living material, the cells in cereal grains used for flour contain enzymes. The most important enzymes in flour are the amylases and proteases. However, the quantities of these enzymes are not always ideal for baking purposes and supplementary enzymes often need to be added.

Bread-making

Bread is the most common and traditional foods around the world. But bread actually has close links with enzymes. For years, enzymes such as malt and fungal alpha-amylase have been used in bread making. Due to the changes in the baking industry and the ever-increasing demand for more natural products, enzymes have gained real importance in bread-making.

The dough for bread, rolls, buns, etc. consists of flour, water, yeast, salt and other ingredients such as sugar and fat. Flour consists of gluten, starch, non-starch polysaccharides, lipids, etc. When the dough is made, the yeast starts to work on the fermentable sugars, transforming them into alcohol and carbon dioxide, thus rising the dough.

In the beginning, the fermentation goes smoothly whether sugar has been added or not, because flour always contains a certain amount of fermentable sugar. But when this has been used up, the fermentation process will cease unless new supplies of sugar are made available to the yeast.

Amylases degrade starch and produce small dextrins for the yeast to act. Gluten is a combination of proteins, which form a large network during dough formation. This network holds the gas in dough proofing and baking. The strength of this network is very important for the quality of all bread raised by yeast. Enzymes such as proteases, xylanases and lipases directly or indirectly improve the strength of the gluten network and so improve the quality the bread.

Dough Improvement

A small percentage of pentosans (non-starch polysaccharides) are present in flour. Pentosans have an important role in bread quality due to their water absorption capability and interaction with gluten, which is vital for the formation of the loaf structure. By hydrolysing the pentosans using some enzymes like hemicellulase, pentosanase or xylanase, the dough becomes easier to handle and the resulting bread has a bigger loaf volume and an improved crumb structure

Maps offers a range of amylases, proteases and xylanases for bread-making and dough improvement

Palkoamylo	Fungal alpha amylase
Palkobake X	Fungal xylanase
Palkotase ACP	Fungal protease

Flour Supplementation

Alpha amylases have significant effects on baked goods. If the content is low, this leads to low dextrin production and poor gas production. This in turn results in inferior quality bread with reduced size and poor crust colour.

To compensate for the deficiencies of the grain, it is necessary to add either sugar or alpha amylase. The addition of enzymes offers certain advantages over sugar. At a flour mill, it is possible to standardize the enzyme content of the flour so that a uniform commodity can be supplied. Furthermore, enzymes bring about a gradual formation of sugar, which matches the needs of the yeast. When the dough is placed in the oven, the steadily increasing temperature leads to an increase in the enzymes' rate of reaction and more sugar is produced. Malt flour and malt extract can be used as enzyme supplements as malt is rich in alpha amylases. However, it is better to use a fungal alpha amylase.

The alpha-amylases degrade the damaged starch in wheat flour into small dextrins, thus allowing yeast to work continuously during dough fermentation, proofing and the early stage of baking. This result in improved bread volume and crumb texture. In addition, the small oligosaccharides and sugars such as glucose and maltose produced by these enzymes enhance the reactions for the browning of the crust and baked flavour.

Maps offers a range of amylases and xylanases for flour supplementation, each with its own special properties which work to obtain specific needs of wheat flour.

Palkoflour	Fungal alpha amylase
Palkoamylo	Fungal alpha amylase
Palkoflour AX	A mixture of fungal alpha amylase and xylanase

Production of Biscuit and Crackers

Another application of enzymes in baking is in the production of biscuits and crackers. The requirements of the flour are altogether different from those in bread-making; a 'soft flour' which produces a dough with pronounced plastic properties is preferred. For this purpose, flour with relatively low protein content is desirable. The gluten protein structure should not be too strong, otherwise the dough will be too difficult to handle.

Unless flour with these properties is available, it is necessary to add an agent to weaken the gluten. Reducing agents (substances which have the opposite effect to oxidizing agents) have been used for this purpose, in particular sodium bisulphite. The bisulphite has the desired effect on the gluten, but unfortunately it affects other substances in the flour, including the content of vitamin B1 (thiamine). This vitamin is completely or partially destroyed. Sodium bisulphite has been banned in certain countries and is becoming less popular due to health risks.

An alternative is the application of a protein-degrading enzyme. This softens the gluten without affecting the other constituents of the dough. Several fungal and bacterial proteases can be used for this purpose. Proteases can also be used when making bread with 'hard flour' i.e. flour high in gluten protein.

Enzymes for Alcohol

The production of fermented alcoholic drinks from starch-based raw materials has been practised since long. The raw material for this differs from, maize (corn) and rye to barley and wheat for whisky and other cereals for grain spirits.

But whatever the raw materials are, starch is still the basic ingredient. Starch is composed of long chains of glucose molecules and these have to be broken down into smaller molecules, which the yeast can transform into alcohol. Enzymes can carry out this process in two stages-liquefaction and saccharification.

Traditionally, enzymes have been provided by adding malt. However, since quite some time, there has been a change and in many countries malt has been totally replaced from distilling operations by the use of enzymes. There are a lot of advantages using industrial enzymes instead of malt, like small quantities of enzyme addition can replace large quantities of malt, resulting easy handling and storage. Even raw material costs are reduced by nearly 30% when switching to commercial enzymes.

Starch Liquefaction

Before enzymes can attack the starch, it must be gelatinised. This is usually done by pressure-cooking. Now-a-days, the old no pressure cooking method has been gaining in popularity, especially in smaller distilleries. Instead of temperatures around 150?C, the maximum temperature ranges from 60?C to 95?C. This results in energy savings and also capital savings since pressure vessels are not required. Whatever the processing technique, alpha amylases are used to break down the gelatinised starch to shorter molecules (dextrins).

Starch Saccharification

Saccharification is the second step in the process. A glucoamylase enzyme is used to break down the starch molecules and the dextrins. This enzyme is capable of achieving the complete degradation of the starch to fermentable sugars (glucose). During fermentation, these sugars are converted into alcohol by yeast in simultaneous saccharification and fermentation process.

Fermentation Improvement

Cereals, particularly maize have low soluble nitrogen. This results in poor yeast growth and longer fermentation time. By adding a small amount of protease enzyme helps in yeast growth and reduce fermentation time. During distillation process it may be necessary to reduce viscosity of the fermentation broth. To facilitate this beta glucanase/pentosanase enzymes are added.

Enzymes for Brewing

Traditionally production of beer is by mixing crushed barley malt and hot water in large vessel known as mash copper. This process is known as 'mashing'. Apart from malt, other starch-based cereals such as maize (corn), sorghum, rice and barley, or pure starch itself, are added to the mash.

These are known as adjuncts. After mashing, the mash is filtered in a lauter tun and the liquid, known as 'sweet wort', is then run off to the copper, where it is boiled with hops. The 'hopped wort' is cooled and transferred to the fermentation vessels where yeast is added. After the fermentation, the so-called 'green beer' is matured before the final filtration and bottling.

Liquefying Adjuncts

Adjuncts are used as an extra source of starch in brewing and usually consist of cereals. Protein also plays a vital role in fermentation by providing soluble nitrogen compounds that the yeast needs. Since the protein proportion in some types of adjunct is very small, it may be necessary to provide extra protein in the wort. This can be done by utilising the prctein in the malt more efficiently. By adding a protein splitting (protease) enzyme, more protein can be solubilised and made available to the yeast.

In their natural form, starch-based cereals such as maize grits and rice are boiled before being added to the malt mash. The boiled (gelatinised) cereals are very viscous and need to be thinned (liquefied) before they can be pumped into the mash copper. This is done using an alpha amylase. Heat-stable alpha amylases are also available that are effective even at 100?C to ensure full liquefaction of the starch.

Filtration

Slow rate of filtration is generally a problem when the mash is run off in the lauter tun and also during the final filtration of the beer. This is due to the presence of certain polysaccharides, mainly beta-glucans and pentosans, which are found in barley and poorly modified barley malt. The solution is to break down the beta-glucans using a beta-glucanase added during mashing or at the start of the fermentation process.

Enzymes for Detergent

For most people, the most popular known application of enzymes is in the manufacture of enzymatic washing agents (detergents). Since last 40 years, the use of enzymes in detergents has been the largest of all enzyme applications. Consumers of detergents are actual users of an enzymatic product. In majority of other applications, enzymes are used as auxiliary agents at some point in the manufacturing process and are not, as a rule, present in the finished product-not at any rate in an active form.

Proteases

Proteases are the most widely used enzymes in the detergent industry. They remove protein stains such as grass, blood, egg and human sweat. These organic stains have a tendency to adhere strongly to textile fibres. The proteins act as glues, preventing the water-borne detergent systems from removing some of the other components of the soiling, such as pigments and street dirt.

The inefficiency of non-enzymatic detergents at removing proteins can result in permanent stains due to oxidation and denaturing caused by bleaching and drying. Blood, for example, will leave a rust-coloured spot unless it is removed before bleaching. Proteases hydrolyse proteins and break them down into more soluble polypeptides or free amino acids. As a result of the combined effect of surfactants and enzymes, stubborn stains can be removed from fibres.

Lipases

Though enzymes can easily digest protein stains, oily and fatty stains have always been troublesome to remove. The trend towards lower washing temperatures has made the removal of grease spots an even bigger problem. This applies particularly to materials made up of a blend of cotton and polyester. The lipase is capable of removing fatty stains such as fats, butter, salad oil, sauces and the tough stains on collars and cuffs.

Amylases

Amylases are used to remove residues of starch-based foods like potatoes, spaghetti, custards, gravies and chocolate. This type of enzyme can be used in laundry detergents as well as in dishwashing detergents.

Celluloses

The development of detergent enzymes has mainly focused on enzymes capable of removing stains. However, a cellulase enzyme has properties enabling it to modify the structure of cellulose fibre on cotton and cotton blends. When it is added to a detergent, it results into the following effects:

Colour brightening-When garments made of cotton or cotton blends have been washed several times, they tend to get a 'fluffy' look and the colours become duller. This effect is due to the formation of microfibrils that become partly detached from the main fibres. The light falling on the garment is reflected back to a greater extent giving the impression that the colour is duller. These fibrils, however, can be degraded by the cellulase enzyme, restoring a smooth surface to the fibre and restoring the garment to its original colour.

Softening-The enzyme also has a significant softening effect on the fabric, probably due to the removal of the microfibrils.

Soil removal-Some dirt particles are trapped in the network of microfibrils and are released when the microfibrils are removed by the cellulase enzyme.

Enzymes for Starch

In the early 19th century, it was discovered that by boiling starch with acid it could be converted into a sweet-tasting substance, which consisted mainly of glucose. Since then acids have been used widely for breaking down starch into glucose. This technique does have a number of drawbacks.

The DE (dextrose equivalent) value is used as an indication of the degree of hydrolysis of syrup. The DE value of starch is zero and that of dextrose is 100. In the last 25 years, as new enzymes are available, starch hydrolysis technology has move away from acids to enzymes.

Enzymatic starch conversion, depending on the enzymes used, syrups with different compositions and physical properties of starch. There are three basic steps in enzymatic starch conversion: liquefaction, saccharification and isomerisation.

Starch Liquefaction

Firstly, there is a liquefaction process. A starch suspension containing 30-40% dry matter is first gelatinised and liquefied. By using heat-stable bacterial alpha amylase, 'maltodextrin' is obtained which contains mainly different oligosaccharides and dextrins. Maltodextrins are only slightly sweet and they usually undergo further conversion.

In most starch conversion plants, starch liquefaction takes place in a jet-cooking process. The heat stable alpha amylase is added to the starch slurry after pH adjustment, and the slurry is pumped through a jet cooker. Live steam is injected here to raise the temperature to 105°C, and the slurry is then passed through a series of holding tubes for 5-7 minutes, which is necessary to gelatinise the starch fully.

Then the temperature of the partially liquefied starch is reduced to 90-100°C by flashing, and the enzyme is allowed to react further at this temperature for 1-2 hours until the required DE (Dextrose Equivalent) is obtained.

Starch Saccharification

Saccharification is the second step in the process. Depending of the desired end product, a glucoamylase or a fungal alpha amylase is used further break down the Maltodextrins. The glucoamylase can hydrolyse starch completely to glucose along with, a little maltose and isomaltose. A pullulanase is a de-branching enzyme that can also be used to aid saccharification. Fungal alpha amylases can also be added in order to produce syrups with a higher maltose content, which means high fermentability and a relatively high degree of sweetness.

Isomerisation

Further going one step ahead, a proportion of the glucose can be isomerised into fructose, which is about twice as sweet as glucose. An immobilized glucose isomerase is used.

Enzymes for Animal Feed

Animal feed is composed of plant material, cereals and vegetable proteins, which cannot be fully digested and utilised by animals. However, feed utilization and digestion can often be increased by the addition of external enzymes to the feed.

Many cereals have a proportion of their energy in the form of non-starch polysaccharides (NSPs), more commonly known as fibre. Enzymes are to break down these NSPs, which lead in increase of metabolisable energy and protein utilisation.

In some cereals, a large part of the NSP is soluble and causes high viscosity in the small intestine of a monogastric animal. As a result, digestion becomes impaired. Selected microbial enzymes can partially degrade this NSP, lowering viscosity in the intestine and improving feed utilisation. Many vegetable protein sources, such as soybean meal, also contain NSP. The addition of selected microbial enzymes can be used to break down the NSP and make it available to the animal. Just as with cereals, the metabolisable energy and protein utilisation for vegetable protein sources can be improved by using the correct combination of amylases and proteases.

In almost all plant material used for animal feed, a large part of the mineral phosphorus is bound in the form of phytic acid, which cannot be degraded by monogastric animals. Phytase liberates part of the bound phosphorus and makes it possible to reduce the phosphorus content of the feed by 25-30%. The R & D centre at Maps has successful produced phytase, and soon shall be commercially launched.

Enzyme for Poultry Feed

The main cost in the raising of poultry is the feed, which is mainly cereal-based. Maize (corn) has a low content of soluble NSPs and is considered to be an ideal cereal. Other cereals contain higher amounts of NSPs that normally impair feed utilisation. It is possible to partially degrading these NSPs with selected enzymes acting on specific raw materials. For example, when using a standard dose of the Maps xylanase enzymes for poultry feed, the metabolisable energy value for wheat increases.

Some cereals are also avoided in poultry diets due to the adverse effects caused by NSP. Barley is a prime example because inclusion of more than 10% barley in broiler diets gives rise to wet and sticky droppings as well as reduced growth rates.

This is due to beta glucan, a soluble, high viscosity NSP located in the cell walls of the barley grain. By adding microbial enzymes to the barley based feed, the NSP could be degraded giving an improvement in droppings, better feed utilisation and faster growth rate.

How are Enzymes Made?

The starting point for enzyme production is a vial of a selected strain of microorganisms. They will be nurtured and fed until they multiply many thousand times. Then the desired end-product is recovered from the fermentation broth and sold as a standardised product. A single bacteria or fungus is able to produce only a very small portion of the enzyme, but billions microorganisms, however, can produce large amounts of enzyme. The process of multiplying microorganisms by millions is called fermentation. Fermentation to produce industrial enzymes starts with a vial of dried or frozen microorganisms called a production strain. One very important aspect of fermentation is sterilisation. In order to cultivate a particular production strain, it is first necessary to eliminate all the native microorganisms present in the raw materials and equipment. If proper sterilisation is not done, other wild organisms will quickly outnumber the production strain and no production will occur.

The production strain is first cultivated in a small flask containing nutrients. The flask is placed in an incubator, which provides the optimal temperature for the microorganism cells to germinate. Once the flask is ready, the cells are transferred to a seed fermenter, which is a large tank containing previously sterilised raw materials and water known as the medium. Seed fermentation allows the cells to reproduce and adapt to the environment and nutrients that will be encountered later on. After the seed fermentation, the cells are transferred to a larger tank, the main fermenter, where fermentation time, temperature, pH and air are controlled to optimise growth. When this fermentation is complete, the mixture of cells, nutrients and enzymes, called the broth, is ready for filtration and purification.

Filtration and purification termed as downstream processing is done after enzyme fermentation. The enzymes are extracted from the fermentation broth by various chemical treatments to ensure efficient extraction, followed by removal of the broth using either centrifugation or filtration. Followed by a series of other filtration processes, the enzymes are finally separated from the water using an evaporation process. After this the enzymes are formulated and standardised in form of powder, liquid or granules.

Symbiotic Nitrogen Fixation in Stem Nodulating Sesbania Rostrata

Leguminosae

Common Name

Description: Tall annual species, becoming perennial under favourable conditions, 1-3 m high, erect, leafy, growing during the short rainy season.

Habitat: Its ecology is quite similar to that of *Aeschynomene indica.*

Soil: Found on deep clayey alluvial soils, periodically flooded or water-logged.

Distribution: *S. rostrata* is a typical Sahelian species.

Products & uses: Unlike *Aeschynomene,* it is not palatable, its main characteristic is its high nitrogen fixing ability that makes it particularly useful for crop rotations with rice, as a green manure or for making a rich compost as the plant contains some 4 % N_2 (25 % CP). This richness in nitrogen, particularly in a species growing in water-logged soils, is due to the existence of aerial stem nodules of the so-called aeschynomenoid type (not the only such species). The amount of nitrogen added by a good crop of *Sesbania rostrata* is evaluated at 60-240 kg/ha/yr; that is : the increase in rice yield resulting from the utilization of *Sesbania* as green manure corresponds to a dressing of this amount of pure nitrogen. A good crop of *Sesbania,* with 500,000 plants per hectare, produces 4,000-5,000 kg DM in 55-60 days. i.e. 160-200 kg of pure nitrogen equivalent. Specialists in nitrogen fixation research reckon this is a fixing power still higher than in soya bean.

Sesbania species can establish symbiotic interactions with rhizobia fromtwo taxonomically distant genera, including the Sesbania rostratastem-nodulating Azorhizobium sp. and Azorhizobium caulinodans and the newly described Sinorhizobium saheli and Sinorhizobium teranga bv. sesbaniae, isolated from the roots of various Sesbania species. A collection of strains from both groups were analysed for their symbiotic properties with different Sesbania species. S. saheli and S. teranga bv. sesbaniae strains were found to effectively stem nodulate Sesbania rostrata, showing that stem nodulation is not restricted to Azorhizobium. Sinorhizobia and azorhizobia, however, exhibited clear differences in other aspects of symbiosis. Unlike Azorhizobium, S. teranga bv. sesbaniae and S. saheli did not induce effective stem nodules on plants previously inoculated on the roots, although stem nodulation was arrested at different stages.

For Sesbania rostrata root nodulation, Sinorhizobium appeared more sensitive than Azorhizobium to the presence of combined nitrogen. S. saheli and S. teranga bv. sesbaniae were effective symbionts with all Sesbania species tested, while Azorhizobium strains fixed nitrogen only in symbiosis with Sesbania rostrata. In a simple screening test, S. saheli and S. teranga bv. sesbaniae were incapable of asymbiotic nitrogenase activity. Thus, Azorhizobium can easily be distinguished from Sinorhizobium among Sesbania symbionts on the basis of symbiotic and free-living nitrogen fixation. The ability of Azorhizobium to overcome the systemic plant control appears to be a stem adaptation function. This last property, together with its host-specific symbiotic nitrogen fixation, makes Azorhizobium highly specialized for stem nodulation of the aquatic legume Sesbania rostrata.

Biological Nitrogen Fixation

Biological fixation of atmospheric nitrogen was the topic for investigations by many workers. Fred, Baldwin and McCoy in 1932 from the U.S.A. put forth their exhaustive work on root nodule bacteria. At Rothamsted in England, Thornton in 1947 studied nodule bacteria from clovers.

Jensen's method devised in 1942 for studying nodulation of plants on agar in test tubes helped in furthering investigations in this field. The German bacteriologist Bortels demonstrated in 1936 that an adequate supply of molybdenum was essential for accelerating nitrogen fixation by nodulating legumes.

The Japanese scientist Kubo found out in 1939 that root nodules of legumes could be effective only in the presence of a red pigment in them. At Helsinki, Virtanen and his school in 1947 studied leghaemoglobin and the chemistry and mechanism of nitrogen fixation. The discovery of consistent nitrogen fixation in cell-free extracts of Clostridium pasteurianum by Carnahan and others at the Du Pont Laboratory in the U.S.A. in 1960 was a landmark in the field of biological nitrogen fixation. Of equal importance was the development of the isotope method to quantify the amount of nitrogen fixed by the use of 15N and mass spectrometer developed by Burris and Wilson in 1957.

The enzyme nitrogenase was isolated, characterized and most of its biochemical properties understood except the nature of intermediates between nitrogen and ammonia. Applying the knowledge obtained with asymbiotic nitrogen fixers, Bergersen in Australia elaborated the biochemistry of nitrogen fixation in legume root nodules in the sixties. Nutman in England from 1948 onwards gave us an insight into the hereditary mechanisms behind nodulation in legumes. At the same time, he proposed the theory of microinvagination of root hairs as an explanation for the origin of infection threads in root hairs of clovers.

The work on nitrogen fixation in nodulated plants other than legumes was taken up by Bond in the U.K. and Quispel, Silver and Becking in Europe who made considerable headway on this aspect between 1950 and 1970. During the same period, Fogg and Stewart in the U.K. intensified the work on nitrogen fixing blue-green algae.

One of the significant advances with regard to non-leguminous root nodulation has been the isolation of an actinomycetous endophyte (Frankia sp.) from root nodules of Comptonia peregrina in 1978 by Callaham, Del Tredici and Torrey.

The isolated endophyte was slow growing and could re-infect and produce nodules on the roots of the same host plant. Another important observation by Trinick from Australia in 1973 was the isolation of Rhizobium from the root nodules of the genus Trema (Parasponia) which has highlighted a unique association of Rhizohil1m with non-leguminous plants resulting in the formation of root nodules.

Bacteria containing fertilizers such as 'Azotobacterin' and 'Phosphobacterin' were extensively used in Russia to improve soil fertility. Krasilnikov and Mishustin in 1937 were busy in Russia on several problems relating to interactions between plants and soil microorganisms. The English translation of 'Mikrobiologiya' opened the door of Russian microbiology to the English-speaking world.

The Australian group led by Vincent in 1954 was very active in the study of all aspects of nodulation, particularly in. understanding the environmental factors in legume root nodulation as evidenced by the work of Gibson in 1965.

While other Australian workers like Date, Brockwell and Roughley in 1962 were busy in the development of techniques involved in inoculant production and application to seed, Burton in 1950 in the U.S.A., was chiefly responsible for the establishment of legume inoculant research in North America with all its industrial implications.

The extensive use of the assay procedure utilizing the nitrogenase-catalysed reduction of acetylene to ethylene coupled with the sensitive gas chromatographic analysis proposed by Hardy and his associates in 1968 of the Du Pont Laboratory in the U.S.A. may be regarded as a turning point in the field of biological nitrogen fixation.

The ease and rapidity with which this technique could be used gave an impetus to many investigators to measure nitrogenase activity in situ in many natural ecosystems.

The Brazilian group of workers headed by Dobereiner showed how Azotobacter paspali could specifically inhabit the roots of the grass Paspalum notatum and fix abundant amounts of nitrogen, an observation which was later confirmed by the acetylene reduction reaction.

Dobereiner also introduced the concept of 'associative symbiosis' or 'diazotrophic biocoenocis' after the rediscovery of Azospirillum within plant cells of many graminaceous plants. Blue-green algae and Azolla received greater attention in developing countries as a nitrogen supplement in rice cultivation.

Rapid advances have been made in the U.K. and the U.S.A. in our knowledge of the physiology, biochemistry and kinetics of the enzyme nitrogenase. Those active in this field are Eady and his associates in England and Burris and Orme-Johnson in Wisconsin. Pate and his associates in Australia have quantified the interrelationship between photosynthates (energy) and nitrogen fixation in legumes.

The rapid expansion in the field of bacterial genetics had its impact on the study of nitrogen-fixing microorganisms. Presently, our understanding of the genetic loci involved in the process of biological nitrogen tixation in Azotobacter, Azospirillum, Rhizobium and blue-green algae (cyanobacteria) has reached new heights due to the efforts of many groups of workers in Europe, Australia and the U.S.A.

We are now aware that rhizobial genes control infection, nodulation and host range in legumes. Extrachromosomal genetic particles known as plasmids are known to carry nitrogen fixing traits. Other nodulation genes have been identified that are functionally and structurally conserved in rhizobia. In short, it has become clear that establishment of legume nodule symbiosis is the result of a multitude of communications and biochemical interactions between the legume host and the rhizobial symbiont.

This area of plant-microorganisms relationships has been enlarged due to the efforts of several molecular biologists such as Postgate, Beringer, Dixon, Johnston, Kennedy, Brill, Ausubel, Helinksy, Fisher, Long, Denarie, Downie, Kondorosi, Pueppke, Rolfe, Gershoff, Broughton, Spanik, Stacey, Verma, Krishnan, Elmerich, Nuti and others all over the world.

Dommergues and.associates from France and Senegal have discovered the occurrence of nodules on stems of Sesbania rostrata which fix nitrogen by virtue of which, the legume serves as an excellent green manure in low-land rice cultivation. Similarly, the group has also discovered nitrogen-fixing Stem nodules on Casuarina spp. caused by Frankia.

It is now known that rhizobia can fix nitrogen independently in cultures and this fact has been substantiated by 15N as well as acetylene reduction methods. The question whether rhizobia can establish endogenously within the cells of plant roots other than those of legumes and Parasponia, and induce the formation of nodules has been tackled by a team of workers headed by Cocking in England who have attempted to induce nodules on roots of rice, wheat and rape seedlings by inoculating the roots with the stem nodulating bacterial species, Azorhizobium caulinodans.

Nodule-like structures capable of fixing nitrogen have no doubt been demonstrated in rice and rape seedlings under laboratory conditions but field applications of any of these laboratory findings are yet to be realised.

Studies on soil microorganisms in relation to soil fertility were enlarged in several agricultural universities in India through applied research programmes under the All India coordinated research project funded by the Indian Council of Agricultural Research and the Department of Science and Technology.

These projects were related to the field application of nitrogen fixing microorganisms and the role in the decomposition of organic matter. The main emphasis has been to minimize the use of chemical fertilizers in the cultivation of rice, legumes and millets.

Biological Nitrogen Fixation

Biological nitrogen fixation (BNF) is the process whereby atmospheric nitrogen (N=N) is reduced to ammonia in the presence of nitrogenase. Nitrogenase is a biological catalyst found naturally only in certain microorganisms such as the symbiotic *Rhizobium* and *Frankia,* or the free-living *Azospirillum* and *Azotobacter*.

Biological nitrogen fixation is brought about bath by free-living soil microorganisms and by symbiotic associations of microorganisms with higher plants. Our main interest in this paper centres on the legume-*Rhizobium* symbiosis. Leguminous plants fix atmospheric nitrogen by working symbiotically with special bacteria, rhizobia, which live in the root nodules. Rhizobia infect root hairs of the leguminous plants and produce the nodules. The nodules become the home for bacteria where they obtain energy from the host plant and take free nitrogen from the soil air and process it into combined nitrogen. In return, the plant receives the fixed N from nodules and produces food and forage protein.

Mechanism of Biological Nitrogen Fixation

The biochemical mechanism of N_2 fixation can be written in simplified form as follows: nitrogenase.

The above mechanism indicates that N_2-fixing systems can thrive in soils poor in N. that they are a source of proteins, and that they provide N for soil fertility. Adenosine triphosphate (ATP) is the source of energy necessary for the cleavage and reduction of N_2 into ammonia. In rhizobia, for instance, ATP results from oxidative degradation of sugars and related molecules. These sugars are manufactured by the host-plant during photosynthesis and transferred to the nodules. In general, for each gram of N_2 fixed by *Rhizobium,* the plant fixes 1-20 grams carbon (C) through photosynthesis. This is an indication that symbiotic N_2 fixation requires additional energy which, in nitrate-fed plants, can be used to produce more photosynthates (products of photosynthesis). The extra energy cost of N_2 fixation can, however safely be carried by most field-grown legumes with little or no loss of production.

It is usually accepted that N_2 fixing systems require more Phosphorus (P) than non-N_2-fixing systems. Phosphorus is needed for plant growth, nodule formation and development, and ATP synthesis, each process being vital for nitrogen fixation.

Nitrogen fixation, which involves the chemical reduction of N_2 to NH_3 or NH_4, requires a source of electrons. Sources of electrons for the nitrogenase activity vary with the organism. They are all small proteins and highly reductive molecules such as flavodoxin, ferredoxin, nicotinamide, or ademine dinucleotide (phosphate).

Nitrogenase is an oxygen sensitive enzyme. The low oxygen tension condition is realized through compartmentation in cyanobacteria (heterokysts in *Anabaena azollae*), active respiration (in *Azotobacter*), synthesis of leghemoglobin (in *Rhizobium* legume). Leghemoglobin is a macromolecule synthesized by both symbiotic partners, the rhizobia and the host plant. *Rhizobium* synthesizes the heme portion, and the plant the globine. Like human hemoglobin, leghemoglobin fixes O_2. It is responsible for the red or brown colour of active (i.e., N_2-fixing) nodules. Non-N_2-fixing nodules have a white nodule content, or a green content when the globine has degenerated.

Specificity and Effectiveness

There are roughly 1,300 leguminous plant species in the world. Of these, nearly 10% have been examined for nodulation, 87% of which were nodulated. Thus not all legumes are infected by rhizobia. *Gliricidia sepium* and *Vigna unguiculata* (cowpea) nodulate freely but nodules have never been found on roots of *Cassia siamea.* A *Rhizobium* that nodulates cowpea may not nodulate *Leucaena* and vice versa. Leguminous species mutually susceptible to nodulation by a particular group of bacteria constitute a cross-inoculation group. Six cross-inoculation groups were defined in the early days of *Rhizobium* research in addition to the cowpea group. This classification scheme is undergoing modifications based on recent research.

Mechanisms of recognition between the microsymbiont and the host-plant have been suggested to explain specificity. (This topic is beyond the scope of this paper).

Not all symbioses fix N_2 with equal effectiveness. This means that a given legume cultivar nodulated by different strains of the same species of *Rhizobium* would fix different amounts of nitrogen. Selection of elite strains of *Rhizobium is* based on this observation. Similarly, a given strain of *Rhizobium* will nodulate and fix different amount of N_2 in symbiosis with a range of cultivars of the same plant species. Thus, different provenances of a given legume (e.g., *Gliricidia sepium* in ILCA's international testing) can nodulate and fix nitrogen at different levels when they are established in the same field. Also, the free-nodulating *Gliricidia* or promiscuous varieties of soybean can nodulate profusely and fix a great deal of nitrogen depending on the effectiveness of the rhizobial populations present.

Table : A short list of* Rhizobium *species and their corresponding hosts

Rhizobium species	*Host plants*
Bradyrhizobium japanicum	*Glycine max* (soybean)
Rhizobium fredii	*Glycine max* (soybean)
R. phaseoli	*Phaseolus vulgaris* (common bean)
R. meliloti	*Medicago sativa* (alfalfa)
	Melilotus sp. (sweet clovers)
R. trifolii	*Trifolium* sp. (clovers)
R. leguminosarum	*Pisum sativum* (peas)
	Vicia faba (broad bean)
"Cowpea rhizobia" group or *Rhizobium* sp.	*Vigna unguiculata* (cowpea),
	Arachis hypogaea (peanut),
	Vigna subterranea (Bambara groundnut)
	Leucaena sp., *Albizia* sp.,
Azarhizobium caulinodans	*Sesbania* sp. *Sesbania rostrata* (stem nodulating)

Factors Limiting Biological Nitrogen Fixation

Interactions between the microsymbiont and the plant are complicated by edaphic, climatic, and management factors. A legume-*Rhizobium* symbiosis might perform well in a loamy soil but not in a sandy soil, in the subhumid region but not in the Sahel, or under tillage but not in no-till plots. These factors affect either the microsymbiont, the host-plant, or both.

Edaphic Factors

Edaphic factors relate to the soil. The six main edaphic factors limiting biological nitrogen fixation are:

- excessive soil moisture,
- drought,
- soil acidity,
- P deficiency,
- excess mineral N, and
- deficiency of Ca, MO, CO and B.

Excessive moisture and waterlogging prevent the development of root hair and sites of nodulation, and interfere with a normal diffusion of O_2 in the root system of plants. *Sesbania rostrata* and *Aeschynomene* sp. can actively fix N_2 under these conditions because they are located on the plant stems, rather than on the roots.

Drought reduces the number of rhizobia in soils, and inhibits nodulation and N_2 fixation. Prolonged drought will promote nodule decay. Deep-rooted legumes exploiting moisture in lower soil layers can continue fixing N_2 when the soil is drying. Mycorrhizal infection has also been found to improve tolerance of plants to drought (e.g., *Acacia auriculiformis* inoculated with the ectomycorrhizal *Baletus suillus*). Mycorrhiza are symbiotic associations between fungi and plant roots. Some mycorrhizal fungi develop exclusively outside the roots; these are called ectomycorrhiza (e.g., *Baletus suillus*). Others, called endomycorrhiza, grow inside the roots with their vesicles and arbuscules inside the roots and with their fungal filaments extended outside (e.g., *Glamus* sp.). These are the vesicular-arbuscular mycorrhiza, usually referred to as VAM.

Soil acidity and related problems of Ca deficiency and aluminum and manganese toxicity adversely affect nodulation, N_2 fixation and plant growth. Research work on the identification of symbioses adapted to acid soil should focus on the host plant, because effective rhizobia adapted to-soil acidity can be found naturally and can be produced through genetic manipulations.

Phosphorus deficiency is commonplace in tropical Africa and reduces nodulation, N_2 fixation and plant growth. Identification of plant species adapted to low-P soils is a good strategy to overcome this soil constraint. The role of mycorrhizal fungi in increasing plant P uptake with beneficial effects on N_2 fixation has been reported. Dual inoculation with effective rhizobia and mycorrhizal fungi shows synergistic effects on nodulation and N_2 fixation in low P soils*. The use of local rock phosphate has been recommended, particularly in acid soils, as an inexpensive source of P. The addition of P-solubilizing microorganisms, particularly of the general *Psemdamaias, Bacillus, Penicillium,* and *Aspergillus* can solubilize rock phosphate and organically bound soil P (which constitutes 95-99% of the total phosphate in soils). However, the use of these microorganisms is not widespread. Some reports show nodulation response to K under field conditions. However, other investigators consider the K effect to be indirect, acting through the physiology of the plant.

Trees are usually infected by mycorrhizal fungi in natural ecosystems in the tropics. The significance of this symbiosis in nature should be better recognised.

Mineral N inhibits the *Rhizobium* infection process and also inhibits N_2 fixation. The former problem probably results from impairment of the recognition mechanisms by nitrates, while the latter is probably due to diversion of photosynthates toward assimilation of nitrates. Some strains of *Rhizobium*, and particularly stem-nodulating *Azarhizobium caulinodans,* fix N_2 actively even when plants are growing in high-N soils (e.g., in the presence of 200 kg fertilizer N ha^{-1}). Application of large quantities of fertilizer N inhibits N_2 fixation, but low doses (<30 kg N ha^{-1}) of fertilizer N can stimulate early growth of legumes and increase their overall N_2 fixation. The amount of this starter N must be defined in relation to available soil N.

Various microelements (Cu, Mo, Co, B) are necessary for N_2 fixation. Some of these are components of nitrogenase for example Mo.

Climatic Factors

The two important climatic determinants affecting BNF are temperature and light. Extreme temperatures affect N_2 fixation adversely. This is easy to understand because N_2 fixation is an enzymatic process. However, there are differences between symbiotic systems in their ability to tolerate high (>35°C) and low (<25°C) temperatures.

The availability of light regulates photosynthesis, upon which biological nitrogen fixation depends. This is demonstrated by diurnal variations in nitrogenase activity. A very few plants can grow and fix N_2 under shade (e.g., *Flemingia congesta* under plantain canopy). In alley farming if hedgerows are not weeded, or if trees are planted with food crops like cassava, their nitrogen fixation and growth will be reduced due to shading. Early growth of legume trees is slow and they cannot compete successfully for light.

Biotic Factors

Among biotic factors, the absence of the required rhizobia species constitute the major constraint in the nitrogen fixation process. The other limiting biotic factors could be:

- excessive defoliation of host plant,
- crop competition, and
- insects and nematodes.

Inoculation of Legumes

If specific and effective rhizobia are absent in a soil, or if they are present in low numbers, it is necessary to introduce the rhizobia in that soil to ensure proper nodulation and nitrogen fixation. This is called inoculation. If specific and effective rhizobia are present in a sufficient number, there will be no need to inoculate the legume. In agrisystems, whenever one is not sure of the presence and effectiveness of the native rhizobia, it could be necessary to inoculate the legume with an adequate strain of rhizobia.

How one can determine the need for inoculation? There are some simple tests: Are nodules absent or sparse on an uninoculated young plant growing in a low-N soil? (This is normally accompanied by plant N deficiencies). Or, are nodule sections white or green? (This is an indication of poor effectiveness).

A more accurate relative effectiveness trial will provide more precise information. The trial, in a simple term, consists of growing the legume with and without fertilizer N while controlling all other limiting factors. The relative effectiveness ratio (RE) is then calculated. RE is defined as: dry weight of unfertilized plants × 100/dry weight of fertilized plants. If the value of RE is more than 5, the inoculation is not required.

When the rhizobia in a soil are infective (i.e., capable of colonizing and nodulating a legume) but poorly effective, they constitute a barrier to the successful exploitation of *Rhizobium* inoculants. Introduced rhizobia must therefore be more aggressive and competitive as nodulators than the native strains. Inoculant rhizobia usually persist in the soil for long periods, particularly when the host is cultivated frequently or is permanent. Persistence of a strain is desirable because it obviates the need for inoculation in subsequent years, assuming inoculant strains maintain their original effectiveness.

Inoculation with rhizobia is usually recommended for newly introduced legumes. Most positive responses to inoculation are confined to crops which have specific requirements for *Rhizobium*, (e.g., *Leucaena leucocephala*, American varieties of soybean). Indigenous legumes seldom respond to inoculation with introduced rhizobia because they nodulate with resident strains, even if these native rhizobia are not the most effective ones.

Inoculation with rhizobia should be considered as an exceptional farming practice rather than the rule. In Australia and the USA, inoculation has played a vital role in legume production. But in developing countries, the practice is not widespread. The major drawback to inoculation technology is the wide variability in yield responses in time and space for a given *Rhizobium*-legume symbiosis. Responses can vary from no response, and sometimes negative responses, to positive yield increases. Response to inoculation with a strain of *Rhizobium* vary with sites, legume cultivars, and the form of inoculant. Changes in climate, such as Africa's long droughts in recent years, and management factors including cropping systems and inoculant handling will also introduce variability in response to inoculation. Local rhizobia are not necessarily better inoculants than exotic strains. All these considerations call for a substantial research support system capable of defining the most appropriate inoculants and procedures for each site and probably for each cropping season as well. The use of freely nodulating legumes will be much easier in this respect.

Inoculation procedures are detailed in Volume 1 of this training manual.

Defoliation, Crop Competition, and Pests

Defoliation (e.g., pruning and lopping) decreases the photosynthetic ability of legumes. It impairs N_2 fixation and can lead to nodule decay. For perennial legumes, nodule decay sheds a high number of rhizobia in the root zone. When new roots develop in subsequent vegetative cycles, nodulation of the legume is expected to improve. Scientists at IITA have observed that uninoculated *Leucaena leucocephala* nodulated very sparsely the first year and showed nitrogen deficiency symptoms. After a number of years nodulation improved and N deficiency symptoms disappeared.

Intercropping legumes with non-leguminous crops can result in competition for water and nutrients. This competition can affect N_2 fixation negatively. However, it has been shown that when mineral N is depleted in the root zone of the legume component by the non-leguminous intercrops, N_2 fixation of legumes may be promoted.

Insects and nematodes have also been reported to interfere with nodule formation, development, and functions.

Estimation of Nitrogen Fixation

From the biochemical reactions of BNF presented in section 2.1, it is evident that N_2 fixing systems contribute to the quality and quantity of agricultural production. Measurement of BNF can provide information on whether actual N_2 fixation in adequate. We discuss below two simple methods of BNF estimation. Measurement of BNF is a more reliable method than nodule counting, nodule weighing, or assessment of leghemoglobin.

Short-term Estimation of BNF: Acetylene Reduction Assay

Nitrogenase not only catalyses the reduction of atmospheric N_2 to NH_3, but can also reduce acetylene (C_2H_4). The acetylene reduction assay (ARA) is carried out on detached nodules, detopped roots, or whole plants in a closed vessel containing 10% acetylene. A gas chromatograph is used to determine the amount of ethylene formed. Data are usually expressed as nanomoles or micromoles of ethylene produced per hour per plant or per weight unit of nodules. The acetylene reduction assay provides an instant measure of nitrogenase activity (but not necessarily of N_2 fixed) under the experimental conditions.

For long-term estimates, a series of measurements must be performed to include diurnal, daily, and seasonal changes. Variation in light intensity, temperature, and moisture in the field will increase the level of variation of nitrogenase activity and will reduce the significance of integration of short-term assays. A problem that is inherent in ARA is the need to calibrate the rates of ethylene production with the actual rates of N_2 fixation. The commonly used ratio of 3:1 for acetylene reduced per N_2 fixed is not always valid. Also, nitrogenase activity of some legumes declines considerably once nodules or roots are detached from the rest of the plant. For plants with long roots, it is difficult to collect all the nodules. To minimize this limitation, the plants are confined to open ended chambers and ARA is done *in situ*.

Medium-term Estimation of BNF: N-solute Analysis of Xylem Exudate

N-solute analysis of xylem exudate is a medium-term type of estimate because it involves the integration of more than one hour of events. The underlying principle is based on the fact that nitrogen from BNF can be transported to the leaves in the form of (1) ureides, allantoin and allantoic acid, or (2) asparagine and glutamine. In agricultural soils, where nitrate is the most readily available form of N for plant growth, the solutes derived from soil mineral N will contain principally free nitrate and organic products of nitrate reduction in the roots. Correlations can be established between the N_2 fixed nitrogen in forms (1), (2), and soil-derived N. Using these correlations, it should be possible to assess N_2 fixation, or at least to obtain an index of BNF by collecting and analysing plant sap for the above-mentioned N compounds.

The methods are simple and have been used successfully in ureide legumes.

Solute analysis can be used in farmers' fields because it is virtually non-destructive. It is also relatively inexpensive. Repeated measurements are also required to fully integrate measurements of total N fixed over a long period of time.

How to Increase BNF and N_2 Fixing Ability

Biological N_2 fixed represents N gain and determines inorganic N fertilizer savings in cropping systems. Legumes can fix more than 250 kg N ha^{-1}. However, the amounts of N_2 fixed can vary considerably in time and space The nitrogen fixation process is influenced by factors such as:

- presence and effectiveness of rhizobia, pest damage,
- plant genotype and age,
- plant and rhizobia interactions,
- changes in soil physiochemical conditions, and
- various management practices such as tree pruning or pesticide application that can affect both symbiotic partners.

Four common approaches to enhance biological nitrogen fixation are:

- inoculation with proven strains,
- microbial screening for improved strains,
- host-plant screening and breeding, and
- adoption of cropping systems and cultural practices.

Microbial Screening

There are collections of effective rhizobia located at centres around in the world for most, if not all, legumes used in agriculture (Takishima et al, 1989). These strains may be screened to identify the most effective and competitive one(s) for a given agroecosystem Once elite strains have been identified, the legume under consideration is inoculated. Instructions on inoculant use are usually given by the manufacturers. Seed inoculation using peat inoculant is the most commonly used method. However, studies are under way to assess the effectiveness of post planting inoculation as a corrective measure Dual inoculation of rhizobia and mycorrhizal fungi has proven beneficial in some cases.

Host-plant Screening and Breeding

A screening of legume plants, with high N_2-fixing components can be carried out Breeders have developed plant varieties with promiscuous nodulation to obviate the need for inoculation with rhizobia In some laboratories in the USA, plants that do not nodulate with indigenous rhizobia but only with introduced "super" strains are being developed

There are still many unexploited legume-*Rhizobium* symbioses in the world. The potential benefit of screening these symbioses is underscored by the fact that only about 0.5% of existing leguminous species are presently used for agricultural purposes.

Cropping Systems and Cultural Practices

It is evident that inclusion of N_2-fixing components in cropping systems will increase N inputs in agrisystems. Cultural practices can control some of the above-mentioned factors which limit BNF. Mulching, for instance, can control weeds and fluctuations of soil moisture and temperature. Liming can eliminate soil acidity, and Al and Mn toxicities.

Summary

Since nitrogen is commonly the most limiting plant nutrient in arable farming in the tropics and also the most expensive element as a mineral fertilizer, biological nitrogen fixation (BNF) holds great promise for smallholder farmers in sub-Saharan Africa. Alley farming systems which use leguminous woody species in the hedgerows can reduce or eliminate farmers' needs for commercial N fertilizer.

Biological nitrogen fixation is the process of capturing atmospheric nitrogen by biological processes. It is accomplished by certain microorganisms and plant-microbe interactions. Legumes are N-fixing systems that have long been used for biological nitrogen fixation in agriculture.

Biologically fixed nitrogen can be estimated using the acetylene reduction assay method, xylem exudate analysis, or by other methods.

A number of edaphic, climatic, and biotic factors inhibit N_2 fixation Among these, the absence of specific and effective rhizobia in the soil is the most important. The amount of biologically fixed nitrogen can be enhanced by different methods, including inoculation with proven strains, screening for improved microbial and host-plant materials, and introduction of improved cultural practices.

8

Paranodulation and Endophytic Nitrogen Fixation in Rice

Nitrogen fixation is the natural process, either biological or abiotic, by which nitrogen (N_2) in the atmosphere is converted into ammonia (NH_3). This process is essential for life because fixed nitrogen is required to biosynthesize the basic building blocks of life, e.g., nucleotides for DNA and RNA and amino acids for proteins. Nitrogen fixation also refers to other abiological conversions of nitrogen, such as its conversion to nitrogen dioxide.

Nitrogen fixation is utilized by numerous prokaryotes, including bacteria, actinobacteria, and certain types of anaerobic bacteria. Microorganisms that fix nitrogen are called diazotrophs. Some higher plants, and some animals (termites), have formed associations (symbioses) with diazotrophs. Nitrogen fixation also occurs as a result of non-biological processes. These include lightning, industrially through the Haber-Bosch Process, and combustion. Biological nitrogen fixation was discovered by the Dutch microbiologist Martinus Beijerinck.

Biological Nitrogen Fixation

Biological nitrogen fixation (BNF) occurs when atmospheric nitrogen is converted to ammonia by an enzyme called nitrogenase. The formula for BNF is:

$$N_2 + 8\ H^+ + 6\ e^- \rightarrow 2\ NH_3 + H_2$$

The process is coupled to the hydrolysis of 16 equivalents of ATP and is accompanied by the co-formation of one molecule of H_2. In free-living diazotrophs, the nitrogenase-generated ammonium is assimilated into glutamate through the glutamine synthetase/ glutamate synthase pathway.

Enzymes responsible for nitrogenase action are very susceptible to destruction by oxygen. (In fact, many bacteria cease production of the enzyme in the presence of oxygen). Many nitrogen-fixing organisms exist only in anaerobic conditions, respiring to draw down oxygen levels, or binding the oxygen with a protein such as Leghemoglobin.

Plants that contribute to nitrogen fixation include the legume family – Fabaceae – with taxa such as clover, soybeans, alfalfa, lupines, peanuts, and rooibos. They contain symbiotic bacteria called *Rhizobia* within nodules in their root systems, producing nitrogen compounds that help the plant to grow and compete with other plants. When the plant dies, the fixed nitrogen is released, making it available to other plants and this helps to fertilize the soil The great majority of legumes have this association, but a few genera (e.g., *Styphnolobium*) do not.

In many traditional and organic farming practices, fields are rotated through various types of crops, which usually includes one consisting mainly or entirely of clover or buckwheat (family *Polygonaceae*), which were often referred to as "green manure."

Non-leguminous Nitrogen-fixing Plants

Although by far the majority of nitrogen-fixing plants are in the legume family Fabaceae, there are a few non-leguminous plants, such as alder, that can also fix nitrogen. These plants, referred to as "actinorhizal plants", consist of 24 genera of woody shrubs or trees distributed among in 8 plant families. The ability to fix nitrogen is not universally present in these families.

For instance, of 122 genera in the Rosaceae, only 4 genera are capable of fixing nitrogen. All these families belong to the orders Cucurbitales, Fagales, and Rosales, which together with the Fabales form a clade of eurosids. In this clade, Fabales were the first lineage to branch off; thus, the ability to fix nitrogen may be plesiomorphic and subsequently lost in most descendants of the original nitrogen-fixing plant; however, it may be that the basic genetic and physiological requirements were present in an incipient state in the last common ancestors of all these plants.

There are also several nitrogen-fixing symbiotic associations that involve cyanobacteria (such as *Nostoc*):

- Some lichens such as *Lobaria* and *Peltigera*
- Mosquito fern (*Azolla* species)
- Cycads
- *Gunnera*.

Microorganisms that Fix Nitrogen (Diazotrophs)

Diazotrophs are bacteria that fix atmospheric nitrogen gas into a more usable form such as ammonia.

A diazotroph is an organism that is able to grow without external sources of fixed nitrogen. Examples of organisms that do this are rhizobia and *Frankia* (in symbiosis) and *Azospirillum*.

All diazotrophs contain iron-molybdenum nitrogenase systems. Two of the most studied systems are those of *Klebsiella pneumoniae* and *Azotobacter vinelandii*. These systems are used because of their genetic tractability and their fast growth.

Types of Diazotrophs

Diazotrophs are scattered across bacterial taxonomic groups (mostly in the Bacteria but also a couple of Archaea). Even within a species that can fix nitrogen there may be strains that do not fix nitrogen.

Fixation is shut off when other sources of nitrogen are available, and, for many species, when oxygen is at high partial pressure. Bacteria have different ways of dealing with the debilitating effects of oxygen on nitrogenases, listed below.

Free-living Diazotrophs

- Anaerobes—these are obligate anaerobes that cannot tolerate oxygen even if they are not fixing nitrogen. They live in habitats low in oxygen, such as soils and decaying vegetable matter. *Clostridium* is an example. Sulphate-reducing bacteria are important in ocean sediments (e.g. *Desulfovibrio*), and some Archean methanogens fix nitrogen in muds and animal intestines.
- Facultative anaerobes—these species can grow either with or without oxygen, but they only fix nitrogen anaerobically. Often, they respire oxygen as rapidly as it is supplied, keeping the amount of free oxygen low. Examples include *Klebsiella pneumoniae, Bacillus polymyxa, Bacillus macerans*, and *Escherichia intermedia*.
- Aerobes—these species require oxygen to grow, yet their nitrogenase is still debilitated if exposed to oxygen. *Azotobacter vinelandii* is the most studied of these organisms. It uses very high respiration rates, and protective compounds, to prevent oxygen damage. Many other species also reduce the oxygen levels in this way, but with lower respiration rates and lower oxygen tolerance.
- Oxygenic photosynthetic bacteria generate oxygen as a by-product of photosynthesis, yet some are able to fix nitrogen as well. These are colonial bacteria that have specialized cells (heterocysts) that lack the oxygen generating steps of photosynthesis. Examples are *Anabaena cylindrica* and *Nostoc commune*. Other cyanobacteria lack heterocysts and can fix nitrogen only in low light and oxygen levels (e.g. *Plectonema*).
- Anoxygenic photosynthetic bacteria do not generate oxygen during photosynthesis, having only a single photosystem which cannot split water. Nitrogenase is expressed under nitrogen limitation. Normally, the expression is regulated via negative feedback from the produced ammonium ion but in the absence of N2, the product is not formed, and the by-product H2 continues unabated [Biohydrogen]. Example species: *Rhodobacter sphaeroides, Rhodopseudomonas palustris, Rhodobacter capsulatus*.

Symbiotic Diazotrophs

- Rhizobia — these are the species that associate with legumes, plants of the

Fabaceae family. Oxygen is bound to leghemoglobin in the root nodules that house the bacterial symbionts, and supplied at a rate that will not harm the nitrogenase.

- Frankias — much less is known about these 'actinorhizal' nitrogen fixers. The bacteria also infect the roots leading to the formation of nodules. Actinorhizal nodules consist of several lobes, each lobe has a similar structure as a lateral root. Frankia is able to colonize the cortical tissue of nodules where it fixes nitrogen. Actinorhizal plants and Frankias also produce hemoglobins, but their role is less well established than for rhizobia. Although at first it appeared that they inhabit sets of unrelated plants (alders, Australian pines, California lilac, bog myrtle, bitterbrush, *Dryas*), revisions to the phylogeny of angiosperms show a close relatedness of these species and the legumes. These footnotes suggest the ontogeny of these replicates rather than the phylogeny. In other words, an ancient gene (from before the angiosperms and gymnosperms diverged) that is unused in most species was reawakened and reused in these species.
- Cyanobacteria — there are also symbiotic cyanobacteria. Some associate with fungi as lichens, with liverworts, with a fern, and with a cycad. These do not form nodules (indeed most of the plants do not have roots). Heterocysts exclude the oxygen, as discussed above. The fern association is important agriculturally: the water fern *Azolla* harbouring *Anabaena* is an important green manure for rice culture.
- Association with animals—although diazotrophs have been found in many animal guts, there is usually sufficient ammonia present to suppress nitrogen fixation. Termites on a low nitrogen diet allow for some fixation, but the contribution to the termite's nitrogen supply is negligible. Shipworms may be the only species that derive significant benefit from their gut symbionts.

Importance

In terms of generating nitrogen available to all organisms, the symbiotic associations greatly exceed the free-living species with the exception of cyanobacteria.

- Cyanobacteria
- *Azotobacteraceae*
- *Rhizobia*
- *Frankia.*

Nitrogen Fixation by Cyanobacteria

Cyanobacteria inhabit nearly all illuminated environments on Earth and play key roles in the carbon and nitrogen cycle of the biosphere. In general, cyanobacteria are able to utilize a variety of inorganic and organic sources of combined nitrogen, like nitrate, nitrite, ammonium, urea, or some amino acids. Several cyanobacterial strains are also capable of diazotrophic growth. Genome sequencing has provided a large amount of information on the genetic basis of nitrogen metabolism and its control in different cyanobacteria. Comparative genomics, together with functional studies, has led to a significant advance in this field over the past years. 2-Oxoglutarate has turned out to be the central signalling molecule reflecting the carbon/nitrogen balance of cyanobacteria. Central players of nitrogen control are the global transcriptional factor NtcA, which controls the expression of many genes involved in nitrogen metabolism, as well as the P_{II} signalling protein, which fine-tunes cellular activities in response to changing C/N conditions.

These two proteins are sensors of the cellular 2-oxoglutarate level and have been conserved in all cyanobacteria. In contrast, the adaptation to nitrogen starvation involves heterogeneous responses in different strains. Nitrogen fixation by cyanobacteria in coral reefs can fix twice the amount of nitrogen than on land–around 1.8 kg of nitrogen is fixed per hectare per day.

Chemical Nitrogen Fixation

Nitrogen can also be artificially fixed for use in fertilizers, explosives, or in other products. The most common method is the Haber process. Artificial fertilizer production is now the largest source of human-produced fixed nitrogen in the Earth's ecosystem.

The Haber process requires high pressures (around 200 atm) and high temperatures (at least 400 °C), routine conditions for industrial catalysis. This highly efficient process uses natural gas as a hydrogen source and air as a nitrogen source.

Research on Catalytic Nitrogen-fixation

Much research has been conducted on the discovery of catalysts for nitrogen fixation, often with the goal of reducing the energy required for this conversion. However, such research has thus far failed to even approach the efficiency and ease of the Haber process. Many compounds react with atmospheric nitrogen under ambient conditions. For example, lithium metal converts to lithium nitride under an atmosphere of nitrogen. Treatment of the resulting nitride gives ammonia.

The first dinitrogen complex was reported in 1965 based on ammonia coordinated to ruthenium ($[Ru(NH_3)_5(N_2)]$), research in chemical fixation focused on transition metal complexes. Since then, a large number of transition metal compounds that contain dinitrogen as a ligand have been discovered. The dinitrogen ligand can either be bound to a single metal or bridge two (or more) metals. The coordination chemistry of dinitrogen is complex and currently under intense investigation. This research may lead to new ways of using dinitrogen in synthesis and on an industrial scale.

The first example of homolytic cleavage of dinitrogen under mild conditions was published in 1995. Two equivalents of a molybdenum complex reacted with one equivalent of dinitrogen, creating a triple bonded MoN complex. Since then, this triple bounded complex has been used to make nitriles. The first catalytic system converting nitrogen to ammonia at room temperature and pressure was discovered in 2003 and is based on another molybdenum compound, a proton source, and a strong reducing agent. However, this catalytic reduction fixates only a few nitrogen molecules.

Nitrogen nutrition most frequently limits the crop productivity. Chemical nitrogen fertilizers being ecologically and economically expensive, interest in exploitation of alternative or supplementary nitrogen sources has been renewed to encourage sustainable agriculture. Reducing the fertilizer nitrogen use while maintaining the native soil nitrogen resources and enhancing crop nitrogen output by biological nitrogen-fixation systems is desirable from both environmental and economic perspectives. Such a production system would enhance agricultural production in the long term in a way that is economically viable and ecologically sound, as well as socially acceptable. Towards the sustainable rice production, rice cultivar with better nitrogen use efficiency is one of the options. In the case of nitrogen inputs into rice production systems, it may be the judicious use of nitrogen fertilizers and/or nitrogen derived from atmospheric nitrogen via biological nitrogen fixation.

Developing rice with biological nitrogen fixation capacity, i.e. autotrophy for nitrogen, could help farmers overcome the nitrogen nutrition limitations and increase the productivity, reducing dependency on chemical nitrogen fertilizers. The various options and the pros and cons of such options are discussed herein, as a part of an approach to the prospect of achieving nitrogen autotrophic rice.

Cereals are the world's major source of food for human nutrition. Among these, rice is very important and represents the staple diet for more than two-fifths (2.4 billion) of the world's population, making it the most important food crop of the developing world1. The challenge of feeding the growing population has to be met with in a sustainable as well as equitable basis. Sustainable production of rice and hence the global food security, depends on reaching even higher levels of production which is not possible without additional nutritional input. In future, more rice is to be produced from less land, at a minimum cost, with reduced use of environmentally adverse inputs.

Nitrogen (N) is the nutrient that most frequently limits the agricultural production. Global agriculture now relies heavily on N fertilizers derived at the expense of petroleum that is vulnerable to the political and economic fluctuations in the oil markets. N fertilizers, therefore, are an expensive input, costing agriculture more than US $ 45 billion per year. The demand for N fertilizers in world agriculture is increasing at a rate approximately equivalent to the rate of increase in the world population, i.e. about two per cent per annum. In rice, it takes about 1 kg of N to produce 15 ~ 20 kg of grain4. Low input efficiency of N fertilizer, decline in crop yield under continuous cropping, nitrate pollution, acidification of soils and emission of greenhouse gases such as ammonia and nitrous oxide, are some economic and environmental factors that must be considered when using chemical N fertilizers.

In addition, run-down of N supply capacity of agricultural soils, lack of purchase power of impoverished farmers, gap between the researchers and farmers in N management of crops and soils for maximum production efficiency, have confounded the situation still further.

These concerns have renewed interest in exploiting alternative or supplementary N sources to encourage sustainable agriculture.

A fundamental shift has taken place in agricultural research and world food production. In the past, the principal driving force was to increase the yield potential of food crops and to maximize productivity. Today, the drive for productivity is increasingly combined with a desire for environmentally sustainable development, with use of renewable resources. For farming systems to remain productive and to be sustainable, it will be necessary to replenish the resources of nutrients which are removed or lost from the soil. Nowhere is this challenge more intense than in supplying rice with nitrogen nutrition.

Producing rice cultivars with better N-use efficiency is one of the options towards the sustainable rice production.

In the case of nitrogen inputs into rice production, it may be the judicious use of N fertilizers or N derived from atmospheric N via biological nitrogen fixation (BNF).

Diazotrophic enterobacteria: What is their role in the rhizosphere of rice?

The search for more interesting and potentially beneficial diazotrophic enterobacterial endophytes continues to be carried out as part of the "New Frontier Project on Nitrogen Fixation in Rice." Diverse diazotrophic enterobacteria, such as Enterobacter cloacae, Erwinia herbicola/Enterobacter agglomerans (Pantoea agglomerans), Klebsiella planticola, K. oxytoca, and Serratia marcescens inhabit the rhizosphere of rice plants growing in both tropical and temperate climates. The ubiquity of enterobacteria can perhaps be attributed to their great metabolic diversity and fast growth rates. Indeed, the widespread occurrence in the rhizosphere of high populations of r-strategist bacteria, such as free-living Klebsiella and endophytic Enterobacter and Serratia species, indicates that there is sufficient availability of organic carbon in and around roots to meet their metabolic needs.

The possession by enterobacters of diverse traits, such as diazotrophy in the presence and absence of O2 and tolerance for heavy metals, probably also contributes enormously to their ability to survive and grow in diverse habitats, including the rhizosphere. In addition, many enterobacters have the ability to produce gums, siderophores, indole acetic acid, and metabolites active against phytopathogens, and they may also have phosphate-solubilizing, chitinase, cellulase, and pectinase activities. The potential role of these diazotrophs in plant growth is now being increasingly understood.

Retrospective on Biological Nitrogen Fixation

In 1838, Boussingault reported that leguminous plants fixed atmospheric nitrogen, but it was not until the convincing 1888 report of Hellriegel and Wilfarth that the validity of this concept was generally accepted. The biochemistry of the process was not investigated seriously until the 1930s when the influence of nitrogen and oxygen pressures and the inhibitory action of hydrogen were established. The application of 15N as a tracer in the 1940s indicated that ammonia is the key intermediate in fixation. Achievement of consistent cell-free nitrogen fixation in the 1960s substantiated the role of ammonia and showed that both an Fe protein and a MoFe protein are involved in the fixation process with energy derived from ATP. In 1992, the tertiary structure of the MoFe protein was established. Recent investigations have emphasized genetic aspects of the nitrogenase system.

Xylem Colonization of Rice and Arabidopsis by Azorhizobium Caulinodans ORS571

The xylem of the roots of Sesbania rostrata is colonized when inoculated with Azorhizobium caulinodans led us to investigate whether the xylem of rice and other nonlegumes could also be colonized. If so, this might provide a nonnodular niche for endophytic symbiotic nitrogen fixation in rice and other nonlegume crops, somewhat comparable to the naturally occurring nonnodular endophytic nitrogen fixation by diazotrophs in xylem vessels and intercellular spaces that other researchers have detected in sugarcane. Earlier studies indicated the colonization of the xylem of lateral roots of wheat inoculated repeatedly with Azorhizobium caulinodans. It is also observed the crack entry of A. caulinodans and intercellular colonization of the root systems of rice, wheat, and the model nonlegume Arabidopsis thaliana, and stimulation by flavonoids such as naringenin. Recently, using the LacZ reporter gene have shown, for the first time, in rice inoculated with Azorhizobium caulinodans, that the xylem of roots can be colonized by azorhizobia. The xylem of the root system of the model nonlegume dicot Arabidopsis thaliana is extensively colonized, when inoculated with Azorhizobium caulinodans. Using this ability to induce xylem colonization by azorhizobia in rice and Arabidopsis, now investigating factors, including flavonoids, that might influence the extent of xylem colonization, and also the extent to which xylem colonization by azorhizobia will provide a niche for nonnodular symbiotic nitrogen fixation.

The Potential Role of Biological Nitrogen Fixation in Meeting Future Demand for Rice and Fertilizer

Rice is the most important crop in Asia, accounting for nearly half of caloric intake in many countries. Future demand for rice will increase at approximately the rate of population growth in the next 20 years, which is projected to be slightly more than 1% per year in Asia. Nitrogen is a key input in rice farming, but the cost of mineral nitrogen fertilizer relative to the value of rice production is typically low in both irrigated and rainfed systems (just 4-7%). This low share is not due to a preponderance of subsidies, which are much less common today than in the past. More important factors are the sharp decline in inflation-adjusted world prices of urea over the past 40 years and the fact that labour is by far the most important cost in rice farming. The low share of N fertilizer in gross returns implies that farmers will not adopt rice varieties with enhanced nitrogen fixation capacity if there is even a small yield penalty associated with such capacity. It also implies that the environmental benefits from reducing nitrogen runoff are likely to be more important than the financial benefits for farmers due to reduced mineral N consumption. Because a large share of rice is grown under irrigated conditions, or in rainfed environments with continuous flooding, it will be important to develop biological nitrogen fixation that can operate under flooded conditions.

Progress in Multinational Collaborative Studies on the Beneficial Association between Rhizobium Leguminosarum bv. Trifolii and Rice

. Two cycles of field and laboratory studies have indicated that diverse indigenous populations of the clover root-nodule symbiont, Rhizobium leguminosarum bv. trifolii, intimately colonize rice roots in cultivated fields of the Egyptian Nile Delta, where rice has been rotated successfully with berseem clover (Trifolium alexandrinum) since antiquity. Certain strain/variety interactions significantly expand rice root architecture, enhance the uptake of several plant nutrients, and increase plant biomass under laboratory and greenhouse conditions. Preliminary results indicating statistically significant increases in grain yield and agronomic fertilizer N-use efficiency following inoculation have been obtained.

Prospects for Constructing Nitrogen-fixing Cereals

The engineering of autonomous nitrogen-fixing plants is undoubtedly a long-term goal because it requires the assembly of a complex enzyme and provision of appropriate physiological conditions in the absence of the environment normally provided by the prokaryotic cell. Here we briefly review the genetic, biochemical, and physiological requirements for nitrogen fixation in the context of engineering nitrogen-fixing plants and report some of our recent findings on the expression of one of the nitrogenase structural subunits within plastids of the model photosynthetic eukaryote Chlamydomonas reinhardtii.

Frontier Project on Nitrogen Fixation in Rice: Looking Ahead

Rice is the world's most important food, particularly for the poor. The future need for rice (more than a 30% increase by 2020) to be grown with less land in the intensive rice areas, and with new technologies in the rainfed systems, demands that science explore new frontiers. IRRI has begun a set of Frontier Projects that explore the feasibility of applying new knowledge and methods to transfer novel traits from other crops to rice. One of these projects is "Nitrogen Fixation in Rice." In 1992, a think-tank workshop concluded that IRRI, working closely with advanced research laboratories, should explore the feasibility of N_2 fixation in rice in a few well-defined approaches: nonnodular symbiosis, nodular symbiosis, transferring nif genes, and CO_2 fixation and N-use efficiency.

A BNF (biological nitrogen fixation) working group for rice was formed to monitor research progress and the free exchange of information and materials. Significant scientific progress has been made in providing new knowledge. Yet this alone is not enough to maintain the project. The Frontier Projects must be periodically assessed on how such new knowledge has furthered our probability of success in meeting the long-term goal. This chapter describes the rationale for IRRI's support of these projects with a long-term, ambitious, and problematic outcome.

Exploring Genetic Programs for Root Endosymbioses

The study of the interaction of plants with diazotrophic bacteria has clearly shown that it is only in intercellular, endosymbiotic interactions that significant amounts of fixed nitrogen are provided to the host plant. While only a minority of plants can establish such root endosymbioses with diazotrophic bacteria, most plants are able to establish endosymbiotic associations with arbuscular endomycorrhizal fungi. Therefore, a large majority of plants, including rice, wheat, and maize, possess the genetic information required to pave the way for successful symbiotic root infection, involving the reactivation of cortical cells in the presence of a microsymbiont, the accepted penetration of the symbiont in host cells, and the synthesis of a membranous interface with the symbiotic partner to facilitate metabolic exchange.

Cyanobacterial Biofertilizer and Quality Control

Several cyanobacteria such as Anabaena, Nostoc, Cylindrospermum, Aulosira, Tolypothrix, Gloeotrichia, Wollea, etc. are very good free nitrogen fixers. In suitable conditions, for example, on the surface of bare moist earth in the tropics, they may add valuable amounts of nitrogen to the soil. Unlike nitrogen fixing bacteria these organisms can photosynthesize and so need no external supply of carbon compounds. In the tropics where paddy rice is grown, the whole nitrogen requirements of the rice crop may be provided by these cyanobacteria. As many as 23 consecutive rice crops have been successfully grown from soils unfertilized with nitrogen at the International Rice Research Institute in Manila, Philippines. In these soils, the nitrogen came from the cyanobacterial fixation.

The Tamil Nadu state in our country has been particularly successful in propagating and using cyanobacteria as a valuable renewable resource in rice cultivation on commercial basis. In this programme, naturally growing cyanobacteria including species of, Anabaena, Aulosira, Nostoc, Cylindrospermum, Gloeotrichia, Lyngbya, Chroococcus, Wollea, Aphanothece, Calothrix, Gloeocapsa, Plectonema and Microcystis are propagated in the rice fields by adding 50 kg of P_2O_5/ha about a month before the planting of rice. Phosphate stimulates growth of cyanobacteria.

Then, 50 kg of K_2O/ha is applied to the fields and the paddy is planted without any nitrogenous and phosphatic fertilization. These fields are provided a thin sheet of water once in 3 or 4 days. Timely plant protection and weed control measures are taken. With this technology, even the recently evolved high yielding varieties of rice respond well and a cent per cent saving of the nitrogen bill is achieved.

Besides, some cyanobacteria (Anabaena azollae) are known to undergo symbiotic association with grasses and pteridophytes and fix considerable amount of nitrogen. The water-fern Azolla containing nitrogen-fixer Anabaena is being currently used in rice cultivation in China, Korea, Japan, India, and some other countries.

Mass Production and Application

The technology for producing cyanobacterial inoculants on a large scale is different from that used for bacterial inoculants. Two approaches are generally in operation. They are:

(a) Efficient strains of required cyanobacteria are cultured in open tanks with water containing adequate amount of mineral nutrients such as phosphate and molybdenum. After considerable amount of growth is obtained, the cyanobacterial mass is removed from the tank and dried. The dried mass is powdered and mixed with suitable carriers.

(b) Another approach has been developed by Farmers Training Centre at Aduthurai by which even small farmers can prepare their own cyanobacteria inoculum. This approach involve artificial multiplication (of either single species or composite mixtures of various species of cyanobacteria) in shallow metal troughs lined with polythene and containing fine so 10 kg, superphosphate 250 g, and rice bran or sawdust 250 g (source of molybdate). These troughs are filled with water up to 2" to 6". Acidity is corrected by adding about 50 g of lime. 25 g of 3% carbofuran granules are applied and this prevents the attack of daphnids and mosquitoes on multiplying cyanobacteria. These troughs are then inoculated with 250 g of powered soil-based cyanobacterial culture.

After good growth has occurred and the standing water has evaporated (about 7 to 10 days), the dry cyanobacteria flakes can be removed and stored dry. One trough (6' ´ 3' ´ 9") may yield as much as 2.5 kg of cyanobacteria every time. The cost of production of cyanobacteria comes to about 50 paise per kg and as much as 15 tons of wet cyanobacteria per ha can be produced in 3 weeks.

The application of the inoculum (composite mixture) to paddy fields has been demonstrated to eliminate the need for using any nitrogen fertilizer. The growth and yield of rice are not adversely affected even when the cyanobacteria are applied to fields containing high levels of combined nitrogen. There is also a strong residual beneficial effect of cyanobacteria on subsequent plantings.

Nitrogen (N) has been established as one of the most important limiting factors in rice production. The overhead labour cost of tillage, irrigation, water control, and other operations do not vary much, whether 400 kg [ha.sup.-1] is produced without N fertilization or 2000 kg [ha.sup.-1] is produced through N fertilization. This has been found to be true for both the modest and the high-yielding rice varieties, since the modem rice varieties yield only little more than do the traditional varieties without an adequate supply of N fertilizer (Brady, 1979). With this came the realization of the yield potential of modern rice varieties, coupled with the practical experience that inorganic N fertilizers are becoming more expensive because of the large fossil-fuel energy requirement for their production, which necessitated a search for, surveys of, and the development of alternate low-cost, assimilable nitrogen sources; i.e., the natural nitrogen resources havin g the unique inherent property to biologically fix a significant amount of atmospheric nitrogen in the soil for agricultural use.

The flooded rice soil ecosystem is characterized by aerobic and anaerobic zones in three major ecological layers: the floodwater and the surface of the soil; the anaerobic plow layer; and the rhizosphere. The floodwater and the surface of the soil provide the sites for aerobic phototrophic nitrogen fixation by free-living cyanobacteria and the Azolla-Anabaena symbiotic [N.sub.2]-fixing complex, and the reduced zone and the rhizosphere of the rice plants are the sites for heterotrophic nitrogen fixation. The present review is an attempt to summarize some of the significant studies in the area of cyanobacteria-based biofertilizers.

A variety of free-living cyanobacteria are now identified as efficient components of cyanobacterial biofertilizers. In addition, a few cyanobacterial species form symbiotic associations with plants (algae, i.e., diatoms; fungi, i.e., lichens; bryophytes, i.e., liverworts, hornworts, and mosses; pteridophytes, i.e., Azolla; gymnosperms, i.e., cycads; and angiosperms, i.e., Gunnera), animals (marine sponges and achiruoid worms), nonphotosynthetic protists (belonging to the group Glaucophyta), bacteria, and hollow shafts of hairs of polar bears (Rai, 1990). The water fern Azolla, holding the [N.sub.2]-fixing cyanobacterium Anabaena azollae, is another established major cyanobacterial biofertilizer. The first part of this review will deal with the free-living cyanobacteria, and the second will focus on the Azolla-Anabaena symbiosis, in order to present a comprehensive view of the present state of knowledge in the area of cyanobacterial biofertilizers.

Free-Living Cyanobacteria

Cyanobacteria are prokaryotic microorganisms that resemble gram-negative bacteria in structure but possess an [O.sub.2]-evolving photosynthetic system similar to that of eukaryotic algae and higher plants. They belong to ancient group of organisms that are recorded even from pre-Cambrian microfossils (Schopf, 1970) and dominate a wide range of diverse environments characterized by extremes of temperature, desiccation, pH, salinity, light intensity, and nutrients. Rice fields in several countries—e.g., Japan, Thailand, China, the Philippines, Bangladesh, and India—have been investigated and show a dominant presence of these organisms. The majority of cyanobacteria are capable of fixing the atmospheric nitrogen, and their presence in rice fields is thought to maintain the nitrogen level in the soil. Ever since the importance of cyanobacteria was recognized, a considerable amount of research has been carried out to evolve methods and means to effectively utilize these organisms as a biofertilizer.

Nitrogen-fixing Cyanobacteria

Cyanobacteria are of morphologically diverse organization. Generally, they can be grouped into unicellular, colonial, unbranched filamentous, pseudoparenchymatous, heterocystous, and heterotrichous-heterocystous forms. Fogg (1949) demonstrated experimentally that heterocysts are the specialized cells that contain the nitrogenfixing mechanism. A considerable amount of research was carried out on heterocystous forms, and it was proved by immunolabeling studies that heterocysts are the location of the enzyme nitrogenase. The methodology for quantification of nitrogen fixation was also developed, through simple estimation of total nitrogen over a period of time, a mass spectrophotometric technique using isotopic nitrogen ([N.sup.15]), and the acetylene-reduction technique. The acetylene-reduction technique has been in use worldwide for assaying the activity of the enzyme nitrogenase. Based on these techniques, the [N.sub.2]-fixing cyanobacteria known so far have been classified into three groups: unicellular, filamentous nonheterocystous, and filamentous heterocystous.

Taxonomic and/or floristic information on soil cyanobacteria is available from, for example, Argentina, Australia, Bangladesh, Czech Republic, Russia, Greece, India, floristic account of Arunachal Pradesh rice fields: Singh et al., 1997a, floristic account of Nagaland rice fields: Singh et al., 1997b), Iraq, the United Kingdom, and the United States.

In majority of the cases referred to above, the soil-borne cyanobacteria have been found to be [N.sub.2] fixers. In addition to light, the pertinent factors that affect cyanobacterial presence in soil are moisture, pH, mineral nutrients, and combined nitrogen. Moisture was noted to be extremely important for their multiplication in loamy soil. Their ability to fix atmospheric nitrogen has a competitive advantage where combined nitrogen levels are low. The abundance of cyanobacteria in tropical soils is supported by their higher temperature optima. Many cyanobacteria tolerate high levels of ultraviolet irradiation, permitting them to survive at the surface of the soil, whereas others photosynthesize efficiently at low photon fluence rates.

They are the most common nitrogen-fixing organisms in rice fields, mostly occurring as a free-floating water blooms, forming a microbial mat; the photographs show the conspicuous vegetative cells and heterocysts. Other important rice-field cyanobacteria include: Nostoc commune, embedded in a dense matrix of mucilage and forming a ball like structure; Scytonema sp., showing heterocysts and characteristic typical false branching; Calothix sp., showing characteristic terminal heterocysts; Nodularia sp., with vegetative cells and heterocysts; Gloeotrichia sp., characterized by a ball-like circular assembly of filaments resembling radiating rays; and Lyngbya sp., having a typical yellow-brown coloration of the mucilage sheath due to the presence of scytonemin, a UV-absorbing compound.

The larger population of [N.sub.2]-fixing cynaobacteria comprises the filamentous and heterocystous forms. Nostoc commune has been found to be the most important source of biologically fixed N in the Arctic Spitsbergen. Activity of sheets of Nostoc commune assayed in situ showed linear relationships with their moisture content and assay temperature. Similarly, more than 100 strains of heterocystous cyanobacteria, belonging to the genera Anabaena, Nostoc, Nodularia, Cylindrospermum, Scytonema, Calothrix, Anabaenopsis, Mastigocladus, Fischerella, Tolypothrix, Aulosira, Stigonema, Haplosiphon, Chlorogloeopsis, Camptylonma, Gloeotrichia, Nostochopsis, Rivularia, Schytonematopsis, Westiella, Westiellopsis, Wollea, and Chlorogloea, have been found to be efficient [N.sub.2] fixers.

Using the acetylene-reduction technique, Wyatt and Silvey (1969) were the first to report nitrogen fixation by a unicellular cyanobacterium, Gloeocapsa. Another report on nitrogen fixation by a nonheterocystous filamentous form, Plectonema boryanum under microaerophilic conditions, was published by Stewart and Lex (1970). Carpenter and Price (1976) reported nitrogen fixation by Trichodesmium sp., a nonheterocystous filamentous form in the marine environment. Detailed studies were recently conducted on the distribution of the [N.sub.2]-fixing enzyme, nitrogenase, and nitrogen physiology in this organism, suggesting that nitrogenase activity and synthesis exhibit an endogenous rhythm in Triclzodesmium sp. cultures which is affected by the addition of nitrogen, and that the organism can simultaneously fix N2 and take up other forms of nitrogen. Gallon and Chaplin (1982) and Gallon (1992) performed intensive studies on the unicellular cyano bacterium Gloeothece and found that it fixed nitrogen in the dark phase of the light/dark cycle. Among such nonheterocystous [N.sub.2]-fixing cyanobacteria, most soil forms appear to be associated with conditions such as water-logging, which lead to reduced ambient oxygen concentration.

The [N.sub.2]-Fixing Enzyme System

The entire group of [N.sub.2]-fixing cyanobacteria (including the heterocystous and nonheterocystous filamentous/unicellular forms), listed above, are able to use molecular nitrogen as a nitrogen source. Dinitrogen is reduced to ammonia by an enzyme complex known as nitrogenase, a reaction that is dependent on reduced ferredoxin and obligatorily coupled to reduction of protons, resulting in the formation of molecular hydrogen:

[N.sub.2] 8 [Fd.sub.red] 8 [H.sup] 16 ATP [right arrow] [NH.sub.3] [H.sub.2] 8 [Fd.sub.ox] 16 ADP 16 [P.sub.]

Like nitrogenases from other prokaryotes, the cyanobacterial nitrogenase consists of dinitrogenase reductase (the homodimeric Fe protein) and dinitrogenase (a heterotetrameric MoFe protein). Electron flow coupled to ATP hydrolysis proceeds from ferredoxin to the Fe protein and finally to the MoFe protein. The genes coding for the structural subunits of cyanobacterial nitrogenase (nifHDK) have been cloned and sequenced. Nitrogenase activity and synthesis are regulated at the level of both enzyme activity and gene expression. Enzyme activity may be regulated by posttranslational modification of the Fe protein.

Nitrogenase is extremely oxygen sensitive, thus requiring special conditions for enzyme activity. In heterocystous cyanobacteria the enzyme is located exclusively in the heterocysts. Differentiation of vegetative cells into heterocysts occurs in response to lack of utilizable nitrogen and involves sequential structural and physiological modifications necessary to provide an anaerobic environment for efficient nitrogen fixation. In most heterocystous species investigated so far, a genome rearrangement during the late stage of differentiation restores nifHDK by excision of an interrupting 11 kb element. The nitrogenase structural genes are subsequently transcribed from the nifH promoter as a polycistronic message. In nonheterocystous species the oxygen-evolving photosynthesis is separated in time from the oxygen-sensitive nitrogenase activity.

A special ferredoxin (FdxH) is synthesized as the specific electron donor to the nitrogenase complex in all [N.sub.2]-fixing cyanobacteria so far tested. The occurrence and function of this ferredoxin were first described in heterocysts of Anabaena; meanwhile, the corresponding gene (fdxH) was identified and sequenced and was crystallized in order to determine its X-ray structure. Two lysine residues were found to be essential for interaction with the nitrogenase.

Cyanobacterial nitrogenase has been purified from Anabaena. Preparation of the enzyme from Anabaena variabilis (ATCC 29413), as described by Reich and Boger (1989), was performed under strictly anaerobic conditions. After cell breakage with a French pressure cell and subsequent ultracentrifugation, the resultant supernatant was subjected to anion exchange chromatography. The MoFe protein and the Fe protein were separated by DEAE-cellulose columns using Tris/HCl buffers (pH 7.8), with increasing concentrations of [MgCl.sub.2] for elution of the enzyme components. For elution of the Fe protein, 80 mM [MgCl.sub.2] was required (in contrast to 45 mM [MgCl.sub.2] for elution of the MoFe protein). Further purification of the Fe protein involved preparative SDS—PAGE.

Routine assays for determination of nitrogenase activity make use of the fact that, in addition to [N.sub.2] several compounds with a triple bond are reduced by the nitrogenase. So acetylene ([C.sub.2][H.sub.2]) is reduced to ethylene ([C.sub.2][H.sub.2]). Both compounds have the advantage of being detected very easily by gas chromatography. Any gas chromatograph equipped with a Porapak column and a flame-ionizing detector is suitable.

For determination of in vivo nitrogenase activity in heterocystous cyanobacterial [N.sub.2]-fixing cells (cyanobacteria grown in the absence of [NO.sub.3]- or [NH.sub.4.sup.]), the cyanobacterial filaments are suspended in the culture medium in reaction vessels that can be sealed with a gas-tight stopper. Acetylene is added to the gas phase, with a final concentration of about 10% (v/v), and aliquots from the gas phase are withdrawn at appropriate intervals and analysed for ethylene formation. Reactions can be performed in the light or the dark; nitrogenase activity in the dark depends on the amount of carbohydrate storage compounds present in the cells at the start of the experiment. The presence of molecular nitrogen in the reaction vessel does not interfere significantly with the experiment, because acetylene is about 60 times more soluble in aqueous solution than is dinitrogen.

Induction of nitrogenase activity in nonheterocystous cyanobacteria is achieved by growth of cyanobacteria with a limiting amount of nitrate or ammonium in the culture medium (1-2 mM). After depletion of the N-source, cells degrade phycobiliproteins, and their colour gradually turns yellowish green. When these cells are transferred to appropriate assay conditions, nitrogenase is induced by de novo protein synthesis within 1-3 hours. Assays are performed in an anaerobic gas phase consisting of argon 10% (v/v) acetylene under low light conditions (about 20 [micro]E [m.sup.-2] [s.sup. 1]). An equilibrium between photosynthetic oxygen evolution and respiratory oxygen uptake is achieved under these conditions, and nitrogenase activity depends on respiratory activity. If higher irradiances are used, addition of DCMU (final concentration 5 x [10.sup.-6] to 1 x [10.sup.-5]) is advantageous to prevent excess photosynthetic oxygen evolution. Alternatively, assays may also be performed in the dark, with a limiting amount of oxygen present.

Nitrogenase activity in isolated heterocysts or cell-free preparations from cyanobacteria can also be determined by the acetylene-reduction technique. In both cases, however, a suitable electron donor has to be used. Molecular hydrogen has been proved to be the best electron donor in heterocyst preparations if the reaction is performed in the light. In cell-free preparations an ATP-regenerating system has to be used in addition, to meet the ATP requirement of the nitrogenase reaction. Sodium dithionite can be used in these systems as a direct electron donor for nitrogenase.

Alternatively, through the reduction of acetylene, the formation of molecular hydrogen can also be monitored to study nitrogenase activity. For these experiments it is necessary to exclude all possible substrates of nitrogenase from the assay. In the absence of suitable substrates, nitrogenase exclusively reduces protons resulting in the evolution of [H.sub.2]. For determination of hydrogen formation, a gas chromatograph equipped with a thermal conductivity detector and a molecular sieve column is required. Due to counteracting hydrogenase activity that consumes molecular hydrogen, rates of nitrogenase activity in intact cyanobacteria measured as formation of [H.sub.2] have to be analysed in detail and considering all parameters involved.

Cyanobacterial [N.sub.2] Fixers of Rice Fields

The abundance of cyanobacteria in rice fields was first observed by Fritsch (1907). De (1939) realized their importance in maintaining the fertility of rice fields through biological nitrogen fixation, later emphasized by Singh (1950, 1961), Holm-Hanson (1968), Fogg and Stewart (1968), Home (1971), and Alexander (1975). The relative occurrence of cyanobactena in rice fields varies within wide limits (Singh & Singh, 1989). They constitute 86% of the total algal flora in southern Iraq, 75% in Indian rice fields, and 70% in Italian soil. It is important to mention here that the cyanobacteria are scarce in Australian rice fields, possibly due to the higher levels of copper sulfate and combined nitrogen present in the irrigation water. They are more prevalent in tropical and subtropical regions, as compared with the temperate belts. Their occurrence is on record in Russia, Northern and Western Europe, and North America. In Sweden the cyanobacteria were present in 80% of the clay soils and more than 90% of the lime soils.

Rice fields in India, being situated in the tropical belt, are quite rich in cyanobacterial flora, as shown in surveys conducted in the states of Karnataka, Kerala, Madhya Pradesh (Agarkar, 1967), Maharashtra, Tamil Nadu (Chacko, 1972), Uttar Pradesh, Orissa, and West Bengal. Singh (1978) conducted a survey in the farms of the Central Rice Research Institute, India, and reported the dominance of Aphanothece, Anabaena, Aulosira Cylindrospermum, Gloeotrichia, and Nostoc. The species were mostly found to be area specific, as cyanobacterial inocula brought from larger distances—i.e., more than 1500 km—could not be established in the Cuttack area, due to the dominance of indigenous species. Out of 2213 soil samples from ric e fields, 33% were found to harbor [N.sub.2]-fixing cyanobacteria in an all-India survey. Upland soils in arid climates are very inhospitable to many microorganisms because of high temperatures and limited water, yet cyanobacteria are especially resistant to such adverse conditions and thus form the dominant component of the microflora in many cases. Through a quantitative study of algal flora of dried soil samples from upland fields (pH 7.8-8.3) at the Indian Agricultural Research Institute, New Delhi, cyanobacteria were found to dominate in all soil samples. Among 120 soil samples collected from the Gulf of Mexico and areas in Ecuador and Colombia, 62 algal species were found. Of those species, 46 were cyanobacteria; and of the cyanobacteria, 23 were [N.sub.2]-fixing species, including 21%, 13%, and 4% populations of Nostoc muscorum, Nostoc Paludosum, and other [N.sub.2]-fixing cyanophytes.

A high density of cyanobacteria, comprising more than half the population of heterocystous species, are found growing at or floating above the surface. This is particularly evident in wetland rice fields, which supply 86% of the world's rice requirement. Singh (1961) investigated the periodicity of cyanobacteria in rice fields in Uttar Pradesh and Bihar and found three prominent filamentous and heterocystous forms; i.e., Aulosira fertilissima, Anabaena ambigua and Cylindrospermum ghorakhpurease. Several cyanobacterial strains have been screened in laboratories for their pH tolerance, because soils of rice fields in India vary from acidic to highly alkaline. Sardeshpande and Goyal (1982) identified many strains that could adapt to a wide range of pH, grow well, and also perform nitrogen fixation efficiently. Anand et al. (1990) reported that Nostoc calcicola could shift the pH of an external medium from acidic or highly alkaline to support its maximum growth within six days. Anand and Revathi (1992) found that N. calcicola could adapt metabolically to varying pH regimes and fix nitrogen efficiently. Singh et al. (1995) reported that the pH of the habitat from which the cyanobacteria were isolated did not seem to confer pH specificity. However, critical studies are still necessary on changes in cyanobacterial species composition in the soil as the rice is growing. Kulasooriya (1998) has stressed how this makes it difficult to compare results from different sites. However, this difficulty cannot be generalized unless we have data from a wide enough range of sites. A survey of 102 soil samples from four countries has shown an abundance of heterocystous forms, positively correlated with pH and available P content of the soils. In Bangladesh rice fields, too, the abundance of heterocystous species was significantly correlated with available P in the soil. Although it is difficult to assess the impact of P fertilization on cyanobacteria in rice fields, because other fertilizers, particularly K, are added at the same time, the highly significant increase in cyanobacterial biomass of the cyanobacterial genera—i.e., Aulosira, Aphanothece, and Gloeotrichia—was shown specifically to be due to phosphate addition.

Light is another decisive factor that has a significant effect on the relative abundance of dominant cyanobacterial genera, as studied in rice fields near Valencia, Spain: Quesada et al. (1998) found that the abundance of nonheterocystous forms was three times more at high than low (about 52% versus 7%) incident radiation and that the three main heterocystous components (Anabaena. Nostoc, and Calothrix spp.) responded differently to the different levels of irradiation. In this study, a higher abundance of Nostoc coincided with a lower abundance of the two other genera, and nitrogenase activity of all the components was greater at the higher irradiance. Most of the cyanophytes appeared to be different in rain-moistened and flooded rice fields of Bangladesh, though mats of Scytonema mirabilis were common under both conditions.

Mass Culture of [N.sub.2]—Fixing Cyanobacterial Biofertilizer Material for Application to Rice

One of the biotechnological applications that has resulted from development of a cyanobacterial biofertilizer program has been extension work, or the preparation and distribution of biofertilizers to farmers. Earlier methods of supplying farmers with seed-based inocula of cyanobacteria received criticism, because contamination of the field by introduced cyanobacterial mixtures with soil was feared. Now, innovative methods have been developed for the distribution of cyanobacterial biofertilizer to farmers. For raising laboratory culture of cyanobacteria, use of polybag bottles, instead of expensive glass flasks, has been suggested. The method of distribution of liquid cyanobacterial cultures in polyethylene and polypropylene sachets has also been developed.

Nutrients fixed by cyanobacteria are made available, mainly in the form of ammonia, to rice plants through exudation autolysis and microbial decomposition. However, this biofertilizer technique is still limited, chiefly because of the inability to efficiently produce good-quality inocula and the difficulty of reestablishing the inoculated strain in fields. But there have been serious attempts to introduce large-scale cyanobacterial culture and applications. During the last decade some progress has been achieved in the production of rice-field isolates in open raceway reactors and flat, inclined modular photobioreactors (Hu et al., 1996). Inmost of these cases the cultures had to be kept at a relatively low cell density to obtain a high yield. Application of the biomass harvested from the pond to the field required processing—i.e., harvesting and drying—which reduced cell viability and increased the cost of production.

One way to overcome some of the above-mentioned problems is to try to cultivate the cyanobacteria in a reactor that ensures a high yield when grown at a relatively high cell density. In such a case, direct application of the cyanobacteria to the field may be considered. Just a few years back, a new kind of reactor, one that is closed and tubular, was developed by Boussiba (1997) to overcame some of the problems of low productivity caused mainly by suboptimal temperatures in open raceway ponds. The efficiency of this bioreactor for the production of a nitrogen-fixing cyanobacterium (rice-field isolate, Anabaena siamensis), was tested. It was shown that the rate of biomass production was significantly higher in this reactor, as compared with the open pond: 0.31 g [1.sup.-1] vs. 0.09 g [1.sup.-1]. This system was claimed to be a reliable means of supplying desired microorganisms, which are in some cases major components in a food-chain production system. Moreover, substituting nitrogen biofertilizers with nitrogen-fixing cyanobacteria produced in a closed, tubular bioreactor has been shown to have other specific advantages. Direct application of the dense cultures into the field may reduce the cost of production, because no harvesting or drying process will be required. It may also improve cell viability, which drops drastically when these algae are dried before application to the field. Using a closed, tubular bioreactor will also prevent rain from diluting the growing cultures (or even totally washing out cells), which occurs often in shallow, open ponds. Finally, the prolonged period of production possible in the closed bioreactor can provide a continuous supply of a good source of protein to be used for enterprises other than rice cultivation, such as fish or chicken production, and may help reduce the migration of farmers to urban areas when the rice season is over. The production of rice-field nitrogen-fixing cyanobacteria in this kind of tubular reactor for rice farming thus has great potential and may considerably enhance use of this fertilization technique in topical countries where rice is considered a major staple in the human diet.

Venkataraman (1964) introduced the terminology "algalization" of rice-field soils by recommending the use of heterocystous cyanobacteria as fertilizer supplements. Several field trials have been conducted since then in India under the All-India Coordinated Research Projects sponsored by the Indian Council of Agricultural Research, and centres for multiplication of such beneficial cyanobacteria came into existence. Venkataraman (1981) also developed the technology for mass cultivation of cyanobacteria. Bisoy and Singh (1988a) reported that 6 kgN [ha.sup.-1] could be produced in 15 days. Kannaiyan (1985) published a detailed account of developing cyanobacteria in nursery plots in the field itself. Small plots of 10 x 2 [m.sup.2] with 1.5 kg of starter culture could yield, after 15 days of growth, 15-20 kg of dry algal flakes. A systematic compilation of the practically possible and less-expensive methods of large-scale [N.sub.2]-fixing cyanobacterial production for use in rice has been provided by Venkataraman (1993). Two of these methods—trough or tank; pit—are essentially for individual farmers, and two—field; nursery-cum-algal—are for bulk production of the [N.sub.2]-fixing cyanobacteria on a commercial scale.

Trough or Tank Method

Either permanent tanks or shallow trays of galvanized iron sheet (2 m x 3 m x 23 cm) are prepared (the size can be increased depending on the amount of material to be produced). For the size specified, 8-10 kg of soil are placed per tray or tank, mixed well with 200 g superphosphate. The trays are filled with 5-15 cm of water, depending on local conditions and rate of evaporation. The soil pH is kept neutral by adding lime. In order to prevent insect pest infestation, any suitable insecticide, preferably Carbofuran (3% granules) is added to this soil preparation. The [N.sub.2]-fixing cyanobacterial clean culture (to be propagated in mass) is brought from the laboratory and sprinkled over the standing water in the tray after the soil has settled down. The unit is exposed to the sun in the open air. Water is added intermittently to make up the loss due to evaporation. During the summer (March to May) the inoculated cyanobacteria will grow vigorously and form a thick mat in 7-10 days. At this stage, watering is stopped so that the cyanobacterial mats dry in the sun and crack into flakes, which are collected and stored for field use. The flakes are removed, the troughs are again filled with water, and a handful of dried algal flakes are added as an inoculum for further propagation. The process can be repeated three or four times before fresh soil material, superphosphate, and insecticide are placed on the troughs or trays. Two harvests of cyanobacterial flakes, thus produced, give approximately 3-5 kg of active material, sufficient to biofertilize one hectare of rice field.

Pit Method

This technique includes the same method as described above, but it is extremely easy and least expensive for small farmers, who can operate it by themselves in their backyards or farms. The difference is that instead of troughs or tanks, shallow pits are dug in the ground and lined with a thick polyethylene sheet to hold the water. The rest of the procedure is the same as that used in the trough or tank method.

Field Method

This method is a large-scale version of the trough or tank method, used to generate bulk quantities of cyanobacterial biofertilizer material. The production field is divided into 40 [m.sup.2] plots. The soil has to be well puddled, if loamy, to facilitate waterlogging. The plots are flooded to have a thin layer of water above the soil. Superphosphate at 12 kg per plot and an insecticide (either Carbofuran 3% granules or Ekalux 5% granules) at 250 g per plot are applied. BHC or Furadon may also be used as an alternative insecticide. The cyanobacterial inoculant is sprinkled as in the trough or tank method.

The water level is always maintained intermittently. In clear, sunny weather, if the soil is clayey, cyanobacterial growth will be optimal within two weeks; in loamy soil it takes almost double the time to achieve such growth. After reaching the thick, floating mat stage, watering is stopped and sun-dried cyanobacterial flakes are collected and stored in sacks for use at the farmer's field. By reflooding the plots and adding superphosphate and insecticide, the process of cyanobacterial propagation and harvest can be repeated many times without further cyanobacterial inoculation. An average yield of cyanobacterial biofertilizer through this method is 16-30 kg per plot in each harvest. Adoption of this method has provided farmers with a record production of 15.6 tons [ha.sup.-1] of fresh [N.sub.2]-fixing cyanobacterial biofertilizer material between March and May in clear sunny weather in some Indian regions.

Nursery-cum-Cyanobacterial Method

In this method, farmers can produce cyanobacterial biofertilizers along with rice-seedling generation in nurseries. In a 360 [m.sup.2] allotted land area, a rice nursery can be prepared in a 320 [m.sup.2] plot, with the adjoining 40 [m.sup.2] plot to be used for cyanobacterial production using the aforementioned field method. Through this concurrent nursery method, 15-20 kg of efficient cyanobacterial material are easily made available to biofertilize 1.5 ha of rice field. Moreover, transplantation is made in the entire 360 [m.sup.2] nursery, so the land is neither wasted nor locked up exclusively for cyanobacterial biofertilizer production. As early as 1977, the Chinese Agricultural Research Institute, in Nanjing, used a mixture of Anabaena sp. and

Nostoc sp. as inoculants at 750 kg [ha.sup.-1] to produce, within 10-15 days, very efficient cyanobacterial biofertilizer material amounting to 7.5 tons [ha.sup.-1] in normal cool weather and 15 tons [ha.sup.-1] when the temperature rose above 30[degrees]C (FAO, 1977).

Benefits of Cyanobacterial Biofertilizer in Rice Cultivation

Enhanced grain and straw yields resulting from the use of cyanobacterial strains as biofertilizers were reported by De and Mandal (1956). Further evidence came from the work of Subramanyan (1972), based on experiments conducted at the Central Rice Research Institute, in Cuttack. Singh (1961) reported that alkaline usar soils in Uttar Pradesh could be reclaimed by using cyanobacteria to neutralize the pH of the soil. The cyanobacteria are also known to increase soil fertility by enhancing the available N and P levels. The beneficial effects of cyanobacterial biofertilization may be represented under the following headings:

Effect of Cyanobacterial Biofertilization on Rice Plants

Although the most frequently used criterion for assessing the effects of cyanobacterial biofertilization has been enhanced grain yield in rice, beneficial effects have also been noted on plant size, on nitrogen content, and on the number of tillers, ears, spikelets, and filled grains per panicle. The relative increase in grain yield over the control has been an average of 28% in pot experiments and 15% in field experiments in India.

Obviously, plants growing with cyanobacterial flakes are more robust than are those without biofertilization. Fruiting without biofertilization. Vaishampayan for testing the responsiveness of rice varieties to cyanobacterial biofertilization under the All-India Coordinated Biological Nitrogen Fixation Research Project, operative at this centre).

Cyanobacterial application to rice fields has been found to result in increased grain yield not only in India but also in China (Ley, 1959), Egypt, Japan, the Philippines, and other rice-growing tropical countries. Watanabe (1962) used the cyanobacterium Tolypothrix tenuis as a source of green manure in about 40 rice varieties in various parts of Japan and noted an average increase in rice yield by 2% in the first year, 8% in the second year, 15% in the third year, and 19.5% in the fourth year. Using the same organism, Aboul-Fadi et al. (1967) obtained a 16.6% increase in paddy yield in the United Arab Republic. Sharma et al. (1986) observed that plots biofertilized with cyanobacteria (with or without chemical N) supported increases in rice yield of 12.7%, 12.9%, and 13.2% during 1982, 1983, and 1984, respectively. Remarkably, Aulosira fertilissima was found to support an increase in rice yield of 368% in pots and 114% in fields (Singh, 1961).

In Tamil Nadu, beneficial effects of cyanobacterial biofertilization were obtained with supplements of only 0 and 25 kg N [ha.sup.-1], helping to save 25 kg N [ha.sup.-1]. Gaur and Singh (1982) reported that the beneficial effect of cyanobacteria was not obtained beyond 20 kg N [ha.sup.-1] level at one out of three experimental locations in Uttar Pradesh. Contrary to this, Venkataraman (1972), in field experiments conducted in different parts of India, showed that beneficial effects of cyanobacteria on rice yield were evident even up to 150 kg N [ha.sup.-1]. Under field conditions, cyanobacterial biofertilizer applied in combination with 50 kg N [ha.sup.-1] supported a grain yield as good as that with 120 kg N [ha.sup.-1]. Srinivasan and Ponnaya (1978) reported an increase in rice yield of 0.2 tons [ha.sup.-1] at two locations and of 0.6 tons [ha.sup.-1] at a third location with cyanobacterial inoculation in the farmer's field.

In addition to increasing the grain and straw yields, cyanobacterial inoculation has also been reported to support an increase in the N content of grain and straw. It was also reported to support an increase in plant height (Singh, 1961), leaf length, number of tillers, ears (Singh, 1961), number of spikelets per panicle, number of filled grains per panicle, and amount of dry matter (Bisoy, 1982). The cyanobacterial inoculation program is generally more successful in the dry season, during which a saving of 20-30 kg N [ha.sup.-1] is generally reported. The effect of cyanobacterial biofertilization was reported to be statistically erratic and limited (Roger, 1991), and decaying rice straw was found to inhibit cyanobacterial growth and [N.sub.2] fixation. The best results were obtained where mixed inocula were p roduced from local stocks and applied with reduced or no inorganic nitrogenous fertilizers.

Availability of Cyanobacterial Nitrogen to Rice Crops

Cyanobacteria release the fixed nitrogen mainly in the form of polypeptides, with lesser amounts of free amino acids, vitamins, and auxinlike substances, either by exudation or by microbial degradation after cell death. Under field conditions, part of the fixed nitrogen is made available to the rice plants, and the rest is reincorporated into the soil. The transfer of fixed N from the cyanobacteria to the crop and other organisms has been investigated using the [N.sup.15] tracer technique, where the uptake could be both by active assimilation and passive processes. Wilson et al. (1980) reported that 37% and 51% of the nitrogen was recorded in rice from [N.sup.15] labelled Aulosira sp. spread on the soil and incorporated into the soil, respectively. Watanabe et al. (1977) found 40% of the fixed N at maturity in the rice portion not directly exposed to [[blank].sup.15][N.sub.2]. However, using the [N.sup.15]/[N.sup.14] tracer technique, 40% of the cyanobacterial nitrogen was found to be recovered by the rice plants, and it has been shown that fixed nitrogen from cyanobacteria is directly transferred to rice plants; addition of ammonium chloride equivalent to 100 kg N [ha.sup.-1] had no adverse effect on the recovery of cyanobacterial nitrogen. Using similar techniques, release of dissolved organic nitrogen—50% of [N.sup.2] fixation—has been shown in a marine cyanobacterium. Tirol et al. (1982) reported that the availability [N.sup.15] from cyanobacteria incorporated into the soil ranged from 23% to 28% for the first crop and 27% to 36% for the second crop. These authors further demonstrated in pot experiments that for the first crop, cyanobacterial [N.sup.15] was less available than was [([NH.sub.4]).sub.2][SO.sub.4] b ut that for the second crop its availability was very similar. After two crops, 57% [N.sup.15] from cyanobacteria and 30-40% from [([NH.sub.4]).sub.2][SO.sub.4] remained in the soil, which indicates the slow-release nature of cyanobacterial nitrogen and thus a cumulative effect of cyanobacterial inoculation. The [N.sup.15] balance in plants and soil after two crops—a result of pot experiments with dried cyanobacterial flakes—showed that losses of [N.sup.15] from [([NH.sub.4]).sub.2][SO.sub.4] were more than those from cyanobacteria irrespective of the mode of application. These observations suggest that cyanobacterial material, because of its organic nature, is less susceptible to N losses, as compared with inorganic N fertilizers and that its low C/N ratio (5-7) provides it with a better N availability than does an organic fertilizer; e.g., green manure. The relative availability of cyanobacterial nitrogen to rice depends on the sus ceptibility to decomposition of the cyanobacterial material, which varies not only with the strain but also with the physiological state, as demonstrated by the discrepancy between the values.

Using similar [N.sup.15] tracer techniques, release of dissolved organic nitrogen (50% of [N.sub.2] fixation) has been shown in a marine cyanobacterium. Cyanobacteria are sometimes conspicuous on the surface of aquatic roots of deepwater rice, and there is evidence from [[blank].sup.15][N.sub.2] studies that part of this nitrogen reaches the rice plant either as the result of release of extracellular combined N or indirectly following grazing and parasitism.

The addition of N by cyanobacterial inoculation was found to increase in the presence of phosphate fertilizers, and both total and organic N levels were maintained beyond the tilling stage. The cyanobacterial ability to mobilize insoluble forms of inorganic phosphate is evident from the findings of Kleiner and Harper (1977), who reported more extractable P in soils with cyanobacterial cover than in nearby soils without cover, Of the 18 strains tested, 17 solubilized tricalcium phosphate. The cyanobacterial P-solubilizing activity also acts on hydroxyapatite and Mussorie rock phosphate. Whitton et al. (1991) assayed 50 cyanobacterial strains, including some soil isolates, for their ability to grow with organic phosphates as their sole P source and found that all strains used the monoesters, that almost all used the diesters, and that just a few used phytic acid. Of these, most strains released part of their phosph omonoesterase extracellularly, but not the phosphodiesterase. Some evidence exists, awaiting necessary scientific interpretation, about cyanobacterial effectiveness at increasing P availability in saline soils.

In a long-term experiment there was a gradual increase in organic carbon due to cyanobacterial inoculation, but the amount remained steady at the end of three years. Bisoyi (1982) also reported that the influence of cyanobacterial inoculation on soil properties was not significant in the first and second rice-cropping seasons but responded well in the third season, showing 2-11% increases in soil organic carbon as a result of cyanobacterial inoculation. Cyanobacterial growth has been noted to initially cause an increase in the soil pH that later declined to the original value. It was also shown in these experiments that P content decreased up to 90 days of cyanobacterial growth and began to increase toward the later period of incubation. A significant reduction in pH, electrical conductivity, exchangeable sodium, hydrolic conductivity, and aggregation status of the soil, as well as a significant increase in the total N content of the soil concomitant with cyanobacterial growth, were noted by Subhashini and Kaushik (1981), who also reported that cyanobacterial inoculation in combination with gypsum had an appreciable reclamation property applicable to sodic soils. In a soil rendered saline due to bad farm management, a 25-30% decrease in salinity was observed through the repeated cultivation of Anabczena torulosa. Cyanobacteria from saline rice fields were studied, and a halotolerant Anabaena sp. was found to be a good nitrogen fixer.

The forms in which Fe, Mn, and also possibly Zn occur in soil are reported to be influenced by cyanobacterial growth in flooded rice fields. Their presence led to adecrease in ammonium acetate—extractable forms of Fe and Mn and an increase in other forms of these elements. These changes were considered to result from the release of oxygen and the addition of organic matter, especially extracellular material. Further changes in the various fractions were noted as a result of the decomposition of the cyanobacterial biomass, which were ascribed to the development of reducing conditions and the formation of organic acids. A decreased content of readily available Fe may help to minimize Zn deficiency in rice. In many cyanobacterial species, the gelatinous sheath was able to chelate Fe, Cu, Mo, Zn, Co, Mn, and other elements essential for their growth. The sheath was also thought to influence the availability of elements to other organisms. These authors ex amined whether N, P, K, Mg, Ca, and Fe were present in significantly larger concentrations in Festuca octoflora growing on soils heavily encrusted with cyanobacteria/cynolichens than in plants on the same soil where the crust had been destroyed. In the case of Mentzelia multiflora, N, Mg, and Fe were present at significantly higher concentrations, but P was present at a significantly lower concentration. In fact, cyanobacterial [N.sub.2] fixation is likely to increase the availability of N to adjacent plants, and the reduced P uptake may have been due to direct competition of its near-surface root layer with the surface crust. A cyanobacterial sheath reduces particle erosion and may adsorb charged nutrient cations. Chelating compounds present in the sheath may also be responsible for the enhanced uptake of Fe. The inorganic and organic nutritional aspects of cyanobacteria are described by Vaishampayan.

9

Status of Inoculant Production and Quality Control

Inoculation of plants with beneficial bacteria can be traced back for centuries. From experience, farmers knew that when they mixed soil taken from a previous legume crop with soil in which nonlegumes were to be grown, yields often improved. By the end of the 19th century, the practice of mixing "naturally inoculated" soil with seeds became a recommended method of legume inoculation in the USA. A decade later, the first patent ("Nitragin") was registered for plant inoculation with *Rhizobium* sp. Eventually, the practice of legume inoculation with rhizobia became common.

For almost 100 years, *Rhizobium* inoculants have been produced around the world, primarily by small companies. Some legumes, like the soybean *(Glycine max* (Merr.)L.) in Brazil, are not fertilized with nitrogen, but are only inoculated. Apart from soybean inoculation, which has made a major agricultural impact in the USA, Brazil, and Argentina, significant contributions to the production of other legumes were made in Australia, North America, Eastern Europe, Egypt, Israel, South Africa, New Zealand, and, to a lesser extent, Southeast Asia.

For the large majority of less developed countries in Asia, Africa, and Central and South America, inoculant technology has had no impact on productivity of the family farm, because inoculants are not used or are of poor quality.

Inoculation with nonsymbiontic, associative rhizosphere bacteria, like *Azotobacter*, was used on a large scale in Russia in the 1930s and 1940s. The practice had inconclusive results and was later abandoned. Interest in *Azotobacter* as an inoculant for agriculture has only recently been revived.

An attempt to use *Bacillus megaterium* for phosphate solubilization in the 1930s on large scale in Eastern Europe apparently failed.

Two major breakthroughs in plant inoculation technology occurred in the late 1970s: (i) *Azospirillum* was found to enhance nonlegume plant growth, by directly affecting plant metabolism, and (ii) biocontrol agents, mainly of the *Pseudomonas fluorescens* and *P. putida* groups, began to be intensively investigated. In recent, years, various other bacterial genera, such as *Bacillus, Flavobacterium, Acetobacter,* and several *Azospirillum*-related microorganisms have also been evaluated.

The immediate response to soil inoculation with associative, nonsymbiotic PGPB (but also for rhizobia) varies considerably depending on the bacteria, plant species, soil type, inoculant density and environmental conditions. In general, shortly after the bacteria are introduced into the soil, the bacterial population declines progressively. This phenomenon (together with bacterial biomass production and the physiological state of the bacteria in the inoculant, discussed below) may prevent the buildup of a sufficiently larg e PGPB population in the rhizosphere to obtain the intended plant response. The key obstacle is that soil is a heterogeneous and unpredictable environment, even on a small scale. The inoculated bacteria sometimes cannot find an empty niche in the soil for survival except in sterilized soil, a condition which does not exist in large-scale agriculture. They must compete with the often better-adapted native microflora and withstand predation by protozoans. A major role of inoculant formulation is to provide a more suitable microenvironment (even temporarily) to prevent the rapid decline of introduced bacteria in the soil.

Although much is known about the survival of bacteria within the protective environment of an inoculant carrier, little is known about the stresses that bacteria must endure upon transfer to the competitive and often harsh soil environment. Inoculants have to be designed to provide a dependable source of beneficial bacteria that survive in the soil and become available to the plant. Although *Rhizobium* inoculants have been in the marketplace for a century, it is only recently that the first commercial preparations of PGPB have appeared on the market. Kenney (1997) noted that "Biological products have had a less than spectacular penetration of the chemical pesticide market. Although great promises have been made, the fulfillment of those promises has not met expectations".

The aims of this review, apart from providing a comprehensive compendium of scattered, hard-to-find, practical information are: (i) to point out present and possible future trends in inoculant technology, and (ii) to mark possible future research pathways for the construction of better inoculants. No attempts have been made to collect technical or marketing data on the numerous inoculants currently available in the market.

Terms and Definitions

"Bacterial inoculant"-*A* formulation containing one or more beneficial bacterial strains (or species) in an easy-to-use and economical carrier material, either organic, inorganic, or synthesized from defined molecules. The inoculant is the means of bacterial transport from the factory to the living plant. The desired effects of the inoculant on plant growth can include nitrogen fixation in legumes, biocontrol of (mainly) soil-borne diseases, the enhancement of mineral uptake, weathering of soil minerals, and nutritional or hormonal effects. Bacterial inoculants may require lengthy and expensive registration procedures in some countries.

"Biofertilizer"-A misleading but widely used term meaning "bacterial inoculant". Usually it refers to preparations of microorganism(s) that may be a partial or complete substitute for chemical fertilization (like rhizobial inoculants). However, other bacterial effects on plant growth are largely ignored. The reason for using the word "fertilizer" is that in some countries it allows easier registration for commercial use. This term, although is appropriate for rhizobia, should be abandoned.

The Scientific Rationale for Plant Inoculation with Emphasis on Developing Countries and Semiarid Agriculture Practices

The first objective when considering inoculation with beneficial bacteria is to find the best bacteria available. Next, a study of the specific inoculant formulation is generally undertaken. In practical terms, the chosen formulation determines the potential success of the inoculant (Fages, 1992). Many potentially useful bacteria reported in the scientific literature never appear on the commercial market, perhaps because of inappropriate formulation.

The technical optimization of an inoculant formulation is independent of the strain used because most of the strains of the same bacterial species share many physiological properties. It may be assumed that a technological process developed for a particular strain is readily adaptable to another strain of the same species with only minor modifications. This normally does not apply to different bacterial genera. Apart from this "rule of thumb", strain-formulation combinations should be optimized in preparations causing stress for the bacteria like dry powders or liquids. Formulation options are usually considered at the inoculant producer's Research and Development facility, located primarily in developed countries where the target market exists. Often, the unique problems of applying these inoculants in developing countries are not considered, since these countries represent only a small share of the market. To highlight this point, I will briefly summarize the main difficulties facing the application of microbial inoculants in some developing countries.

Agricultural practices in developing countries or under semiarid conditions are two examples in which bacterial inoculants may find their greatest challenges. Developing countries practice mainly low-input agriculture in which fertilizers, pesticides, and agrotechnical machinery are scarce. For example, about 80% of Mexico's farm and grazing land is owned, rented, or worked by poor families. Many farmers are only part-time agricultural labourers who receive a daily wage or a percentage of the crop. Yields of legumes grown in family farms in developing countries are often meager and fall far short of those that can be obtained under experimental conditions. It is irrelevant whether or not responses to bacterial inoculation can be demonstrated under fertile, high-yield conditions in experimental stations in these countries or in developed countries, since this has little relevance to subsistence farming. Naturally, this type of farming does not have the resources to invest in improved agricultural techniques. Artificial inoculation, in particular, requires an infrastructure to store and transport biological products in large quantities into rural areas, and this infrastructure is not available. In developing countries, even the wealthier growers lack sufficient knowledge of modern agricultural techniques. Governmental extension services (there are almost no private consultancies) and the growers' formal agricultural education are poor. Most growers tend to practice traditional methods or copy methods from more developed countries without being aware of the deficiencies of such practices in their own particular region or without knowing the "cost" to the environment. In most cases, fertilizers are too expensive or the crop's value does not justify the expense. In places where fertilizers are available, overfertilization is common, and this practice may contaminate deep water reservoirs, produce significant health hazards to the nearby population, and disrupt the local environment. In several Latin American, African, and Pacific Island countries, a shortage of foreign currency to purchase large quantities of imported fertilizers is the main limiting factor. Consequently, most yields are low and much of the agricultural production is destined for self-consumption or local markets. The contribution of such agriculture to the world's food supply remains well below its agronomic potential.

The use of beneficial microorganisms in these areas is almost unheard of, except for rhizobia, which were introduced by many governments in the past. Many growers, when asked to replace available nitrogen fertilizer with rhizobial inoculants, are reluctant to do so for fear of reducing soil fertility. Except for soybeans, inoculants have had very little impact on legume production in developing countries. Furthermore, in many developing countries, there are no inoculant industries (as in all of Central America) or usable peat sources (as in India) for traditional inoculant formulations (described later) (Eaglesham, 1988), which increases the difficulty of popularizing the inoculation concept. Finally, in many developing countries, there is a basic reluctance to use bacteria and fungi as beneficial organisms, because in these rural cultures, microbes, are perceived to be associated with human and animal diseases (Anonymous, 1995).

Semiarid conditions, make survival difficult for the introduced bacteria even in developed countries like Israel, Australia, and parts of the USA. Harsh conditions, including frequent droughts, lack of sufficient irrigation, high salinity and soil erosion, may quickly diminish the population of any bacteria introduced into the soil unless precautions are taken to select the proper inoculant and provide irrigation concomitant with inoculation. Inoculation should be timed to coincide with sowing into moist soil or be delivered quickly with irrigation to assure rapid colonization of the target plants, as practiced in developed countries. Semiarid agriculture in developing countries, however, is a greater challenge, and technologically speaking, it seems destined for failure. However, in this type of agriculture, beneficial microorganisms may make the greatest contribution, if inexpensive and easy-to-use formulations can be developed. There, the inoculated bacteria would be subject to few, if any, side effects from pesticides. If properly instructed, the farmer, by virtue of cultivating small family plots, can apply inoculants at the right time and at the right dosage. Since the yields are low in the first place, any enhancement will be greater than in developed agriculture (Hegazi,1988) in which plants are already grown almost at their maximum genetic potential. Finally, most of the world's farmers are located in developing countries, although their productivity is presently only a small fraction of that of developed countries. Since microbial inoculants can be produced and marketed inexpensively, with some organized financial aid and better information transfer, these farmers are potential clients for microbial technology (Anonymous, 1995).

In summary, microbial inoculants can play an increasing role in the agriculture of developing countries, but field procedures involving their use should first be conceived in developed countries where the capacity for R&D is greater.

Mixed Bacterial Inoculants: Concept and Potential for the Future

Numerous recent studies show a promising trend in the field of inoculation technology. Mixed inoculants (combinations of microorganisms) that interact synergistically are currently being devised. Microbial studies performed without plants indicate that some mixtures allow the bacteria to interact wit h each other synergistically, providing nutrients, removing inhibitory products, and stimulating each other through physical or biochemical activities that may enhance some beneficial aspects of their physiology, like nitrogen fixation. It still has to be demonstrated that these bacterial synergistic effects also benefit plant growth. An example of this is *Azospirillum*, one of the most studied bacteria that associates with plants. It may associate with sugar- or polysaccharide-degrading bacteria (PDB), establishing a metabolic association where the sugar-degrading bacteria produce degradation and fermentation products used by *Azospirillum* as a carbon source, which in turn provides PDB with nitrogen.

Other examples are the association between *Azospirillum* and *Bacillus* that degrades pectin, *Azospirillum* and *Cellulomonas* that degrades cellulose, and *Azospirillum* and Emerobacter cloacae that ferments glucose.

Plant studies have shown that the beneficial effects of *Azospirillum* on plants can be enhanced by co-inoculation with other microorganisms. Co-inoculation, frequently, increased growth and yield, compared to single inoculation, provided the plants with more balanced nutrition, and improved absorption of nitrogen, phosphorus, and mineral nutrients. Thus, plant growth can be increased by dual inoculation with *Azospirillum* and phosphate-solubilizing bacteria. *Azospirillum* is also considered to be a *Rhizobium*-"helper" stimulating nodulation, nodule activity, and plant metabolism, all of which stimulate many plant growth variables and plant resistance to unfavourable conditions. Other successful combinations include *Azospirillum* or *Azotobacter* mixed with Streplomyces, and *Azospirillum* with the fungal biocontrol agent, Phialophora radicola.

Mixed inoculation with diazotrophic bacteria and arbuscular-mycorrhizal fungi creates synergistic interactions that may result in a significant increase in growth, in the phosphorus content in plants, enhanced mycorrhizal infection, and an enhancement in the uptake of mineral nutrients such as phosphorus, nitrogen, zinc, copper, and iron. These studies point out further advantages of mixed cultures over single strains: i) in vitro studies have shown that *Azospirillum* can produce more phytohormones when grown in mixed culture mixed cultures provide conditions more suitable for nitrogen fixation than pure cultures, and iii) mixed inoculation of biocontrol microorganisms is more efficient in controlling pathogens than the use of single-strain inoculants, e.g., combinations of *Pseudomonas* with *Serratia* and *Pseudomonas* with a nonpathogenic *Fusarium*. Despite progress in research on mixed inoculants, they are not yet produced commercially. In Canada, for example, apart from rhizobial inoculants, there are only two registered (single) microbial commercial inoculants Polonenko (1994), and more single strains of microbial inoculants must be registered before the inoculation industry can contemplate the development and commercialization of multibacterial inoculants.

Inoculant Formulations

Optimal Charasterics of a Carrier for Inoculants

The carrier is the delivery vehicle of live microorganisms from the factory to the field; however, no universal carrier or formulation is presently available for the release of microorganisms into soil. The carrier is the major portion (by volume or weight) of the inoculant.

The materials of which the carrier is composed and the type of formulation vary. The carrier can be a slurry or a powder. A good carrier should have one essential characteristic : the capacity to deliver the right number of *viable* cells in good physiological condition at the right time. Additional desirable characteristics for a good inoculant should be as follows:

(i) *Chemical and physical characteristics.* The inoculants should be nearly sterile or easily sterilized, and as chemically and physically uniform as possible. They should also be of consistent quality, high water-holding capacity (for wet carriers) and suitable for as many bacterial species and strains as possible.

(ii) *Manufacturing qualities.* The inoculant should be easily manufactured and mixed by existing industry, it should allow for the addition of nutrients, have an easily adjustable pH, and be made of a reasonably priced raw material in adequate supply.

(iii) *Farm handling qualities.* A good inoculant allows for ease of handling (a major concern for the farmer), provides rapid and controlled release of bacteria into the soil, a nd can be applied with standard agrotechnical machinery.

(iv) *Environmental characteristics.* The inoculant should be nontoxic, biodegradable and nonpolluting, and should minimize environmental risks such as the dispersal of cells to the atmosphere or to the ground water.

(v) *Storage qualities.* The inoculant should have sufficient shelf life (one or two years at room temperature is often necessary for successful integration into the agricultural distribution system in some countries).

Naturally, no single carrier can have all these qualities, but a good one should have as many as possible. A "super-inoculant" such as the one described above is theoretically possible. To date, no effort to synthesize a carrier with predefined superior characteristics has been reported, probably because of the cost involved. The raw materials of most commercial carriers are cheap and naturally abundant (peat and soil fractions).

Types of Existing Carriers For Inoculants

The almost universal carrier for rhizobia (which is the only inoculant being sold in large volume today) is peat. Despite its popularity, it has several drawbacks (discussed below), and many alternative materials have been evaluated.

Carriers can be divided into four basic categories:

(i) Soils: peat, coal, clays, and inorganic soil.

(ii) Plant waste materials: composts, farmyard manure, soybean meal, soybean and peanut oil, wheat bran, "press-mud", agricultural waste material, spent mushroom compost, and plant debris.

(iii) Inert malerials: vermiculite, perlite, ground rock phosphate, calc ium sulfate, Polyacrylamide gels, and alginate beads.

(iv) Plain lyophilized microbial cultures and oil-dried bacteria (Johnston, 1962). These preparations can later be incorporated into a solid carrier or used as they are.

To produce an inoculant, the target microorganism can be introduced into a sterile or non sterile carrier. From a purely microbiological point of view, the sterile c arrier has significant advantages but has not usually been cost effective from a commercial standpoint. Nonetheless, sterile-originated inoculants have been successfully marketed even with their higher price tag.

Formulation is the industrial "art" of converting a promising laboratory-proven bacterium into a commercial field product. Chemical formulations of agroproducts set high standards for long shelf life, ease of use, and resistance to abuse by the farmers.

Microbial inoculant formulations are expected to match the above characteristics and overcome two major problems for living organisms: (i) loss of viability during short storage in the grower's warehouse (which in developing countries usually lack refrigeration), and (ii) long shelf life and stability over the range of-5° to 30°C within the marketing distribution systems. Products lacking this range of temperature tolerance will be unacceptable in the agricultural market.

Inoculants come in four basic dispersal forms:

(i) Powders. This form is used as a seed coating before planting. The smaller the particle size, the better the inoculant will adhere to the seeds. Standard sizes vary from 0.075 to 0.25 mm, and the amount of inoculant used is around 200 to 300 g/ha. These inoculants are the most common both in developed (Smith, 1997) and developing countries.

(ii) Slurries. This inoculant is based on powder-type inoculants suspended in liquid (usually water). The suspension is directly applied to the furrow or alternatively, the seeds are dipped just prior to sowing.

(iii) *Granulars.* These inoculants are applied directly to the furrow together with the seeds. Size ranges are from 0.35 to 1.18 mm. Rhizobia inoculant is used at a rate of 5 to 30 Kg/ha. These inoculants are popular and have been successfully commercialized since 1975. Bead-like forms are synthetic variations of granular forms. These can be in macro sizes (1 to 3 mm in diameter) used as granules form, or in micro size (100 to 200 μm) used as a powder for seed coating. These inoculants are a new, as yet unproven, possibility in inoculation technology, and their features will be described later in detail.

(iv) *Liquids.* These inoculants use broth cultures or liquid formulations, mainly in water, but also in mineral or organic oils.

The seeds are either dipped into the inoculant before sowing, or an applicator evenly sprays the liquid inoculant on the seeds. After drying, the seeds are sown. This method ensures even coverage of the seeds without interference with the seed monitoring system of the planters or inoculum loss when dried (Smith, 1995). These inoculants are currently popular in the USA, Canada, Argentina, and Brazil, mainly for soybeans, but also for lentils, peas, and peanuts. For biocontrol agents of leaf diseases, the inoculant c an be diluted in water and sprayed for better coverage of the leaves. Alternatively, the suspension can be sprayed directly into the furrow or on the seeds before sowing. The in-furrow inoculant provides a larger amount of bacteria to the plant than seed inoculation. In rhizobia, this improves plant nodulation (Smith, 1995). For bacteria with poor survival in the soil, like *Azospirillum* sp., these formulations are largely useless since they do not provide a protective environment for the bacteria. Furthermore, in some plant species, these formulations should be applied several days after sowing at seedling germination, causing extra work and cost for the farmer.

The microbial inoculant is not merely a suitable carrier containing the bacteria. Other materials might be involved in the final formulation. For example: an *Azospirillum lipoferum* inoculant for corn, developed in France, was based on 1% alginate containing the bacterial cells and 99% inert calcium carbonate "diluent", which allowed for the right bacterial concentration, because the alginate contained too many cells of *Azospirillum* for optimal inoculation. Apparently this alginate formulation was never commercialized, perhaps due to its high cost, and *A. lipoferum* was commercialized in a sterile peat inoculant instead.

The use of each type of inoculant depends upon market availability, cost, and the needs of a particular crop under specific environmental conditions. For example, the granular form is better than powder inoculants for rhizobia, under stressful planting conditions, but since more is required, it is costlier.

Traditional Peat Formulations

Peat formulations have been the carriers of choice, and are the most commonly used in the rhizobia inoculation industry. Since it was adopted decades ago, farmers are by now quite comfortable with peat, and governmental agencies usually know how to monitor its quality. Its popularity is primarily due to successful field results obtained under commercial cultivation. In this form, the bacteria are metabolically active, and in some inoculants, bacterial multiplication continues during the storage period, as long as sufficient nutrients, moisture, and the correct temperature are maintained.

Recipes for peat inoculants are widely known and will not be repeated here. Instead, the major drawbacks of such inoculants and why other types of carriers are being sought are discussed. The principle drawbacks originate from the great variability in peat quality (which is source dependent), and because peat is an undefined, complex organic material. This greatly affects the final product and may cause difficulties in inoculant dosage, storage conditions, and variation of inoculant effectiveness between different manufacturers and between different batches from the same manufacturer. Heat sterilization of some peats may release compounds toxic to the bacteria resulting in low bacterial counts. In peat inoculants, bacteria have a lower tolerance for physical stress during storage, in particular for temperature variations. Some types of peat can even reduce plant growth. Finally, peat formulations are prone to contamination that can reduce the shelf life of the inoculant. From the delivery standpoint, peat powder is easily blown away from the seeds by the commonly used seed air-delivery system used by the planter. Peat also interferes with the seed monitoring mechanism of the planters. Addition of adhesives to the inoculant during its application to the seeds or slurry application will improve its adhesion, but that requires additional time and labour for a process that is already labour-intensive (Smith, 1995).

New Trends in Formulations Using Unconventional Synthetic Materials

During the last decade, several experimental formulations based on polymers have been evaluated. These polymers have demonstrated potential as bacterial carriers (Jung et al. 1982) that offered substantial advantages over peat. These formulations encapsulate the living cells, protect the microorganisms against many environmental stresses, and release them to the soil, gradually but in large quantities, when the polymers are degraded by soil microorganisms, usually at the time of seed germination and seedling e mergence. They can be stored dried at ambient temperatures for prolonged periods, offer a consistent batch quality and a better defined environment for the bacteria, and can be manipulated easily according to the needs of specific bacteria. These inoculants can be amended with nutrients to improve the short-term survival of the bacteria upon inoculation, which is essential to the success of the inoculation process, especially with associative PGPR. However, *a major constraint* for the inoculation industry is that polymers are expensive compared to peat-based inoculants and require more handling by the industry (Fages, 1992). Thus, even relatively large inoculant manufacturers have currently abandoned the technique.

Encapsulated Formulations

The encapsulation of microorganisms into a polymer matrix is still experimental in the field of bacterial-inoculation technology. At present there is no commercial bacterial product using this technology. The concept underlying immobilized microbial cells, is to entrap beneficial microorganisms into a matrix. The formulation (bacteria-matrix) is then fermented in a bacterial growth medium. These formulations can produce many useful compounds for industrial and environmental applications (such as organic acids, amino acids, enzymes) and biodegrade toxic materials (bioremedation) over extended periods of time. The bacterial products are then extracted from the bioreactor while fermentation continues. An excellent review of environmental applications of immobilized microbial cells was recently published.

Immobilized microbial cells are easy to produce, store, and handle during industrial operations. The main goal of these industrial formulations is to maintain the cells entrapped in an active form for as long as possible. Any premature release of the microorganisms from these encapsulated forms is undesirable. Encapsulated bacterial formulations in agriculture have at least two distinctly different goals from those of the fermentation industry: (i) to temporarily protect the encapsulated microorganisms from the soil environment and microbial competition, and (ii) to release them gradually for the colonization of plant roots.

Macro and Micro Formulations of Alginate

Alginate is the material most commonly used for encapsulation of microorganisms. The resulting inocula are used for various purposes: the immobilization of cell organelles and enzymes, the application of biological control agents and mycoherbicides, to increase the stability of recombinant plasmids in the host cells, in bacterial chemotaxis research, and mushroom cultivation. Alginate is a naturally occurring polymer composed of f1-1,4-linked D-mannuronic acid and L-glucuronic acid. It is available from different macroalgae as well as several bacteria. Because of the massive production of alginate in the far east, alginate cost has recently dropped making it potentially more attractive to the inoculant industry.

The preparation of beads containing bacteria is fairly easy and involves a multistep procedure. In cases where the biomass of the entrapped strain is low, a secondary multiplication of the entrapped bacteria in the already formed beads is required (Bashan, 1986a Several other alginate-based preparations have been tried for the encapsulation of VAM fungi, ectomycorrhizal fungi, *Frankia* inoculation, and fungi used as biocontrol agents against soil-borne pathogens.

The main advantages of alginate preparations are their nontoxic nature, biodegradability, and their slow release of microorganisms into the soil. This technology was used to encapsulate the plant-beneficial bacteria *A. brasilense* and *P. fluorescens*, which were later successfully used to inoculate wheat plants under field conditions. The bacteria survived in the field long enough and their populations were comparable to the survival of bacteria originating from peat-based inoculants. The encapsulation of genetically engineered *P. fluorescens,* which was released into the soil, showed significantly improved survival rates over nonencapsulated cells after 3 months. Furthermore, the addition of clay and skim milk to the beads significantly increased bacterial survival over alginate beads alone. Colonization of wheat roots by beneficial cells released from the beads was superior to that achieved by direct soil inoculation. These studies provide clear evidence that alginate beads are efficient slow-release carriers for plant inoculants, providing a protective environment in the soil. *P. fluorescens* numbers decreased only moderately in the soil when the cells were encapsulated, whereas a large reduction in bacterial numbers was observed when free, nonencapsulated inoculum was used. When the bacteria were encapsulated, they were washed down only slightly compared to. free inoculated bacteria. Most of them remained in the root zone. Another recent example is the colonization of tomato roots with *P. fluorescens* embedded in alginate beads to produce a biocontrol preparation against bacterial wilt of tomato. Trevors et al. (1992) proposed that good rhizosphere colonization and survival, a low wash-down rate, and resistance to drying show that the use of alginate in bacterial inoculation, has potential.

The use of macro alginate beads has two major disadvantages: (i) the need for an additional treatment during sowing, and (ii) the need for the bacteria to move through the soil towards the plants. In developed countries, a farmer who is already busy during sowing might be pressed for time and reluctant to incur the additional expense (Anonymous, 1995). In developing countries, the farmer might not inoculate the seeds at all. This is because of insufficient agricultural education and conservative traditions that make them suspicious of new technologies, especially those involving live bacteria. Under agricultural practices, when beads are loosely mixed with seeds and sown together, the beads containing the bacteria might fall far from the seeds (up to several cm).

The bacteria released from the beads must migrate through the soil, facing competition from the native microflora and sometimes the absence of a continuous film of water needed for their movement. These distances might prove prohibitive for many beneficial bacteria, even those with a proven motility in soil like *Azospirillum*.

To overcome these difficulties, the microbead concept was conceived. If the beads are small enough but still capable of encapsulating a sufficient number of bacteria, it would be possible to produce a powder-like formulation. The seeds would be coated with this "bead dust" at the seed handling facility and sold to the farmer in developing countries as "improved seeds". Coated seeds (but with fertilizers or fungicides) are common and accepted in most rural areas of Mexico, for example. In developed countries, in large scale agricultural practice such as in North America, precoated seeds will eliminate the need for an additional expensive field treatment and provide the ultimate convenience for the grower. To be sure, precoating seeds with bacteria is not an easy industrial task, however, a similar idea, but with peat inoculant, has already been applied for preinoculation of forage legumes such as alfalfa. The peat is applied to the seeds as a slurry using an adhesive and the inoculated seeds are covered with finely ground calcium carbonate.

The production of alginate microbeads is simple and involves a "mist-like" extrusion via a thin nozzle. This technology produced alginate beads in sizes of 50 to 200 μm, which entrapped a significant number of bacteria (approx. 108 to 109 cfu/g), similar to the level obtained in alginate macrobeads (Bashan, 1986a). Application of this novel formulation to the soil is in progress. An alternative to production of microbeads directly is the mechanical crushing of dehydrated macrobeads to any desired size and their use to inoculate seeds as microbeads.

So far, it appears that alginate is the most promising of the encapsulating materials tested. However, because of the limited published research on alginate beads in agriculture and because of their possible deficiencies, especially their higher price when compared to peat, it is premature to predict whether alginate will displace peat in the inoculation industry or will remain only in the domain of industrial and environmental microbiology.

Encapsulation with other Materials

Although commercial alginate preparations are not yet available for bacterial plant inoculation, several other materials, which are used in industrial and environmental microbiology, may be considered a s substitutes when the microorganism fails to adapt to alginate preparations. To the best of our knowledge, almost none have been tested in soil or in the field. However, this review should point out their existence to promote further research with these carriers.

Dried Synthetic Carriers

The main objective of encapsulating bacteria is to increase their survival time during storage (not bacterial number, however, which usually decreases during the process). Until now, the most common solutions to this problem of survival time have been air-dried and lyophilized preparations. The lowered water content in the final product is responsible for long-term survival during storage. In this way, the bacteria in the formulation remain inactive, resistant to environmental stresses, insensitive to contamination, and are more compatible with fertilizer application (Paau, 1988).

The dehydration phase is perhaps the *most critical* of the entire formulation process especially for nonspore-forming bacteria. Bacterial survival is affected by several variables: the culture medium used for bacterial cultivation, the physiological state of the bacteria when harvested from the medium, the process of cell encapsulation, the use of protective materials, the type of drying technology used, and the rate of dehydration. If properly dehydrated, the shelf life of the dried formulation is much longer than that of any peat-based product.

From both the commercial and the agricultural point of view, the extremely long survival of bacteria in these preparations makes the dry formulations very attractive. By studying the effect of dehydration on encapsulated *Azospirillum* cells, Paul et al. (1993) demonstrated that a large proportion of the cells are destroyed during dehydration. Howevert, when properly dehydrated, the surviving cells are sufficient for inoculation. An additional benefit is that the bacteria survive for almost a year without a drop in population. Dry alginate beads containing *A. brasilense* and *P. fuorescens* produced in 1983, and which were preserved for archival purposes, maintained their bacterial population for 14 years, although at lower level than at encapsulation time.

Alternatively to dried synthetic carrier inoculants, air-dried clay powder with features and advantages similar to these of synthetic carriers can be used because, with current technology, it is more cost effective.

Regulation and Control of Contamination of Commercial Inoculants

Naturally, an inoculant should contain a level of bacteria sufficient to inoculate plants and produce an economic gain. The required level of bacteria cannot be established as a general standard because it varies from one bacterial species to another. Only rhizobial inoculants have legally established standards. Since this is a new research field, standards for PGPB numbers in inoculants do not yet exist, and every manufacturer can claim whatever he deems appropriate for his product. In fact, some commercial preparations do not even contain any viable cells.

Quality control methods to determine the number of bacteria within the inoculant are not standardized either. To measure the bacterial number, commonly known methods in microbiology are used; the traditional Plate Count methods, Most Problable Number.

Many developed countries (but not the USA) have regulations for inoculant quality, but in almost all countries in Latin America, inoculant quality is not regulated, nor are the existing regulations well enforced. The level of rhizobia required in the inoculant varies worldwide (between 107 and 4 x 109 cfu/g inoculant) and no set of common international standards exists. As far as I know, "Agriculture Canada" is the only governmental agency in North America that has continuously regulated inoculants during the last 20 years. This was after an internal survey by "Agriculture Canada" during the early 1970s which found low levels of *Rhizobium* in commercial preparations. The latest standards in Canada for rhizobia are 103 to 105 nodulating rhizobia per seed (size-dependent) or 1011, rhizobia per hectare. Currently, there are only two research stations in Canada which routinely evaluate inoculant quality and publish the results annually. The INRA Laboratory of Soil Microbiology in Dijon plays a similar role in France.

Since the introduction of governmental regulations, there has been an improvement in the quality of commercial inoculants in several countries, including Australia, Canada, and the UK. In most countries, there are no regulations of the level of contaminants in the most commonly used nonsterile peat preparations. A Canadian survey showed the contamination level sometimes exceeded, by several orders of magnitude, the level of rhizobia in the inoculant, indicating poor quality inoculants. In at least one study, it was shown that the contaminant species had detrimental effects on the rhizobia. It is still unknown whether these bacterial contaminants p resent any health hazards to humans, animals, and plants. These potential hazards should not be ignored as long as nonsterile-carrier inoculants are widely used. These are likely to stay in the market in the future since sterile-carrier inoculants are usually 5 to 10 times higher in price (R.S. Smith, 1995 personal communication). However, one should note that the use of non-sterile carrier inoculants has caused no reported health hazards in decades of usage. France, which has the highest standards for inoculant quality and mandated field testing of new formulations, has a strict requirement that prohibits contaminants in rhizobia inoculants (Catroux, 1991). Australia permits low levels of contaminants (0.1 % of the total bacterial populatio n), but at the same time requires high population levels of rhizobia (Thompson, 1991). Even some developing countries have very high standards for inoculants. In Rwanda, high rhizobia counts and no more than 0.001% contaminants are allowed (Scaglia, 1991), but it is doubtful whether these high standards are enforced. Surprisingly, the USA and UK have no regulations, perhaps because there have been no reported adverse effects.

Inoculant quality control is left to market forces and the manufacturers' discretion (Smith, 1992). The self-imposed standards for rhizobia in American industry are also lower than most European standards. This lack of standard regulation makes it particularly risky for small and medium size growers who tend to purchase inoculants from only one source and may end up with noneffective inoculants. Olsen et al (1994a) noted that Canadian regulations sometimes allow even low levels of rhizobia to be legally acceptable, perhaps because the cost of regulation is too high, compared to the risk of misuse. Olsen et al. (1994a) concluded that increased standards not only ensure that the farmer is provided with effective inoculants but are also in the best interest of the inoculation industry. Outlawing low quality inoculants from the market (especially new and lesser known PGPB) will help to prevent a bad public image for the industry and will facilitate the introduction and acceptance of inoculants. It should be noted that the percentage of substandard inoculants in the market is not known, and perhaps the problem is just hypothetical.

Examples of New Commercial Microbial Inoculants

The inoculant industry is not standing still. Its main future prospect is the incorporation of PGPB into products. Apart from *Rhizobium* inocula nts, the market for all other bacterial products is smaller than 1% of the respective agrochemical industry. In biologicals, bioinsecticides based on *Bacillus thuringiensis* have a 90% share of the market (and are not within the scope of this review). *Rhizobium* inoculants have the same current dollar value as the bioinsecticides. Numerous manufactures, mainly small to medium size, exist worldwide and have been producing products similar in quality and quantity for decades. For the purpose of illustration only, this review will cite a few examples of these commercial inoculants. A commercial example of a peat-based inoculant is the nonsterile peat granule Soil implant® that still comprises a large portion of the sales of one company. Despite the wide use of peat-based inoculants, other types of preparations for rhizobia recently appeared on the market, such as the vermiculite-based Gold CoatTM *Rhizobium* inoculant, a liquid seed-applied soybean inoculant, Cell-Tech® (Smith, 1995), a liquid in-furrow inoculant LIFT (Smith, 1995) and an air-dried clay powder for alfalfa, Nitragin® Gold (Smith 1995).

Commercial microbial inoculants of other beneficial microorganisms have begun to appear on the market on a small scale. These include "Azogreen", a French-approved *Azospirillum* inoculant (Fages 1992), Mycostop from Finland, one of the first biofungicides based on *Streptomyces* sp., and the biofungicide "Kodiak" based on *Bacillus subtilis*. Although few strains of PGPB were officially registered in China, many others are being sold off the shelves of research institutes and a few have commercial names [YIB (*Bacillus cerus*); 5406-antagonist (*Streptomyces jingyangensis*)], identifying them as biocontrol agents and as growth promoters. In Japan, a product based on *Agrobacterium radiobacter* K84 (Bakuterohzu) to control crown gall in rose was registered (Tsuchiya, 1997). Its Australian predecessor, NoGall, has been on the market for years (Kerr, 1989). BioCoatTm, a product of *P. fuorescens* for suppression of *Fusarium* sp. in radish, was phased out in 1996 after several years of use, because of low disease suppression and short shelf-life. Several other PGPB products are currently in the pipe-line for final release.

Time and Application Methods for Bacterial Inoculants

A major advantage of formulated bacteria is that they are usually readily applied using standard farming machinery. It is unlikely, even in developed countries, let alone in developing countries, that farmers will purchase specialized equipment to test an unproven product. Additionally, ease of use and the absence of time-consuming application procedures are also major factors. There are several common methods of inoculant application, yet farmers are not always willing to practice them. The primary limiting factors in developed countries are the cost of labour and time pressures at planting (Smith 1992). In developing countries, there is insufficient knowledge, lack of adequate machinery, and improper distribution and importation laws for live inoculants that can lose their viability and effectiveness (Anonymous, 1995). In North America, for example, Smith (1992) wrote that farmers follow recommended procedures *only* if they have plenty of time for planting. This is particularly difficult for large growers in temperate zones who need to sow in a limited period of time. Relatively laborious methods such as the recommended slurry method are phased out for less laborious ones like the common dry method. Farmers are discouraged from inoculant use if they have to make additional passes over the sown field. Two main methods of inoculation are currently being used: seed inoculation and soil inoculation (Hegde, 1992). The latter is done by delivering the inoculant directly into the sowing furrow with the seeds (Gault, 1982). Seed inoculation is the most popular method worldwide, as long as the farmer is willing to take the extra step of mixing the inoculant with the seeds immediately before sowing. The less common method, soil inoculation, is now being used successfully for rhizobia inoculation, but has several disadvantages which limit its future for the application of *Azospirillum*, which survives poorly in many soils.

By analysing the advantages and disadvantages of the two methods, it can be concluded that while there is currently room for both methods, the future probably lies with seed inoculation. For inoculation of soybeans, currently the major inoculated crop, changing management practices (conservation tillage and narrow rows) also limit the use of the granular form (R.S. Smith, 1995 personal communication).

Microbial inoculants can be applied during three possible phases: (i) at the seed processing plant as a seed coating, months before the actual sowing, (ii) "on site", as a seed application just before sowing, or by inoculant delivery directly onto the seeds in the furrow, and (iii) after seedlings emerge (Bashan, 1986b). The most popular method to date with peatbased inoculants is the "on site" method, primarily because of lower costs. Fages (1992) pointed out some major drawbacks for "on-site" seed inoculation: (i) additional work is required during sowing, which is time restricted, (ii) the seed germination rate may decrease if some seeds are damaged during the mixing step with the inoculant, (iii) since the bacteria in the inoculant are alive, they may be subjected to UV irradiation which can reduce their population during the field mixing operation, and (iv) the bacterial population may be reduced when the wet inoculant is attached to the chemically coated seeds. All these drawbacks also exist when seeds are precoated by industry, albeit to a lesser extent.

Although it requires a more expensive technology, the precoating of seeds merits more research to eliminate all or most of the above mentioned difficulties (Paul et al. 1991), primarily because it is most convenient for the farmer. Of particular interest is the interaction between the seed and the bacterial cells. Polysaccharide (either natural or synthetic) matrixes, which can be attached harmlessly and easily to the seeds, are the most promising materials to be evaluated.

Soil inoculation is an alternative to seed inoculation. It is more convenient for the farmer than seed inoculation, but is sometimes not as effective. It is also more expensive because more inoculant is required. Soil inoculation can be done either with peat-based granules or with microgranulated forms of inert materials; sand, calcium carbonate, or marble powder. These materials have been previously mixed with the inoculum in the factory or can be mixed with the seeds by the farmer prior to sowing. The technique uses a specific granular applicator which makes use of insecticide applicators farmers already. have. Fages (1994) reported that a commercial product for *A. lipoferum* has successfully used this technology with corn for four years in 13 locations in France.

Alternatively, liquids applied in furrow are also available for legume inoculation.

It is clear that some applications use a large amount of inoculant (by weight or volume), up to several tons of inoculant per hectare. This makes them uneconomical for many crops. It is expected that such bulky inoculants will be phased out of the market soon. In the agrochemical industry, application levels (of active ingredient and water) are low and decreasing. The new bacterial inoculants must meet these standards if they are to compete with chemicals on the farmer's list.

Cost of Development and Marketing

The cost of developing a new product by the agrochemical industry has been estimated at over $80 million US and rising. The development of resistance to pesticides may shorten the commercial life of these products and thus their potential return. The development of bacterial inoculants is claimed to be cheaper than that of agrochemicals, although the large scale screening of strains with biological activity is still required (comparable to more than 1:20,000 screened molecules for a new chemical product). However, having a predetermined pathogen as a research goal will dictate where to look for its antagonist, making the initial screening easier. In the case of a biocontrol PGPB, it has been shown that it is rather easy to detect a potential antagonist in vitro. The following are some factors that reduce the costs of development of bacterial inoculants which makes them attractive to the agrochemical industry: (i) reduced registration costs compared to those of chemical-product test programs that are well-established and costly. The cost of registration of naturally-occurring bacteria is relatively low (Markle, 1983) because the registration agencies worldwide expect a lower environmental impact from an indigenous microorganism, even in large quantities, than from a man-made, and possibly toxic, molecule. This does not consider the cost of registration of genetically engineered bacteria since those regulations are still being formulated by agencies in each country; (ii) reduced registration time decreases the time span from first screening to market, thus increasing revenues; (iii) the possibility of developing bacterial products for small markets. Since the cost involved in bringing a new chemical to the marketplace is so large, the product must be targeted to a market large enough to have a good return on investment. This limits the choice of crops to the major crops only. With a cheaper bacterial inoculant, the smaller markets of site-specific agriculture become available.

Yet, one should bear in mind that this type of future agriculture will probably increase the cost of development of bacterial inoculants, since many types of inoculants should be developed; and (iv) although fermentation is costlier than chemical production, the fermentation plant is more versatile.

It can produce, with minor modifications, different products rather then specializing in a single chemical formulation.

Other incentives for the agrochemical industry to develop bacterial inoculants might be: (i) It is less likely that pathogens will develop resistance as fast as they do to chemical products. However, some bacterial inoculants, especially those that use an organism employing a single me chanism against the pathogen, can also develop resistance. (ii) They are "environmentfriendly". The "natural" tag of bacterial inoculants (especially those that are nonengineered and indigenous) make them more acceptable in the public eye, and especially to the "Green movement" pressure groups, than chemicals. Although theoretically true, this is not always necessarily so, i.e., the acceptance of *B. thuringiensis* as safe for humans and the environment occurred only after the product's effectiveness was confirmed at a cost equivalent to the chemical alternative. On the other hand, a "natural" label by itself will not guarantee acceptance of a mildly effective product in the market place.

Regarding the commercialization of inoculant products, any inoculant R&D laboratory should consider the guidelines of Lethbridge (1988), Fages (1992) and Kenney (1997). First, all the considerations mentioned above (efficient strains, optimized formulations, cost-effective production, and good and practical inoculation techniques) are not sufficient to launch a new product on the market nor guarantee its success. The following practical variables should be considered: (i) the product must be efficient and reliable in large-scale field trials and especially under "real life" conditions; the value to the grower is the number one factor of success of any new inoculant, (ii) registration costs and delays that vary from one country to another, (iii) obviously, patents on industrial processes and registration of biological products must be secured (patent registration is time consuming and expensive in most countries albeit less tedious than the registration of chemicals), (iv) for every potential customer country, a market survey must be done which examines customer demand, market size, and expected selling price, (v) safety concern and environmental consciousness are markedly more important today than they were in the past, (vi) the ease of handling of an inoculant is a major issue for the end users; the need to keep an inoculant refrigerated has an adverse effect on its commercialization, and (vii) the biological product must have the ability to be compatible with chemicals used in Integrated Pest Management which today is the common state of the art of pest control.

10

Integrated Pest Management (IPM)

Integrated pest management (IPM) is an integrated approach of crop management to solve ecological problems when applied in agriculture.

These methods are performed in three stages: *prevention*, *observation*, and *intervention*. It is an ecological approach with a main goal of significantly reducing or eliminating the use of pesticides while at the same time managing pest populations at an acceptable level. For their leadership in developing and spreading IPM worldwide, Dr. Perry Adkisson and Dr. Ray F. Smith received the 1997 World Food Prize.

History

Shortly after World War II, when synthetic insecticides became widely available, entomologists in California developed the concept of "supervised insect control". Around the same time, entomologists in cotton-belt states such as Arkansas were advocating a similar approach. Under this scheme, insect control was "supervised" by qualified entomologists, and insecticide applications were based on conclusions reached from periodic monitoring of pest and natural-enemy populations. This was viewed as an alternative to calendar-based insecticide programs. Supervised control was based on a sound knowledge of the ecology and analysis of projected trends in pest and natural-enemy populations.

Supervised control formed much of the conceptual basis for the "integrated control" that University of California entomologists articulated in the 1950s. Integrated control sought to identify the best mix of chemical and biological controls for a given insect pest. Chemical insecticides were to be used in manner least disruptive to biological control. The term "integrated" was thus synonymous with "compatible." Chemical controls were to be applied only after regular monitoring indicated that a pest population had reached a level (the economic threshold) that required treatment to prevent the population from reaching a level (the economic injury level) at which economic losses would exceed the cost of the artificial control measures.

IPM extended the concept of integrated control to all classes of pests and was expanded to include tactics other than just chemical and biological controls. Artificial controls such as pesticides were to be applied as in integrated control, but these now had to be compatible with control tactics for all classes of pests. Other tactics, such as host-plant resistance and cultural manipulations, became part of the IPM arsenal. IPM added the multidisciplinary element, involving entomologists, plant pathologists, nematologists, and weed scientists.

In the United States, IPM was formulated into national policy in February 1972 when President Nixon directed federal agencies to take steps to advance the concept and application of IPM in all relevant sectors. In 1979, President Carter established an interagency IPM Coordinating Committee to ensure development and implementation of IPM practices.

Principles

An IPM system is designed around six basic components: The US Environmental Protection Agency has a useful set of IPM principles.

1. Acceptable pest levels: The emphasis is on *control,* not *eradication.* IPM holds that wiping out an entire pest population is often impossible, and the attempt can be more costly, environmentally unsafe, and frequently unachievable. IPM programs first work to establish acceptable pest levels, called action thresholds, and apply controls if those thresholds are crossed. These thresholds are pest and site specific, meaning that it may be acceptable at one site to have a weed such as white clover, but at another site it may not be acceptable. This stops the pest gaining resistance to chemicals produced by the plant or applied to the crops. If many of the pests are killed then any that have resistance to the chemical will rapidly reproduce forming a resistant population. By not killing all the pests there are some un-resistant pests left that will dilute any resistant genes that appear.
2. Preventive cultural practices: Selecting varieties best for local growing conditions, and maintaining healthy crops, is the first line of defence, together with plant quarantine and 'cultural techniques' such as crop sanitation (*e.g.* removal of diseased plants to prevent spread of infection).
3. Monitoring: Regular observation is the cornerstone of IPM. Observation is broken into two steps, first; inspection and second; identification. Visual inspection, insect and spore traps, and other measurement methods and monitoring tools are used to monitor pest levels. Accurate pest identification is critical to a successful IPM program. Record-keeping is essential, as is a thorough knowledge of the Behaviour and reproductive cycles of target pests. Since insects are cold-blooded, their physical development is dependent on the temperature of their environment. Many insects have had their development cycles modelled in terms of degree days. Monitor the degree days of an environment to determine when is the optimal time for a specific insect's outbreak.
4. Mechanical controls: Should a pest reach an unacceptable level, mechanical

methods are the first options to consider. They include simple hand-picking, erecting insect barriers, using traps, vacuuming, and tillage to disrupt breeding.

5. Biological controls: Natural biological processes and materials can provide control, with minimal environmental impact, and often at low cost. The main focus here is on promoting beneficial insects that eat target pests. Biological insecticides, derived from naturally occurring microorganisms (*e.g.*: *Bt*, entomopathogenic fungi and entomopathogenic nematodes), also fit in this category.
6. Chemical controls: Synthetic pesticides are generally only used as required and often only at specific times in a pests life cycle. Many of the newer pesticide groups are derived from plants or naturally occurring substances (*e.g.*: nicotine, pyrethrum and insect juvenile hormone analogues), and further 'biology-based' or 'ecological' techniques are under evaluation.

An IPM regime can be quite simple or sophisticated. Historically, the main focus of IPM programs was on agricultural insect pests. Although originally developed for agricultural pest management, IPM programs are now developed to encompass diseases, weeds, and other pests that interfere with the management objectives of sites such as residential and commercial structures, lawn and turf areas, and home and community gardens.

Process

IPM is applicable to all types of agriculture and sites such as residential and commercial structures, lawn and turf areas, and home and community gardens. Reliance on knowledge, experience, observation, and integration of multiple techniques makes IPM a perfect fit for organic farming (sans artificial pesticide application). For large-scale, chemical-based farms, IPM can reduce human and environmental exposure to hazardous chemicals, and potentially lower overall costs of pesticide application material and labour.

1. Proper identification of pest-What is it? Cases of mistaken identity may result in ineffective actions. If plant damage due to over-watering are mistaken for fungal infection, spray costs can be incurred, and the plant is no better off.
2. Learn pest and host life cycle and biology. At the time you see a pest, it may be too late to do much about it except maybe spray with a pesticide. Often, there is another stage of the life cycle that is susceptible to preventative actions. For example, weeds reproducing from last year's seed can be prevented with mulches. Also, learning what a pest needs to survive allows you to remove these.
3. Monitor or sample environment for pest population-How many are here? Preventative actions must be taken at the correct time if they are to be effective. For this reason, once the pest is correctly identified, monitoring must begin *before* it becomes a problem. For example, in school cafeterias where roaches may be expected to appear, sticky traps are set out before school

starts. Traps are checked at regular intervals so populations can be monitored and controlled before they get out of hand. Some factors to consider and monitor include: Is the pest present/absent? What is the distribution-all over or only in certain spots? Is the pest population increasing or decreasing?

4. Establish action threshold (economic, health or aesthetic)- How many are too many? In some cases, a certain number of pests can be tolerated. Soybeans are quite tolerant of defoliation, so if there are a few caterpillars in the field and their population is not increasing dramatically, there is not necessarily any action necessary. Conversely, there is a point at which action *must* be taken to control cost. For the farmer, that point is the one at which the cost of damage by the pest is *more* than the cost of control. This is an economic threshold. Tolerance of pests varies also by whether or not they are a health hazard (low tolerance) or merely a cosmetic damage (high tolerance in a non-commercial situation).

Different sites may also have varying requirements based on specific areas. White clover may be perfectly acceptable on the sides of a tee box on a golf course, but unacceptable in the fairway where it could cause confusion in the field of play.

Choose an appropriate combination of management tactics For any pest situation, there will be several options to consider. Options include, mechanical or physical control, cultural controls, biological controls and chemical controls. Mechanical or physical controls include picking pests off plants, or using netting or other material to exclude pests such as birds from grapes or rodents from structures. Cultural controls include keeping an area free of conducive conditions by removing or storing waste properly, removing diseased areas of plants properly. Biological controls can be support either through conservation of natural predators or augmentation of natural predators.

Augmentative control includes the introduction of naturally occurring predators at either an inundative or inoculative level. An inundative release would be one that seeks to inundate a site with a pest's predator to impact the pest population. An inoculative release would be a smaller number of pest predators to supplement the natural population and provide ongoing control. Chemical controls would include horticultural oils or the application of pesticides such as insecticides and herbicides. A Green Pest Management IPM program would use pesticides derived from plants, such as botanicals, or other naturally occurring materials.

Evaluate results- How did it work? Evaluation is often one of the most important steps. This is the process to review an IPM program and the results it generated. Asking the following questions is useful: Did actions have the desired effect? Was the pest prevented or managed to farmer satisfaction? Was the method itself satisfactory? Were there any unintended side effects? What can be done in the future for this pest situation? Understanding the effectiveness of the IPM program allows the site manager to make modifications to the IPM plan prior to pests reaching the action threshold and requiring action again.

Association of Natural Biocontrol Producers

The Association of Natural Biocontrol Producers (ANBP) is an international not-for-profit professional association representing the biological pest management industry.

Biocontrol refers to the use of beneficial animals (especially insects, such as beetles and bumblebees) for controlling pests such as aphids, thrips and mites that are their natural prey.

Members in the association include commercial producers, university researchers, regulatory agencies, and others.

Biodynamic Agriculture

Biodynamic agriculture is a method of organic farming that treats farms as unified and individual organisms, emphasizing balancing the holistic development and interrelationship of the soil, plants, animals as a self-nourishing system without external inputs insofar as this is possible given the loss of nutrients due to the export of food.

Regarded by some as the first modern ecological farming system and one of the most sustainable, biodynamic farming has much in common with other organic approaches, such as emphasizing the use of manures and composts and excluding of the use of artificial chemicals on soil and plants. Methods unique to the biodynamic approach include the use of fermented herbal and mineral preparations as compost additives and field sprays and the use of an astronomical sowing and planting calendar. Biodynamics originated out of the work of Rudolf Steiner, the founder of the spiritual philosophy anthroposophy.

History

The development of biodynamic agriculture began in 1924 with a series of eight lectures on agriculture given by Rudolf Steiner at Schloss Koberwitz in what was then Silesia, Germany. The course was held in response to a request by farmers who noticed degraded soil conditions and a deterioration in the health and quality of crops and livestock resulting from the use of chemical fertilizers. An agricultural research group was subsequently formed to test the effects of biodynamic methods on the life and health of soil, plants and animals.

Today biodynamics is practiced in more than 50 countries worldwide. Demeter International is the primary certification agency for farms and gardens using the methods.

Geographic Developments

- In the United States, the Biodynamic Farming & Gardening Association was founded in 1938 as a New York state corporation.
- In Australia the first biodynamic preparations were made by Ernesto Genoni in Melbourne in 1927 and by Bob Williams in Sydney in 1939. Since the 1950s

research work has continued at the Biodynamic Research Institute (BDRI) in Powelltown, near Melbourne Australia under the direction of Alexei Podolinsky. In 1989 Biodynamic Agriculture Australia was established, as a not for profit association.

- The University of Kassel has a dedicated Department of Biodynamic Agriculture.

Biodynamic Method of Farming

Biodynamic agriculturalists conceive of the farm as an organically self-contained entity with its own individuality, within which organisms are interdependent. "Emphasis is placed on the integration of crops and livestock, recycling of nutrients, maintenance of soil, and the health and well being of crops and animals; the farmer too is part of the whole." Cover crops, green manures and crop rotations are used extensively and the farms foster bio-diversity.

Biodynamic farms often have a cultural component and encourage local community. Some biodynamic farms use the Community Supported Agriculture model, which has connections with social threefolding.

Biodynamic Preparations

Steiner prescribed nine different preparations to aid fertilization which are the cornerstone of biodynamic agriculture, and described how these were to be prepared. Steiner believed that these preparations transferred supernatural terrestrial and cosmic "forces" into the soil. The prepared substances are numbered 500 through 508, where the first two are used for preparing fields whereas the latter seven are used for making compost. A long term trial (DOK experiment) evaluating the biodynamic farming system in comparison with organic and conventional farming systems, found that preparations have influence on soil structure and micro-organisms enhancing soil fertility and increasing biodiversity. Regarding compost development beyond accelerating the initial phase of composting, some positive effects have been noted:

- The field sprays contain substances that stimulate plant growth include cytokinins.
- Some improvement in nutrient content of compost.

Field Preparations

Field preparations, for stimulating humus formation:

- 500: (horn-manure) a humus mixture prepared by filling the horn of a cow with cow manure and burying it in the ground (40–60 cm below the surface) in the autumn. It is left to decompose during the winter and recovered for use the following spring.
- 501: Crushed powdered quartz prepared by stuffing it into a horn of a cow and buried into the ground in spring and taken out in autumn. It can be mixed with 500 but usually prepared on its own (mixture of 1 tablespoon of quartz

powder to 250 litres of water) The mixture is sprayed under very low pressure over the crop during the wet season, in an attempt to prevent fungal diseases. It should be sprayed on an overcast day or early in the morning to prevent burning of the leaves.

Both 500 and 501 are used on fields by stirring about one teaspoon of the contents of a horn in 40–60 litres of water for an hour and whirling it in different directions every second minute. Although some biodynamic beliefs refer to buried quartz "fermenting", a 2004 review commented that it is unclear what this actually means, as rock does not ferment.

Compost Preparations

Compost preparations, used for preparing compost, employ herbs which are frequently used in medicinal remedies:

- 502: Yarrow blossoms (*Achillea millefolium*) are stuffed into urinary bladders from Red Deer (*Cervus elaphus*), placed in the sun during summer, buried in earth during winter and retrieved in the spring.
- 503: Chamomile blossoms (*Matricaria recutita*) are stuffed into small intestines from cattle buried in humus-rich earth in the autumn and retrieved in the spring.
- 504: Stinging nettle (*Urtica dioica*) plants in full bloom are stuffed together underground surrounded on all sides by peat for a year.
- 505: Oak bark (*Quercus robur*) is chopped in small pieces, placed inside the skull of a domesticated animal, surrounded by peat and buried in earth in a place where lots of rain water runs past.
- 506: Dandelion flowers (*Taraxacum officinale*) is stuffed into the peritoneum of cattle and buried in earth during winter and retrieved in the spring.
- 507: Valerian flowers (*Valeriana officinalis*) are extracted into water.
- 508: Horsetail (*Equisetum*).

One to three grams (a teaspoon) of each preparation is added to a dung heap by digging 50 cm deep holes with a distance of 2 meters from each other, except for the 507 preparation, which is stirred into 5 litres of water and sprayed over the entire compost surface. All preparations are thus used in homeopathic quantities. Each compost preparation is designed to guide a particular decomposition process in the composting mass.

One study found that the oak bark preparation improved disease resistance in zucchini.

Astronomical Planting Calendar

The approach considers that there are astronomical influences on soil and plant development, specifying, for example, what phase of the moon is most appropriate for planting, cultivating or harvesting various kinds of crops. This aspect of biodynamics has been termed "astrological" in nature.

Treatment of Pests and Weeds

Biodynamic agriculture sees the basis of pest and disease control arising from a strong healthy balanced farm organism. Where this is not yet achieved it uses techniques analogous to fertilization for pest control and weed control. Most of these techniques include using the ashes of a pest or weed that has been trapped or picked from the fields and burnt. A biodynamic farmer perceives weeds and plant vulnerability to pests as a result of imbalances in the soil.

- Pests such as insects or field mice (*Apodemus*) have more complex processes associated with them, depending on what pest is to be targeted. For example field mice are to be countered by deploying ashes prepared from field mice skin when Venus is in the Scorpius constellation.
- Weeds are combated (besides the usual mechanical methods) by collecting seeds from the weeds and burning them above a wooden flame that was kindled by the weeds. The ashes from the seeds are then spread on the fields, then lightly sprayed with the clear urine of a sterile cow (the urine should be exposed to the full moon for six hours), this is intended to block the influence from the full moon on the particular weed and make it infertile.

Seed Production

Biodynamic agriculture has focused on open pollination of seeds (permitting farmers to grow their own seed) and the development of locally adapted varieties. The seed stock is not controlled by large, multinational seed companies.

Trademark Protection of Term Biodynamic

The term *Biodynamic* is a trademark held by the Demeter association of biodynamic farmers for the purpose of maintaining production standards used both in farming and processing foodstuffs.(This is not a trademark held privately in New Zealand) The trademark is intended to protect both the consumer and the producers of biodynamic produce. Demeter International is an organization of member countries; each country has its own Demeter organization which is required to meet international production standards (but can also exceed them). The original Demeter organization was founded in 1928; the U.S. Demeter Association was formed in the 1980s and certified its first farm in 1982. In France, Biodivin certifies biodynamic wine. In Egypt, SEKEM has created the Egyptian Biodynamic Association (EBDA), an association that provides training for farmers to become certified.

Studies of Efficacy

Studies have compared biodynamic farming methods to both other organic methods and to conventional methods. Most studies have found that biodynamic farms have soil quality significantly better than conventionally farmed soils but comparable to the soil quality achieved by other organic methods; the decisive factor is likely to be the use of compost. Studies of yields differ in their conclusions.

- A 1993 study compared soil quality and financial performance of Biodynamic and conventional farms in New Zealand. The study reported that, "The Biodynamic farms proved in most enterprises to have soils of higher biological and physical quality: significantly greater in organic matter, content and microbial activity, more earthworms, better soil structure, lower bulk density, easier penetrability, and thicker topsoil." The biodynamic farms were just as financially viable on a per hectare basis. The study compared biodynamic farms with adjacent conventional farms, but didn't attempt to compare farms of similar size, or with similar crops.

- A further study investigated whether biodynamic preparations had any effect on the yield and growth of lentil and wheat crops, weed populations and soil fertility in the short term. The study found that "[i]n general, soils and crops treated with biodynamic preparations showed few differences from those not treated". Plots tended with biodynamically treated compost produced results for yield, crop quality and soil fertility that were similar to those tended with non-biodynamic composts and NPK fertilizers. Some alteration was observed in the nitrogenous chemistry of the soil and grain where biodynamic field sprays were applied, however the study did not ascribe or discern any biological significance to the difference. Among the variables considered by the study, some measured outcomes correlated with biodynamic field spray usage, including a higher per-unit biomass yield ratio for lentils and a lowering of carbon and crude protein contents in wheat grains. The study's conclusion remarked that "any additional short-term benefits from biodynamic preparations remain questionable."
- A long-term study conducted at a commercial vineyard in California compared vineyard blocks treated with biodynamic preparations alongside those tended with general organic farming methods, to examine effects upon soil and crop quality. "No differences were found in soil quality" during the first six years of the study, and analyses of other indicators including the yield per vine, clusters per vine, cluster and berry weight also showed there were no differences. The study did find a statistically significant (p-value < 0.05) difference in the yield-to-pruning weight ratio, indicating an "ideal vine balance for producing high-quality wine grapes" for the biodynamically treated crop, but noted the control vines had been "slightly overcropped". In one particular year of the study the biodynamically treated wine grapes had significantly higher Brix and notably higher total phenols and anthocyanins. In conclusion, the study found that biodynamic preparations "may affect" the vine canopy and chemistry, but showed no effects on the soil and tissue nutrient parameters measured in the study.
- A 21-year study by the FiBL Institute in Switzerland compared the agronomic and ecological performance of biodynamic, organic and two conventional systems. The study found that nutrient input in the biodynamic and organic systems was 34 to 51% lower than in the conventional systems but crop yield was only 20% lower on average, indicating more efficient production. The total

energy (for fuel, production of mineral fertilizer and pesticides, etc.) to produce a dry-matter unit of crop was 20 to 56% lower for the biodynamic and organic systems, and pesticide input was reduced by 97% (by 100% for the biodynamic system). In regard to soil aggregate stability, soil pH, humus formation, soil calcium, microbial biomass, and faunal biomass (earthworms and arthropods), the biodynamic system was superior even to the organic system, which in turn had superior results over the conventional systems. With the significant increase in microbial diversity in the biodynamic and organic systems, there was a significant associated decrease in metabolic quotient, indicating a greater ability to use organic material for plant growth.

Related

Ehrenfried Pfeiffer, a biochemist prominent in the early development of biodynamic preparations, developed a process for the bacterial conversion of municipal waste into compost usable in agriculture. The process was first used on a commercial scale in Oakland, California in the early 1950s.

Chitosan (Natural Biocontrol for Agricultural & Horticultural Use)

Chitosan is a linear polysaccharide composed of randomly distributed β-(1-4)-linked D-glucosamine (deacetylated unit) and N-acetyl-D-glucosamine (acetylated unit). It has a number of commercial and possible biomedical uses.

Manufacture and Properties

Chitosan is produced commercially by deacetylation of chitin, which is the structural element in the exoskeleton of crustaceans (crabs, shrimp, etc.) and cell walls of fungi. The degree of deacetylation can be determined by NMR spectroscopy, and the %DD in commercial chitosans is in the range 60-100 %. On average, the molecular weight of commercially produced chitosan is between 3800 to 20,000 daltons. A common method for the synthesis of chitosan is the deacetylation of chitin using sodium hydroxide in excess as a reagent and water as a solvent. This reaction pathway, when allowed to go to completion (complete deacetylation) yields up to 98% product.

The amino group in chitosan has a pKa value of ~6.5, which leads to a protonation in acidic to neutral solution with a charge density dependent on pH and the %DA-value. This makes chitosan water soluble and a bioadhesive which readily binds to negatively charged surfaces such as mucosal membranes. Chitosan enhances the transport of polar drugs across epithelial surfaces, and is biocompatible and biodegradable. Purified qualities of chitosans are available for biomedical applications.

Chitosan and its derivatives such as trimethylchitosan (where the amino group has been trimethylated) have been used in non-viral gene delivery. Trimethylchitosan, or quaternised chitosan, has been shown to transfect breast cancer cells; with increased degree of trimethylation increasing the cytotoxicity and at approximately 50% trimethylation the derivative is the most efficient at gene delivery. Oligomeric derivatives (3-6 kDa) are relatively non-toxic and have good gene delivery properties.

Usage

Agricultural and Horticultural Use

In agriculture, chitosan is used primarily as a natural seed treatment and plant growth enhancer, and as a ecologically friendly biopesticide substance that boosts the innate ability of plants to defend themselves against fungal infections. The natural biocontrol active ingredients, chitin/chitosan, are found in the shells of crustaceans, such as lobsters, crabs, and shrimp, and many other organisms, including insects and fungi. It is one of the most abundant biodegradable materials in the world. Degraded molecules of chitin/chitosan exist in soil and water. Chitosan applications for plants and crops are regulated by the EPA, and the USDA National Organic Program regulates its use on organic certified farms and crops. EPA approved biodegradable chitosan products are allowed for use outdoors and indoors on plants and crops grown commercially and by consumers. The natural biocontrol ability of chitosan should not be confused with the effects of fertilizers or pesticides upon plants or the environment. Chitosan active biopesticides represent a new tier of cost effective biological control of crops for agriculture and horticulture. The biocontrol mode of action of chitosan elicits natural innate defence responses within plant to resist insects, pathogens, and soil-borne diseases when applied to foliage or the soil. Chitosan increases photosynthesis, promotes and enhances plant growth, stimulates nutrient uptake, increases germination and sprouting, and boosts plant vigor. When used as seed treatment or seed coating on cotton, corn, seed potatoes, soybeans, sugar beets, tomatoes, wheat and many other seeds, it elicits an innate immunity response in developing roots which destroys parasitic cyst nematodes without harming beneficial nematodes and organisms. Agricultural applications of chitosan can reduce environmental stress due to drought and soil deficiencies, strengthen seed vitality, improve stand quality, increase yields, and reduce fruit decay of vegetables, fruits and citrus crops. Horticultural applications of chitosan increases blooms and extends the life of cut flowers and Christmas trees. The US Forest Service has conducted research on chitosan to control pathogens in pine trees. and chitosan's ability to increase pine tree resin pitch outflow by 40% to resist pine beetle infestation.

Chitosan has a rich history of being researched for applications in agriculture and horticulture dating back to the 1980s. By 1989, Bentech Labs patented chitosan salt solutions applied to crops for improved freeze protection or to crop seed for seed priming. Shortly thereafter, Bentech's chitosan salt received the first ever biopesticide label from the EPA. Numerous other chitosan patents for plants soon followed. Chitosan applications to protect plants have been used in space as well. NASA first flew a chitosan experiment to protect adzuki beans grown aboard the space shuttle and Mir space station in 1997. NASA results revealed chitosan induces increased growth (biomass) and pathogen resistance due to elevated levels of beta 1-3 glucanase enzymes within plant cells. NASA confirmed chitosan elicits the same effect in plants on earth. Over 20 years of R&D by DuPont/ConAgra Ventures (DCV) and AgriHouse Inc have gone into developing non-toxic low molecular weight chitosan polymer solutions safe enough for broad spectrum agricultural and horticultural use. In 2008,

AgriHouse Inc, Denver (Berthoud), Colorado, was granted EPA natural broad spectrum elicitor status for YEA! Yield Enhancing Agent, a liquid solution containing an ultra low molecular active ingredient of 0.25% chitosan. YEA! is a next generation natural chitosan elicitor solution for agriculture and horticultural uses, and was granted an amended label for foliar and irrigation applications by the EPA in June, 2009. A milliliter of YEA! contains over 14.4 X 10^{13} bioactive low molecular weight chitosan molecules, and it is 600 times more effective than common chitosan. Given its low potential for toxicity and its abundance in the natural environment, chitosan does not harm people, pets, wildlife, or the environment when used according to label directions. Agricultural chitosan facts are located on USDA and EPA web sites.

The US Forest Service tested chitosan as an eco-friendly biopesticide, to pre-arm pine trees to defend themselves against mountain pine beetle.

Water Filtration

Chitosan can also be used in water processing engineering as a part of a filtration process. Chitosan causes the fine sediment particles to bind together, and is subsequently removed with the sediment during sand filtration. Chitosan also removes phosphorus, heavy minerals, and oils from the water. Chitosan is an important additive in the filtration process. Sand filtration apparently can remove up to 50% of the turbidity alone, while the chitosan with sand filtration removes up to 99% turbidity. Chitosan has been used to precipitate caseins from bovine milk and cheese making.

Chitosan is also useful in other filtration situations, where one may need to remove suspended particles from a liquid. Chitosan, in combination with bentonite, gelatin, silica gel, isinglass, or other fining agents is used to clarify wine, mead, and beer. Added late in the brewing process, chitosan improves flocculation, and removes yeast cells, fruit particles, and other detritus that cause hazy wine. Chitosan combined with colloidal silica is becoming a popular fining agent for white wines, because chitosan does not require acidic tannins (found primarily in red wines) with which to flocculate.

Industrial Use

Scientists have recently developed a polyurethane coating that heals its own scratches when exposed to sunlight, offering the promise of scratch-free cars and other products. The self-healing coating uses chitosan incorporated into traditional polymer materials, such as those used in coatings on cars to protect paint. When a scratch damages the chemical structure, the chitosan responds to ultraviolet light by forming chemical chains that begin bonding with other materials in the substance, eventually smoothing the scratch. The process can take less than an hour.

Marek W. Urban, a scientist working on this project, said the polymer can only repair itself in the same spot once, and would not work after repeated scratches.

Biomedical Uses

Chitosan's properties allow it to rapidly clot blood, and has recently gained approval in the United States and Europe for use in bandages and other hemostatic agents. Chitosan hemostatic products have been shown in testing by the U.S. Marine Corps to quickly stop bleeding, and result in 100% survival of otherwise lethal arterial wounds in swine and to reduce blood loss. Chitosan hemostatic products reduce blood loss in comparison to gauze dressings and increase patient survival. Chitosan hemostatic products have been sold to the U.S. Army and are currently used by the UK military. Both the US and UK have already used the bandages on the battlefields of Iraq and Afghanistan. Chitosan is hypoallergenic and has natural antibacterial properties, which further support its use in field bandages.

Chitosan's properties also allow it to be used in trans-dermal drug delivery. One of these properties is that it is mucoadhesive in nature. Another property that is important to chitosan's ability to aid in drug delivery is that it is fairly reactive so it can be produced in many different forms. The most important property of chitosan with regards to drug delivery is its positive charge under acidic conditions. This positive charge comes from protonation of its free amino groups. Lack of a positive charge means chitosan is insoluble in neutral and basic environments. However, in acidic environments, protonation of the amino groups leads to an increase in solubility. The implications of this are very important to biomedical applications. This is a molecule that will maintain its structure in a neutral environment but will solubilize and degrade in an acidic environment. This means that chitosan can be used to transport a drug to an acidic environment, where the chitosan packaging will then degrade, releasing the drug to the desired environment. One example of this drug delivery has been seen in the transport of insulin.

Claimed Health Benefits

Chitosan is frequently sold in tablet form at health stores as a "fat binder": It is supposed to have the capability to interact with lipids (fat) from the digestive system and limit their absorption in the body. Therefore, chitosan can be an effective complement to help lose weight during diet period or to stabilise one's weight. In the 2007 Cochrane meta-analysis which evaluated all available clinical trials performed with chitosan on the subject of weight loss, it was concluded that body weight and all parameters related to cholesterol changed in favour of chitosan compared to placebo. The mean difference in body weight was "1.7 kg (range: –2.1 to –1.3) in favour of chitosan. This change in body weight was statistically significant. There was no difference between chitosan and placebo concerning side effects. The various qualities (in terms of duration, sample size, doses, subject characteristics, type of diet, chitosan quality and characteristics, etc.) of the clinical trials performed to evaluate the effect of chitosan on body weight might account for some of the disparities observed in clinical trial results and the subsequent critics regarding the real efficacy of chitosan. In an experimental model of the stomach and duodenum tract, chitosan has shown to interact with oil, which inhibited duodenal absorption and enhanced lipid excretion. However, the mechanism of interaction between chitosan and fat is not very well understood and has not been really proved clinically yet. This is certainly the reason why the FDA has issued in 2004 Warning Letters to two companies who made inappropriate claims according to the regulator. Although detractors claim that using chitosan may have the deleterious effect of rendering ineffective certain minerals found in foodstuffs, several animal studies contradicted this statement by showing no or little effect. In mice, dietary ingestion of chitosan did not depress the level of iron, zinc or copper. Moreover, there is no proof of any adverse events, in particular regarding nutrient absorption, in humans.

Medical Research

Chitosan is currently the focus of much medical research, as it is a polyglucosamine (the second-most-common dietary fibre, after cellulose). Studies have shown chitosan has the following properties:

- As a soluble dietary fibre, it increases gastrointestinal lumen viscosity and slows down the emptying of the stomach.
- It alters bile acid composition, increasing the excretion of sterols and reducing the digestibility of ileal fats. It is unclear how chitosan does this, but the currently favored hypotheses involve the increase of intestinal viscosity or bile acid-binding capacity.
  - o Chitosan is relatively insoluble in water, but can be dissolved by dilute acids, which would make it a highly-viscous dietary fibre. Such fibres might inhibit the uptake of dietary lipids by increasing the thickness of the boundary layer of the intestinal lumen, which has been observed in animal experiments.
  - o Having very few acetyl groups, chitosan contains cationic groups. This may cause chitosan to have bile acid-binding capacity, which causes mixed micelles to be entrapped or disintegrated in the duodenum and ileum. This would interrupt bile acid circulation, causing reduced lipid absorption and increased sterol excretion, which has also been observed in animal experiments.

Chitin

Chitin is a long-chain polymer of a *N*-acetylglucosamine, a derivative of glucose, and is found in many places throughout the natural world. It is the main component of the cell walls of fungi, the exoskeletons of arthropods such as crustaceans (e.g. crabs, lobsters and shrimps) and insects, the radulas of mollusks and the beaks of cephalopods, including squid and octopuses. Chitin may be compared structurally to the polysaccharide cellulose and functionally to the protein keratin. Chitin has also proven useful for several medical and industrial purposes.

Etymology

The English word "chitin" comes from the French word "chitine", which first appeared in 1836. These words were derived from the Greek word "chitôn", meaning mollusk. That is either influenced by, or related to the Greek word *khitôn*, meaning "tunic" or "frock", the Central Semitic word "*kittan", the Akkadian words "kitu" or "kita'um", meaning flax or linen, and the Sumerian word "gada" or "gida".

Chemistry, Physical Properties and Biological Function

Chitin is a modified polysaccharide which contains nitrogen; it is synthesized from units of *N*-acetylglucosamine (more precisely, 2-(acetylamino)-2-deoxy-D-glucose). These units form covalent β-1,4 linkages (similar to the linkages between glucose units forming cellulose). Chitin may therefore be described as cellulose with one hydroxyl group on each monomer substituted with an acetyl amine group. This allows for increased hydrogen bonding between adjacent polymers, giving the chitin-polymer matrix increased strength.

In its unmodified form, chitin is translucent, pliable, resilient and quite tough. In arthropods, however, it is often modified, becoming embedded in a hardened proteinaceous matrix, which forms much of the exoskeleton. In its pure form it is leathery, but when encrusted in calcium carbonate it becomes much harder. The difference between the unmodified and modified forms can be seen by comparing the body wall of a caterpillar (unmodified) to a beetle (modified).

Fossil Record

Chitin first appeared in the exoskeletons of Cambrian arthropods, e.g. trilobites. The oldest preserved chitin dates to the Oligocene, about 25 million years ago.

Uses

Agriculture

Most recent studies point out that chitin is a good inducer for defence mechanisms in plants. It has also been assessed as a fertilizer that can improve overall crop yields yields. The EPA regulates chitin for agricultural use within the USA. Chitosan is derived from chitin, which is used as a biocontrol elicitor in agriculture and horticulture.

Industrial

Chitin is used industrially in many processes. It is used as an additive to thicken and stabilize foods and pharmaceuticals. It also acts as a binder in dyes, fabrics, and adhesives. Industrial separation membranes and ion-exchange resins can be made from chitin. Processes to size and strengthen paper employ chitin.

Medicine

Chitin's properties as a flexible and strong material make it favourable as surgical thread. Its biodegradibility means it wears away with time as the wound heals. Moreover, chitin has some unusual properties that accelerate healing of wounds in humans.

Occupations associated with high environmental chitin levels, such as shellfish processors, are prone to high incidences of asthma. Recent studies have suggested that chitin may play a role in a possible pathway in human allergic disease. Specifically, mice treated with chitin develop an allergic response, characterized by a build-up of interleukin-4 expressing innate immune cells. In these treated mice, additional treatment with a chitinase enzyme abolishes the response.

Biological Pest Control

Biological control of pests in agriculture is a method of controlling pests (including insects, mites, weeds and plant diseases) that relies on predation, parasitism, herbivory, or other natural mechanisms. It can be an important component of integrated pest management (IPM) programs.

Biological Control is defined as the reduction of pest populations by natural enemies and typically involves an active human role. Natural enemies of insect pests, also known as biological log control agents; and include predators, parasitoids, and pathogens. Biological control agents of plant diseases are most often referred to as antagonists. Biological control agents of weeds include herbivores and plant pathogens. Predators, such as birds, lady beetles and lacewings, are mainly free-living species that consume a large number of prey during their whole lifetime. Parasitoids are species whose immature develops on paper or within a single insect host, ultimately killing the host. Most have a very narrow host range. Many species of wasps and some flies are parasitoids. Pathogens are disease-causing organisms including bacteria, fungi, and viruses. They kill or debilitate their own host and are relatively specific. There are three basic types of biological control strategies; conservation, classical biological control, and augmentation.

Conservation

The conservation of natural enemies is probably the most important and readily available biological control practice available to homeowners and gardeners. Natural enemies occur in all areas, from the private garden to the open field. They are adapted to the habitat and to the target pest, and their conservation is generally simple and cost-effective. Lacewings, lady beetles, hover fly larvae, and parasitized aphid mummies are almost always present in aphid colonies. Fungus-infected adult flies are often common following periods of high humidity. These naturally occurring biological controls are often susceptible to the same pesticides used to target their pest hosts. Preventing the accidental eradication of natural enemies is termed "simple conservation."

To conserve and encourage pest insect eating birds, including native plants and ornamental plants that supply berries, acorns, nuts, seeds, nectar, and other vegetative foods, and also bird nest building materials, encourages their presence, health, and new generations. These qualities can also increase the visible population to enjoy in a garden. Using companion planting and the birds' insect cuisine habits is a traditional method for biological control agent pest control in an organic garden and any landscape, and in organic farming and sustainable agriculture. Installing specified Nest boxes for mosquito eating Bats reduces a pest and increases endangered species conservation.

Effects of Biological Control

Effects on Native Biodiversity

Biological control can potentially have positive and negative effects on biodiversity. Usually a biological control is introduced to an area to protect a native species from an invasive or exotic species that has moved into its area. The control is introduced to lessen the competition between native and introduced species. However, the introduced control does not always target only the intended species. It can also target native species.

When introducing a high biological control to a new area, a primary concern is the host- or prey-specificity of the control agent. Generalist feeders (control agents that are not restricted to a single species or a small range of species) often make poor biological control agents, and may become invasive species themselves. For this reason, potential biological control agents should be subject to extensive testing and quarantine before release into any new environment. If a species is introduced and attacks a native species, the biodiversity in that area can change dramatically. When one native species is removed from an area, it may have filled an essential ecological niche. When this niche is absent it may directly affect the entire ecosystem.

Because they tend to be generalist feeders, vertebrate animals seldom make good biological control agents, and many of the classic cases of "biocontrol gone awry" involve vertebrates. For example, the cane toad, *Bufo marinus*, was introduced as a biological control and had significant negative impact on biodiversity. The cane toad was intentionally introduced to Australia to control the cane beetle. When introduced, the cane toad thrived very well and did not only feed on cane beetles but other insects as well. The cane toad soon spread very rapidly, thus taking over native amphibian habitat. The introduction of the cane toad also brought foreign disease to native reptiles. This drastically reduced the population of native toads and frogs. "The cane toad also exudes and can squirt poison from the parotid glands on their shoulders when threatened or handled. This toxin contains a cocktail of chemicals that can kill animals that eat it. Freshwater crocodiles, goannas, tiger snakes, dingos and northern quolls have all died after eating cane toads, as have pet dogs. This example shows how a small mis-introduced organism can alter the native biodiversity in a large ecosystem in a very expedient manner. A pyramid effect can take place if native species are reduced or eradicated. The domino effect keeps on going and can potentially exude on other bordering ecosystems until an equilibrium is reached.

A second example of a biological control agent that subsequently crossed over to native species is the *Rhinocyllus conicus*. The seed feeding weevil was introduced to North America to control exotic thistles (Musk and Canadian). However, the weevil did not target only the exotic thistles, it also targeted native thistles that are essential to various native insects. The native insects rely solely on native thistles and do not adapt to other plant species. Therefore, they cannot survive. Biological controls do not always have negative impacts on biodiversity (Corry 2000).

Successful biological control reduces the density of the target species over several years, thus providing the potential for native species to re-establish. In addition, regeneration and reestablishment programs can aid to the recovery of native species. Native species can be affected in a positive way as well. To develop or find a biological control that exerts control only on the targeted species is a very lengthy process of research and experiments. In the late 19th century, the citrus industry was in great fear when the cottony cushion scale was discovered. This organism could cause a great deal of economic loss to the industry. However, a biological control was introduced. The vedalia beetle and a parasitoid fly were introduced to control the pest. Within a few years time, the cottony cushion scale was controlled by the natural enemies and the citrus industry suffered little financial loss. Many exotic or invasive species can suppress the development of native species. The introduction of an effective biological control that reduces the population of the invasive species allows the rejuvenation of the native species. Biological controls can reduce competition for biotic and abiotic factors which can result in the re-establishment of the once over ran native species.

Effects on Invasive Species

Invasive species are closely associated with biological controls because the environment in which they are invasive most likely does not contain their natural enemies. If invasive species are not controlled, biodiversity may be at great threat in the affected area. An example of an invasive species is the alligator weed. This plant was introduced to the United States from South America. This aquatic weed spreads rapidly and causes many problems in lakes and rivers. The weed takes root in shallow water causing major problems for navigation, irrigation, and flood control. The alligator weed flea beetle and two other biological controls were released in Florida. Because of their success, Florida banned the use of herbicides to control alligator weed three years after the controls were introduced (Cofrancesco 2007). Biological controls for invasive species also can have a negative impact on biodiversity.

The cane toad, as mentioned previously, is an example of trying to control an invasive species. The cane toad was introduced to eradicate an invasive species. It became invasive, thus altering the biodiversity. The introduction of the cane toad could have caused more of a disturbance in biodiversity than the targeted species did.

Classical Biological Control

Classical biological control is the introduction of natural enemies to a new locale where they did not originate or do not occur naturally. This is usually done by government authorities. In many instances the complex of natural enemies associated with an insect pest may be inadequate. This is especially evident when an insect pest is accidentally introduced into a new geographic area without its associated natural enemies. These introduced pests are referred to as exotic pests and comprise about 40% of the insect pests in the United States. Examples of introduced vegetable pests include the European corn borer (*Ostrinia nubilalis*), one of the most destructive insects in North America. To obtain the needed natural enemies, scientists turned to classical biological control. This is the practice of importing, and releasing for establishment, natural enemies to control an introduced (exotic) pest, although it is also practiced against native insect pests. The first step in the process is to determine the origin of the introduced pest and then collect appropriate natural enemies associated with the pest or closely related species. The natural enemy is then passed through a rigorous quarantine process, to ensure that no unwanted organisms (such as hyperparasitoids) are introduced, then they are mass produced, and released. Follow-up studies are conducted to determine if the natural enemy becomes successfully established at the site of release, and to assess the long-term benefit of its presence.

There are many examples of successful classical biological control programs.

Joseph Needham noted a Chinese text dating from 304AD, *Records of the Plants and Trees of the Southern Regions*, by Hsi Han, which describes mandarin oranges protected by biological pest control techniques that are still in use today.

One of the earliest successes in the west was in controlling *Icerya purchasi*, the cottony cushion scale, a pest that was devastating the California citrus industry in the late 19th century. A predatory insect *Rodolia cardinalis* (the Vedalia Beetle), and a parasitoid fly were introduced from Australia by Charles Valentine Riley. Within a few years the cottony cushion scale was completely controlled by these introduced natural enemies.

Damage from *Hypera postica* Gyllenhal, the alfalfa weevil, a serious introduced pest of forage, was substantially reduced by the introduction of several natural enemies. 20 years after their introduction the population of weevils in the alfalfa area treated for alfalfa weevil in the Northeastern United States was reduced by 75 percent. A small wasp, *Trichogramma ostriniae*, was introduced from China to help control the European corn borer making it a recent example of a long history of classical biological control efforts for this major pest. Many classical biological control programs for insect pests and weeds are under way across the United States and Canada. The population of *Levuana irridescens* (the Levuana moth), a serious coconut pest in Fiji, was brought under control by a classical biological control program in the 1920s.

Classical biological control is long lasting and inexpensive. Other than the initial costs of collection, importation, and rearing, little expense is incurred. When a natural enemy is successfully established it rarely requires additional input and it continues to kill the pest with no direct help from humans and at no cost. Classical biological control does not always work. It is usually most effective against exotic pests and less so against native insect pests. The reasons for failure are not often known but may include the release of too few individuals, poor adaptation of the natural enemy to environmental conditions at the release location, and lack of synchrony between the life cycle of the natural enemy and host pest.

Augmentation

This third type of biological control involves the supplemental release of natural enemies. Relatively few natural enemies may be released at a critical time of the season (inoculative release) or literally millions may be released (inundative release). Additionally, the cropping system may be modified to favour or augment the natural enemies. This latter practice is frequently referred to as habitat manipulation. An example of inoculative release occurs in greenhouse production of several crops. Periodic releases of the parasitoid, *Encarsia formosa*, are used to control greenhouse whitefly, and the predaceous mite, *Phytoseiulus persimilis,* is used for control of the two-spotted spider mite.

Lady beetles, lacewings, or parasitoids such as those from the genus *Trichogramma* are frequently released in large numbers (inundative release). Recommended release rates for Trichogramma in vegetable or field crops range from 5,000 to 200,000 per acre (1 to 50 per square metre) per week depending on level of pest infestation. Similarly, entomopathogenic nematodes are released at rates of millions and even billions per acre for control of certain soil-dwelling insect pests. The spraying of octopamine analogues (such as 3-FMC) has been suggested as a way to boost the effectiveness of augmentation. Octopamine, regarded as the invertebrate counterpart of dopamine plays a role in activating the insects' flight-or-fight response. The idea behind using octopamine analogues to augment biological control is that natural enemies will be more effective in their eradication of the pest, since the pest will be behaving in an unnatural way because its flight-or-fight mechanism has been activated. Octopamine analogues are purported to have two desirable characteristics for this type of application: (1) they affect insects at very low dosages (2) they do not have a physiological effect in humans (or other vertebrates).

Habitat or environmental manipulation is another form of augmentation. This tactic involves altering the cropping system to augment or enhance the effectiveness of a natural enemy. Many adult parasitoids and predators benefit from sources of nectar and the protection provided by refuges such as hedgerows, cover crops, and weedy borders. Also, the provisioning of natural shelters in the form of wooden caskets, boxes or (turnaround) flowerpots is a form of this. For example, the stimulation of the natural predator *Dermaptera* is done in gardens by hanging up turnaround flowerpots with straw or wood wool. Mixed plantings and the provision of flowering borders can increase the diversity of habitats and provide shelter and alternative food sources. They are easily incorporated into home gardens and even small-scale commercial plantings, but are more difficult to accommodate in large-scale crop production. There may also be some conflict with pest control for the large producer because of the difficulty of targeting the pest species and the use of refuges by the pest insects as well as natural enemies.

Examples of habitat manipulation include growing flowering plants (pollen and nectar sources) such as Buckwheat near crops to attract and maintain populations of natural enemies. For example, hover fly adults can be attracted to umbelliferous plants in bloom.

Biological control experts in California have demonstrated that planting prune trees in grape vineyards provides an improved overwintering habitat or refuge for a key grape pest parasitoid. The prune trees harbor an alternate host for the parasitoid, which could previously overwinter only at great distances from most vineyards. Caution should be used with this tactic because some plants attractive to natural enemies may also be hosts for certain plant diseases, especially plant viruses that could be vectored by insect pests to the crop. Although the tactic appears to hold much promise, only a few examples have been adequately researched and developed.

Examples of Predators

Ladybugs, and in particular their larvae which are active between May and July in the northern hemisphere, are voracious predators of aphids such as greenfly and blackfly, and will also consume mites, scale insects and small caterpillars. The ladybug is a very familiar beetle with various coloured markings, whilst its larvae are initially small and spidery, growing up to 17 mm long. The larvae have a tapering segmented grey/black body with orange/yellow markings and ferocious mouthparts. They can be encouraged by cultivating a patch of nettles in the garden and by leaving hollow stems and some plant debris over winter so that they can hibernate.

Hoverflies resemble slightly darker bees or wasps and they have characteristic hovering, darting flight patterns. There are over 100 species of hoverfly whose larvae principally feed upon greenfly, one larva devouring up to fifty a day, or 1000 in its lifetime. They also eat fruit tree spider mites and small caterpillars. Adults feed on nectar and pollen, which they require for egg production. Eggs are minute (1 mm), pale yellow white and laid singly near greenfly colonies. Larvae are 8–17 mm long, disguised to resemble bird droppings, they are legless and have no distinct head. Semi-transparent in a range of colours from green, white, brown and black.

Hoverflies can be encouraged by growing attractant flowers such as the poached egg plant *(Limnanthes douglasii)*, marigolds or phacelia throughout the growing season. Dragonflies are important predators of mosquitoes, both in the water, where the dragonfly naiads eat mosquito larvae, and in the air, where adult dragonflies capture and eat adult mosquitoes. Community-wide mosquito control programs that spray adult mosquitoes also kill dragonflies, thus removing an important biocontrol agent, and can actually increase mosquito populations in the long term. Other useful garden predators include lacewings, pirate bugs, rove and ground beetles, aphid midge, centipedes, spiders, predatory mites, as well as larger fauna such as frogs, toads, lizards, hedgehogs, slow-worms and birds. Cats and rat terriers kill field mice, rats, June bugs, and birds. Dogs chase away many types of pest animals. Dachshunds are bred specifically to fit inside tunnels underground to kill badgers.

More examples:

- *Phytoseiulus persimilis* (against spider mites)
- *Amblyscius californicus* (against spider mites)
- *Amblyseius cucumeris* (against spider mites)
- *Typhlodromips swirskii* (against spider mites, thrips, and white flies)
- *Feltiella acarisuga* (against spider mites)
- *Stethorus punctillum* (against spider mites)
- *Macrolophus caluginosus* (against spider mites).

Parasitoid Insects

Most insect parasitoids are wasps or flies. Parasitoids comprise a diverse range of insects that lay their egg on or in the body of an insect host, which is then used as a food for developing larvae. Parasitic wasps take much longer than predators to consume their victims, for if the larvae were to eat too fast they would run out of food before they became adults. Such parasites are very useful in the organic garden, for they are very efficient hunters, always at work searching for pest invaders. As adults they require high energy fuel as they fly from place to place, and feed upon nectar, pollen and sap, thereby pollinating plenty of flowering plants, particularly buckwheat, umbellifers, and composites will encourage their presence.

Four of the most important groups are:

- Ichneumonid wasps: (5–10 mm). Prey mainly on caterpillars of butterflies and moths.
- Braconid wasps: Tiny wasps (up to 5 mm) attack caterpillars and a wide range of other insects including greenfly. A common parasite of the cabbage white caterpillar- seen as clusters of sulphur yellow cocoons bursting from collapsed caterpillar skin.
- Chalcid wasps: Among the smallest of insects (<3 mm). Parasitize eggs/larvae of greenfly, whitefly, cabbage caterpillars, scale insects and Strawberry Tortrix Moth (*Acleris comariana*).
- Tachinid flies: Parasitize a wide range of insects including caterpillars, adult and larval beetles, true bugs, and others.

Examples of parasitoids:

- *Encarsia formosa* (against white flies)
- *Eretmocerus* spp. (against white flies)
- *Aphidius colemani* (against aphids).

Biological Control with Micro-organisms

Various microbial insect diseases occur naturally, but may also be used as biological pesticides. When naturally occurring, these outbreaks are density dependent in that they generally only occur as insect populations become denser.

Bacteria and Biological Control

Bacteria used for biological control infect insects via their digestive tracts, so insects with sucking mouth parts like aphids and scale insects are difficult to control with bacterial biological control. *Bacillus thuringiensis* is the most widely applied species of bacteria used for biological control, with at least four sub-species used to control Lepidopteran (moth, butterfly), Coleopteran (beetle) and Dipteran (true flies) insect pests.

Fungi and Biological Control

Fungi that cause disease in insects are known as entomopathogenic fungi, including at least fourteen species of entomophthoraceous fungi attack aphids. Species in the genus *Trichoderma* are used to manage some soilborne plant pathogens. *Beauveria bassiana* is used to manage different types of pest such whiteflies, thrips, aphids and weevils.

Examples of entomopathogenic fungi:

- *Beauveria bassiana* (against white flies, thrips, aphids and weevils)
- *Paecilomyces fumosoroseus* (against white flies, thrips and aphids)
- *Metarhizium* spp. (against beetles, locusts, Hemiptera, spider mites and other pests)
- *Lecanicillium lecanii* (against white flies, thrips and aphids)
- *Cordyceps* species (sometines teleomorphs of the above: that infect a wide spectrum of arthropods).

Plants to Regulate Insect Pests

Choosing a diverse range of plants for the garden can help to regulate pests in a variety of ways, including;

- Masking the crop plants from pests, depending on the proximity of the companion or intercrop.
- Producing olfactory inhibitors, odors that confuse and deter pests.
- Acting as trap plants by providing an alluring food that entices pests away from crops.
- Serving as nursery plants, providing breeding grounds for beneficial insects.
- Providing an alternative habitat, usually in a form of a shelterbelt, hedgerow, or beetle bank where beneficial insects can live and reproduce. Nectar-rich plants that bloom for long periods are especially good, as many beneficials are nectivorous during the adult stage, but parasitic or predatory as larvae. A good example of this is the soldier beetle which is frequently found on flowers as an adult, but whose larvae eat aphids, caterpillars, grasshopper eggs, and other beetles.
- Some plants have chemical defences in order to regulate pests. The geranium has developed such a defence against Japanese beetles, one of the most damaging and expensive pests to control when it comes to ornamental and turf plants. The geranium's petals contain a chemical compound that paralyzes the beetle within 30 minutes of ingestion. The beetle will remain paralyzed for several hours and will typically regain movement within 24 hours. However, while paralyzed the beetle is very vulnerable to its predators and is usually hunted before the paralysis subsides. Agricultural Research Service (ARS) scientists are working to isolate the chemical compound in geraniums that causes the paralysis in the beetles. Scientists hope to one day use this natural pesticide to control the population of beetles. In addition to this research ARS scientists are studying ways to help geranium leaves better hold on to pesticide chemicals that are sprayed on them, that way less pesticides will have to be applied to the leaves.

Plants to Regulate Plants

The legume vine *Mucuna pruriens* is used in the countries of Benin and Vietnam as a biological control for problematic *Imperata cylindrica* grass. *Mucuna pruriens* is said not to be invasive outside its cultivated area. *Desmodium uncinatum* can be used in push-pull farming to stop the parasitic plant, *Striga*.

Directly Introducing Biological Controls

Most of the biological controls listed above depend on providing incentives in order to 'naturally' attract beneficial insects to the garden. However there are occasions when biological controls can be directly introduced. Common biocontrol agents include parasitoids, predators, pathogens or weed feeders. This is particularly appropriate in situations such as the greenhouse, a largely artificial environment, and are usually purchased by mail order.

Some biocontrol agents that can be introduced include;

- *Encarsia formosa*. This is a small predatory chalcid wasp which is parasitical on whitefly, a sap-feeding insect which can cause wilting and black sooty moulds. It is most effective when dealing with low level infestations, giving protection over a long period of time. The wasp lays its eggs in young whitefly 'scales', turning them black as the parasite larvae pupates. It should be introduced as soon as possible after the first adult whitefly are seen. Should be used in conjunction with insecticidal soap.
- Red spider mite, another pest found in the greenhouse, can be controlled with the predatory mite *Phytoseilus persimilis*. This is slightly larger than its prey and has an orange body. It develops from egg to adult twice as fast as the red spider mite and once established quickly overcomes infestation.
- A fairly recent development in the control of slugs is the introduction of 'Nemaslug', a microscopic nematode (*Phasmarhabditis hermaphrodita*) which will seek out and parasitize slugs, reproducing inside them and killing them. The nematode is applied by watering onto moist soil, and gives protection for up to six weeks in optimum conditions, Nemaslug nematodes are mainly effective with small and young slugs under the soil surface.
- A bacterial biological control which can be introduced in order to control butterfly caterpillars is *Bacillus thuringiensis*. This available in sachets of dried spores which are mixed with water and sprayed onto vulnerable plants such as brassicas and fruit trees. The bacterial disease will kill the caterpillars, but leave other insects unharmed. There are strains of *Bt* that are effective against other insect larvae. *Bt israelensis* is effective against mosquito larvae and some midges.
- A viral biological control which can be introduced in order to control the overpopulation of European rabbit in Australia is the rabbit haemorrhagic disease virus that causes the rabbit haemorrhagic disease.
- A biological control being developed for use in the treatment of plant disease is the fungus *Trichoderma viride*. This has been used against Dutch Elm disease, and to treat the spread of fungal and bacterial growth on tree wounds. It may also have potential as a means of combating silver leaf disease.

- Several species of dung beetle were introduced to Australia from South Africa and Europe during the Australian Dung Beetle Project (1965–1985) led by Dr. George Bornemissza of the Commonwealth Scientific and Industrial Research Organisation in order to biologically control the population of pestilent bush flies and parasitic worms.
- The parasitoid *Gonatocerus ashmeadi* (Hymenoptera: Mymaridae) has been introduced to control the glassy-winged sharpshooter *Homalodisca vitripennis* (Hemipterae: Cicadellidae) in French Polynesia and has successfully controlled ~95% of the pest density

Negative Results of Biological Pest Control

In some cases, biological pest control can have unforeseen negative results that could outweigh all benefits. For example, when the mongoose was introduced to Hawaii in order to control the rat population, it preyed on the endemic birds of Hawaii, especially their eggs, more often than it ate the rats. (Note, however, that the introduction of the mongoose was not undertaken based on scientific—or perhaps any—understanding of the consequences of such an action. The introduction of a generalist mammal for biocontrol of anything would be unthinkable by any reasonable standards today.)

Cane toads (*Bufo marinus*) were introduced to Australia in the 1930s in a failed attempt to control the cane beetle, a pest of sugar cane crops. 102 toads were obtained from Hawaii and bred in captivity to increase their numbers until they were released into the sugar cane fields of the tropic north in 1935. It was later discovered that the toads can't jump very high so they did not eat the cane beetles which stayed up on the upper stalks of the cane plants. The toads soon became very numerous and out-competed native species and became very harmful to the Australian environment, including being very toxic to would-be predators such as native snakes.

Living organisms evolve and increase their resistance to biological, chemical, physical or any other form of control. Unless the target population is completely exterminated or is rendered incapable of reproduction, the surviving population will inevitably acquire a tolerance of whatever pressures are brought to bear- this results in an evolutionary arms race.

Biological Pesticide

The term biopesticide is used for microbial biological pest control agents that are applied in a similar manner to chemical pesticides. Commonly these are bacterial, but there are also examples of control agents based on fungi, viruses and nematodes. Weeds and rodents have also been controlled with microbial agents.

One well-known insecticide example is *Bacillus thuringiensis*, a bacterial disease of Lepidoptera, Coleoptera and Diptera. Because it has little effect on other organisms, it is considered more environmentally friendly than synthetic pesticides. The toxin from *Bacillus thuringiensis* (Bt toxin) has been incorporated directly into plants through the use of genetic engineering.

Other microbial control agents include products based on:

- entomopathogenic fungi (*e.g.Beauveria bassiana, Lecanicillium* spp., *Metarhizium* spp.),
- plant disease control agents: include *Trichoderma* spp. and *Ampelomyces quisqualis* (a hyper-parasite of grape powdery mildew); *Bacillus subtilis* is also used to control plant pathogens.
- beneficial nematodes attacking insect (*e.g. Steinernema feltiae*) or slug (*e.g. Phasmarhabditis hermaphrodita*) pests
- entomopathogenic viruses (*e.g.. Cydia pomonella* granulovirus).

Biopesticides, key components of integrated pest management (IPM) programmes, are receiving much practical attention as a means to reduce the load of synthetic chemical products being used to control plant diseases. In most cropping systems, biological pesticides should not necessarily be viewed as wholesale replacements for chemical control of plant pests and diseases, but rather as a growing category of efficacious supplements that can be used as rotation agents to retard the onset of resistance to chemical pesticides and improve sustainability. In organic cropping systems, biopesticides can represent valuable tools that further supplement the rich collection of cultural practices that ensure against crop loss to diseases.

Biopesticides for use against crop diseases have already established themselves on a variety of crops. For example, biopesticides already play an important role in controlling downy mildew diseases. Their benefits include: a 0-Day Pre-Harvest Interval, the ability to use under moderate to severe disease pressure, and the ability to use as a tank mix or in a rotational program with other registered fungicides. Because some market studies estimate that as much as 20% of global fungicide sales are directed at downy mildew diseases, the integration of biofungicides into grape production has substantial benefits in terms of extending the useful life of other fungicides, especially those in the reduced-risk category.

A major growth area for biopesticides is in the area of seed treatments and soil amendments. Fungicidal and biofungicidal seed treatments are used to control soil borne fungal pathogens that cause seed rots, damping-off, root rot and seedling blights. They can also be used to control internal seed–borne fungal pathogens as well as fungal pathogens that are on the surface of the seed. Many biofungicidal products also show capacities to stimulate plant host defences and other physiological processes that can make treated crops more resistant to a variety of biotic and abiotic stresses.

The Manual of Biocontrol Agents gives a review of the available biological insecticide (and other biology-based control) products. In order to implement these environmentally-friendly pest control agents, it is often especially important to pay attention to their formulation and application.

Perceived Advantages

- do not leave harmful residues

- substantially reduced impact on non-target species
- when locally produced, may be cheaper than chemical pesticides
- in the long-term may be more effective than chemical pesticides (*e.g.* as demonstrated by the LUBILOSA Programme).

Perceived Disadvantages

- high specificity, which will require an exact identification of the pest/pathogen and may require multiple pesticides to be used
- often slow speed of action (thus making them unsuitable if a pest outbreak is an immediate threat to a crop)
- often variable efficacy due to the influences of various biotic and abiotic factors (since biopesticides are usually living organisms, which bring about pest/ pathogen control by multiplying within the target insect pest/pathogen)
- living organisms evolve and increase their resistance to biological, chemical, physical or any other form of control. Unless the target population is completely exterminated or is rendered incapable of reproduction, the surviving population will inevitably acquire a tolerance of whatever pressures are brought to bear- this results in an evolutionary arms race.

Bacillus Thuringiensis a Potential Biocontrol Agent for Pest(s) Control

Bacillus thuringiensis (or Bt) is a Gram-positive, soil-dwelling bacterium, commonly used as a biological alternative to a pesticide; alternatively, the Cry toxin may be extracted and used as a pesticide. *B. thuringiensis* also occurs naturally in the gut of caterpillars of various types of moths and butterflies, as well as on the dark surface of plants.

During sporulation many Bt strains produce crystal proteins (proteinaceous inclusions), called δ-endotoxins, that have insecticidal action. This has led to their use as insecticides, and more recently to genetically modified crops using Bt genes. There are however many crystal-producing Bt strains that do not have insecticidal properties.

Discovery and Study

B. thuringiensis was first discovered in 1902 by Japanese biologist Shigetane Ishiwatari. In 1911, *B. thuringiensis* was rediscovered in Germany by Ernst Berliner, who isolated it as the cause of a disease called *Schlaffsucht* in flour moth caterpillars. In 1976, Zakharyan reported the presence of a plasmid in a strain of *B. thuringiensis* and suggested the plasmid's involvement in endospore and crystal formation. *B. thuringiensis* is closely related to *B.cereus*, a soil bacterium, and *B.anthracis*, the cause of anthrax: the three organisms differ mainly in their plasmids. Like other members of the genus, all three are aerobes capable of producing endospores. Upon sporulation, *B. thuringiensis* forms crystals of proteinaceous insecticidal δ-endotoxins (called crystal proteins or Cry proteins), which are encoded by *cry* genes. In most strains of *B. thuringiensis* the *cry* genes are located on the plasmid.

Cry toxins have specific activities against insect species of the orders Lepidoptera (moths and butterflies), Diptera (flies and mosquitoes), Coleoptera (beetles), hymenoptera (wasps, bees, ants and sawflies) and nematodes. Thus, *B. thuringiensis* serves as an important reservoir of Cry toxins for production of biological insecticides and insect-resistant genetically modified crops. When insects ingest toxin crystals, the alkaline pH of their digestive tract activates the toxin. Cry inserts into the insect gut cell membrane, forming a pore. The pore results cell lysis and eventual death of the insect.

Use in Pest Control

Spores and crystalline insecticidal proteins produced by *B. thuringiensis* have been used to control insect pests since the 1920s. They are now used as specific insecticides under trade names such as Dipel and Thuricide. Because of their specificity, these pesticides are regarded as environmentally friendly, with little or no effect on humans, wildlife, pollinators, and most other beneficial insects. The Belgian company Plant Genetic Systems was the first company (in 1985) to develop genetically engineered (tobacco) plants with insect tolerance by expressing *cry* genes from *B. thuringiensis.*

B. thuringiensis-based insecticides are often applied as liquid sprays on crop plants, where the insecticide must be ingested to be effective. It is thought that the solubilized toxins form pores in the midgut epithelium of susceptible larvae. Recent research has suggested that the midgut bacteria of susceptible larvae are required for *B. thuringiensis* insecticidal activity.

Bacillus thuringiensis serovar *israelensis*, a strain of *B. thuringiensis* is widely used as a larvicide against mosquito larvae, where it is also considered an environmentally friendly method of mosquito control.

Genetic Engineering for Pest Control

Usage

In 1995, potato plants producing Bt toxin were approved safe by the Environmental Protection Agency, making it the first pesticide producing crop to be approved in the USA. By 1996, Bt Maize, Bt Potato and Bt cotton were being grown by farmers in the USA.

Bt crops (in corn and cotton) were planted on 281,500 km^2 in 2006 (165,600 km^2 of Bt corn and 115900 km^2 of Bt cotton). This was equivalent to 11.1% and 33.6% respectively of global plantings of corn and cotton in 2006. Claims of major benefits to farmers, including poor farmers in developing countries, have been made by advocates of the technology, and have been challenged by opponents. The task of isolating impacts of the technology is complicated by the prevalence of biased observers, and by the rarity of controlled comparisons (such as identical seeds, differing only in the presence or absence of the Bt trait, being grown in identical situations). The main Bt crop being grown by small farmers in developing countries is cotton, and a recent exhaustive review of findings on Bt cotton by respected and unbiased agricultural economists concluded that "the overall balance sheet, though promising, is mixed. Economic returns are highly variable over years, farm type, and geographical location".

Environmental impacts appear to be positive during the first ten years of Bt crop use (1996–2005). One study concluded that insecticide use on cotton and corn during this period fell by 35.6 million kg of insecticide active ingredient which is roughly equal to the amount of pesticide applied to arable crops in the EU in one year. Using the Environmental Impact Quotient (EIQ) measure of the impact of pesticide use on the environment, the adoption of Bt technology over this ten year period resulted in 24.3% and 4.6% reduction respectively in the environmental impact associated with insecticide use on the cotton and corn area using the technology.

Advantages

There are several advantages in expressing Bt toxins in transgenic Bt crops:

- The level of toxin expression can be very high thus delivering sufficient dosage to the pest.
- The toxin expression is contained within the plant system and hence only those insects that feed on the crop perish.
- The toxin expression can be modulated by using tissue-specific promoters, and replaces the use of synthetic pesticides in the environment. The latter observation has been well documented worldwide.

Health and Safety

Overall, Bt-modified crops appear to be safe for farmers and consumers. The proteins produced by Bt have been used in sprays for agricultural weed control in France since 1938 and the USA since 1958 with seemingly no ill effects on the environment or human health.

Bt toxins are considered environmentally friendly by many farmers and may be a potential alternative to broad spectrum insecticides. The toxicity of each Bt type is limited to one or two insect orders, and is nontoxic to vertebrates and many beneficial arthropods. The reason is that Bt works by binding to the appropriate receptor on the surface of midgut epithelial cells. Any organism that lacks the appropriate receptors in its gut cannot be affected by Bt.

There is clear evidence from laboratory settings that Bt toxins can affect non-target organisms. Usually, but not always, affected organisms are closely related to intended targets. Typically, exposure occurs through the consumption of plant parts such as pollen or plant debris, or through Bt ingested by their predatory food choices. Nevertheless, due to significant data gaps, the real-world consequences of Bt transgenics remains unclear.

Not all scientific reports on Bt safety have been positive. A 2007 study funded by the European arm of Greenpeace, suggested the possibility of a slight but statistically meaningful risk of liver damage in rats. While small statistically significant changes may have been observed, statistical differences are both probable and predictable in animal studies of this kind,(known as Type I errors), that is, the probability of finding a false-positive due to chance alone. In this case, the number of positive results was within the statistically predicted range for Type I errors.

The observed changes have been found to be of no biological significance by the European Food Safety Authority. A 2008 Austrian study investigating the usefulness of a long-term reproduction mouse model for GM crop safety reported that Bt-treated corn consumption in mice appeared to be correlated with reduced fertility via an unknown biochemical mechanism.

Limitations of Bt Crops

Constant exposure to a toxin creates evolutionary pressure for pests resistant to that toxin. Already, a Diamondback moth population is known to have acquired resistance to Bt in spray form (i.e., not engineered) when used in organic agriculture. The same researcher has now reported the first documented case of pest resistance to biotech cotton.

One method of reducing resistance is the creation of non-Bt crop refuges to allow some non-resistant insects to survive and maintain a susceptible population. To reduce the chance that an insect would become resistant to a Bt crop, the commercialization of transgenic cotton and maize in 1996 was accompanied with a management strategy to prevent insects from becoming resistant to Bt crops, and insect resistance management plans are mandatory for Bt crops planted in the USA and other countries. The aim is to encourage a large population of pests so that any genes for resistance are greatly diluted. This technique is based on the assumption that resistance genes will be recessive.

This means that with sufficiently high levels of transgene expression, nearly all of the heterozygotes (S/s), the largest segment of the pest population carrying a resistance allele, will be killed before they reach maturity, thus preventing transmission of the resistance gene to their progenies. The planting of refuges (i.e., fields of non-transgenic plants) adjacent to fields of transgenic plants increases the likelihood that homozygous resistant (s/s) individuals and any surviving heterozygotes will mate with susceptible (S/S) individuals from the refuge, instead of with other individuals carrying the resistance allele. As a result, the resistance gene frequency in the population would remain low.

Nevertheless, there are limitations that can affect the success of the high-dose/refuge strategy. For example, expression of the Bt gene can vary. For instance, if the temperature is not ideal this stress can lower the toxin production and make the plant more susceptible. More importantly, reduced late-season expression of toxin has been documented, possibly resulting from DNA methylation of the promoter. So, while the high-dose/refuge strategy has been successful at prolonging the durability of Bt crops, this success has also had much to do with key factors independent of management strategy, including low initial resistance allele frequencies, fitness costs associated with resistance, and the abundance of non-Bt host plants that have supplemented the refuges planted as part of the resistance management strategy.

Insect Resistance

In November 2009, Monsanto scientists found that the pink bollworm had become resistant to Bt cotton in parts of Gujarat, India. In four regions, Amreli, Bhavnagar, Junagarh and Rajkot the crop is no longer effective at killing the pests. This was the first instance of Bt resistance that was confirmed by Monsanto anywhere in the world. Monsanto confirmed field resistance of the worm to the *Cry1Ac* first generation Bollgard cotton, which expresses a single Bt gene.

Secondary Pests

Chinese farmers have found that after seven years of growing BT cotton the populations of other insects other than bollworms, such as mirids, have become significant problems. Similar problems but with mealy bugs have been reported in India.

Possible Problems

Lepidopteran Toxicity

The most publicised problem associated with Bt crops is the claim that pollen from Bt maize could kill the monarch butterfly. This report was puzzling because the pollen from most maize hybrids contains much lower levels of Bt than the rest of the plant and led to multiple follow-up studies.

It appears that the initial study was flawed by faulty pollen-collection procedure; researchers fed non-toxic pollen mixed with anther walls containing Bt toxin. The weight of the evidence is that Bt crops do not pose a risk to the monarch butterfly.

Wild Maize Genetic Contamination

A study in *Nature* reported that Bt-containing maize genes was contaminating maize in its centre of origin. *Nature* later "concluded that the evidence available is not sufficient to justify the publication of the original paper." However, there still remains a controversy over the highly unorthodox retraction on the part of Nature. In 1998, Chapela, one of the original paper's authors spoke out against Berkeley accepting a multi-million dollar research grant from the Swiss pharmaceutical company, Novartis.

A subsequent large-scale study, in 2005, failed to find any evidence of contamination in Oaxaca. However, further research confirmed initial findings concerning contamination of natural maize by transgenic maize.

However, further studies, such as that published in *Molecular Ecology* in 2008, have shown some small-scale (about 1%) genetic contamination (by the 35S promoter) in sampled fields in Mexico. One meta-study has found evidence for and against Bt contamination of maize, concluding that the preponderance of evidence points to Bt maize contamination in Mexico.

Possible Link to Colony Collapse Disorder

As of 2007, a new phenomenon called Colony Collapse Disorder (CCD) is affecting bee hives all over North America. Initial speculation on possible causes ranged from cell phone and pesticide use to the use of Bt resistant transgenic crops. The Mid-Atlantic Apiculture Research and Extension Consortium published a report in March 2007 that found no evidence that pollen from Bt crops is adversely affecting bees. The actual cause of CCD remains unknown, and scientists believe that it may have multiple causes.

Bio-control Agents and Biopesticides in IPM Technology for Sustainable Crop Production

Biological control of pests and diseases has undergone a great deal of development in the last one and a half decade in India. Both Centre and State Governments have been strong supporters of Integrated Pest Management (IPM) with main emphasis on biological control. Realizing the potential of biocontrol and taking cognizance of the need to extend this technology from laboratory to field, Department of Biotechnology (DBT) has gone in a big way to support a major programme in this area since 1989. DBT has made concerted efforts towards the development of biopesticide technology in the country in the last 15 years in a systematic way by launching various programmes and projects. The main objective of the programme is to develop safe and efficient biopesticide formulations, as well as cost effective commercially viable mass production technologies of various biocontrol agents/biopesticides (which include microbial pesticides-baculoviruses, entomopathogenic fungi and nematodes, antagonistic fungi and bacteria, parasitoids and predators; pheromones, other semiochemicals; botanical pesticides and plant growth regulator) for use under Integrated Pest Management and their bioefficacy tested through large scale front line field demonstrations against the key pests and diseases of economically important crops in varied agro-climatic zones.

DBT established a National Biocontrol Network Programme (NBNP) in 1989 to study the management of key insect pests, diseases and weeds of economically important crops viz., cotton, sugarcane, tobacco, oilseeds, pulses, vegetables, fruits, spices and plantation crops. This network is a continued activity of the Department. Almost 200 R&D projects have been implemented so far in ICAR/CSIR Institutes and SAUs throughout the country.

Based on the technology developed through DBT's, NBNP, two model production units for biocontrol agents were set up at TNAU, Coimbatore and Madurai to assess their commercial viability. Already some private individuals, entrepreneurs, progressive farmers, unemployed agricultural graduates and science graduates have started producing biocontrol agents.

During 1993-94, a time bound, goal oriented mission mode programme was launched on the development, production and demonstration of biocontrol agents under "IPM" in 15 States. Under this programme, 20-production units and 2 repository centres were set up in various States with the main objective to revalidate and fine-tune the technology packages best suited for various agro-climatic zones and also to develop farmer's oriented technology packages for their large-scale adaptation and adoption. The cost of crop protection was optimized and income/profit maximized through demonstration trials. Economic benefits have been realized in terms of yield increasé, monetary gain and reduction in the pesticide consumption. Successful demonstration trials under this programme (55000ha) have culminated in the recommendation of specific biocontrol agents as an ecofriendly option in IPM modules by various State Departments of Agriculture Subsequently a major R&D programme on Biological control of crop pests and diseases under IPM as a component of Integrated Nutrient Management (INM) programme was launched in 1998-99 on the role of biological resources for Integrated Nutrient and Pest Management (INPM) for increasing agricultural productivity in existing cropping systems. It was launched at 14 centres in 12 states with the main aim to develop cost effective, sustainable and eco-friendly modules/package of practices in different ecosystems and also to demonstrate the sustained preservation of ecosystem by conducting front line demonstrations in identified clusters/adopted villages by adopting IPM and INM techniques.

Several cost effective, sustainable and eco-friendly IPM-INM modules in different ecosystems have been developed for various crops viz., rice, pigeonpea, chickpea, french beans, soybean, lentil, green pea, mustard, tomato, brinjal, capsicum, cauliflower, cabbage, chilli, coconut etc. The cost-benefit ratio of IPM-INM modules worked out at large scale trials showed significant tilt in favour of adopting the technology by farmers for different crops.

Achievements made in the National Biocontrol R&D Network Programme, mission mode and INM-IPM programme are:

i) Development and standardization of cost

ii) Effective and commercially viable mass production technologies of about 30 candidate biocontrol agents/biopesticides

iii) Demonstration of the field efficacy of biocontrol agents/biopesticides under different ecosystem in various economically important crops covering 1,80,000 ha.

iv) Economic benefits realized in the IPM trials by computing the cost benefit ratio in terms of a) monetary gain; b) yield increase; c) reduction in pesticide consumption

v) Several cost effective, sustainable and eco-friendly IPM-INM modules have been developed for various crops

vi) Sustained preservation of the ecosystem has been successfully demonstrated in IPM-INM adopted villages.

vii) General awareness created about the biocontrol based IPM technology among the extension functionaries and the farming community.

viii) Gap of availability of sufficient quantity of biocontrol agents alongwith the quality control and their adoption and adaptation in farmers' field have been bridged to a considerable extent.

Mass Production Technologies Developed

Standardized cost effective and commercially viable mass production technologies of various candidate biocontrol agents/biopesticides viz. a) Microbials-NPVs of *Helicoverpa armigera, Spodoptera litura, Hyblaea puera, Spilosoma obliqua* and GVs of *Chilo infuscatellus* and *Plutella xylostella, Trichoderma viride, T. harzianum, T. koningii, T. virens, Myrothecium verrucaria, Aspergillus niger, Bacillus subtilis, Pseudomonas fluorescense, Beauveria bassiana, Verticillium lecanii, Nomuraea rileyi, Debaromyces henseni* and *Pochonia chlamydosporia* b) Parasitoids and Predators – *Trichogramma chilonis, T. polae, T. japonicum* and *Chrysoperla carnea* c) Entomopathogenic Nematodes (EPN)-*Steinernema carpocapsae, S. biocornutum, S. seemai* sp. n., and *Heterorhabditis indica.* d) Pheromones – Insects sex pheromone lures for *Chilo sacchariphagus indicus* and *Scirpophaga incertulas.*

Bioefficacy Demonstration

Field efficacy of aforesaid biocontrol agents Biopesticides were tested at multilocation covering various crops viz., cotton, rice, chickpea, tobacco, groundnut, sunflower, sugarcane, sesamum, soyabean, blackgram, greengram, pigeonpea, lentil and other pulses, cauliflower, tomato, chilli and other vegetables, spices, tea, coffee etc. Substantial area of land (1,80,000 ha. approx) was covered in various agro-climatic zones of the country for managing pests and diseases of important crops.

Technologies Developed and their Status of Transfer

Several mass production technologies of biocontrol agents/biopesticides have been developed, standardised and transferred. The fermentation based technologies of 3 biocontrol agents viz., *Trichoderma viride, T. virens, Bacillus* spp. and 3 biofertilizers viz., *Rhizobium* spp, *Azotobacter* spp., and *Azospirillium* spp. have been transferred as an integrated package for biopesticides and biofertilizers to M/s Prathishta Industries Ltd., Secundrabad (A.P.) and M/s Haryana Biotech, Gurgaon. Mass production technology of *Aspergillus Niger* (AN27) was transferred to M/s Cadila Pharma. Mass production technology of *Ha*NPV(Heliokill), *Sl*NPV (Magic), *So*NPV (Spilocide), Phule Trichocards (*Trichogramma chilonis*), Phule *Trichoderma* have been developed, formulated, branded and being produced and supplied to farmers in large quantities for demonstration.

A solid state fermentation technology of *Trichoderma harzianum* has been developed at Indian Institute of Spices Research (IISR) Calicut and transferred to 10 private entrepreneurs in the states of Tamil Nadu, Karnataka and Andhra Pradesh. Mass production technologies of *Ha*NPV(1000 LE/day) and *Sl*NPV (500 LE/day) developed at UAS, Dharwad are ready for transfer. *Ha*NPV(1000 LE/day) has also been developed using local strains/resources at RARS, Lam, Guntur and have been transferred to five industries. A mechanized system of mass production of *Trichogramma chilonis* has been developed at IARI and ready for transfer. At G B Pant University of Agriculture and Technology, Uttaranchal, mass production technologies of two high temperature tolerant strains of *Trichogramma poliae* and *T. japonicum* have been developed and negotiations are on for technology transfer.

Mass production technologies of *Trichoderma viride* and *Pseudomonas fluorescens* developed at Tamil Nadu Agricultural Universisity, Coimbatore have already been transferred to various industries. At Kerala Forest Research Institute (KFRI), Kerala, mass production technology of *Hp*NPV against *Hyblaea puera* (teak defoliator) has been developed for teak forest ecosystem. Techno-economic feasibility has been studied. Seven formulations of this NPV have also been developed and mode of application on teak trees has been standardised and perfected. Technology is ready as a full package for transfer. In the case of *Steinernema carpocapsae* and *Myrothecium verrucaria,* technologies are being fine tuned for upscaling the production and increasing the shelf life for commercialisation.

The pheromone technologies for the management of Internode Borer (*Chilo sacchariphagus indicus*) of sugarcane have been developed at SPIC research foundation and is ready for transfer. Rice yellow stem borer (*Scirpophaga incertulas*) has been successfully managed through sex pheromone technology and demonstrated at multi location trials (300 acre). This technology has been developed at IICT, Hyderabad in collaboration with and DRR.

Popularization

Several extension activities were organized at the instance of DBT for creating awareness among the farmers on the large scale use of biocontrol agents under IPM. About 75000 farmers were benefited by various extension activities which included regional workshops, training programmes, agroclinics, agrifair, farmers ghosthi, krishi melas, farmers field schools etc.

11

Role of Bacteria and Actinomycetes in Biological Control of Pests and Diseases

MICROORGANISMS We find that more life is living below the surface than above in most environments. Soil are put into six groups: earthworms, nematodes, arthropods, bacteria, fungi, and protozoa with each having their own function. They do play an important role in plant health and water.

The cycle of nutrients going through our environment are mainly driven by these microorganisms. They maintain:

1. Decomposition
2. Mineralization
3. Nitrogen cycling
4. Storage and release of nutrients
5. Carbon cycling
6. Take the pollutants out of the water before it reaches underground or surface water.

The Location, Living Conditions, and Functions of Microorganisms

Actinomycetes Effective microorganism, they are responsible for distinctive scent of freshly exposed, moist soil. Prefer neutral to alkaline soils; high oxygen requirement; prevalent in dry regions. They release carbon, nitrogen, and ammonia during decomposition of organic matter. They help form humus. Associate with non-leguminous plants like bitterbrush to fix nitrogen and make it available to other plants in the area. Some nitrogen may be unusable without the bacteria converting it to a form that can be used.

Bacteria Microorganisms found all over the world and even found down in the earth as far as 1 mile. It thrives under most conditions but found near plant roots, an important food source. Feeds on organic matter; encourage organic and inorganic chemical reactions that affect plant growth; and fix nitrogen from air in soil.

Soil Bacteria

What bacteria lack in size, they make up in numbers. They are tiny, one-celled organisms. A teaspoon of productive soil generally contains between 100 million and 1 billion bacteria.

Has Four Functional Groups

1. Most are decomposers that consume simple carbon compounds. By this process, it convert energy in soil organic matter into forms useful to the rest of the organisms. A number of decomposers can break down pesticides and pollutants in soil. Decomposers are especially important in stopping or retaining, nutrients in their cells, thus preventing the loss of nutrients, such as nitrogen, from the rooting zone.
2. A second group are the mutualists that form partnerships with plants. The most well-known of these are the nitrogen-fixing bacteria.
3. The third group is the pathogens.
4. A fourth group, called lithotrophs (literally meaning rock eaters) or chemoautotrophs (which are able to synthesize all of the organic compounds they need from inorganic raw materials in the absence of sunlight.), obtains its energy from compounds of nitrogen, sulfur, iron or hydrogen instead of from carbon compounds.

Functions

- They perform important services related to water dynamics.
- nutrient cycling
- disease suppression.
- many organisms will compete with disease-causing organisms in roots and on above ground surfaces of plants.

Important Bacteria

Nitrogen-fixing The plant supplies simple carbon compounds to the bacteria, and it converts nitrogen (N2) from air into a form the plant host can use. When leaves or roots from the host plant decompose, soil nitrogen increases in the surrounding area.

Nitrifying change ammonium to nitrite then to nitrate – a preferred form of nitrogen for grasses and most row crops. Nitrate is leached more easily from the soil, so some farmers use nitrification inhibitors to reduce the activity of one type of nitrifying (scavenging potentially toxic nitrogen compounds from their surroundings, including: ammonia and nitrite). Nitrifying it are suppressed in forest soils, so that most of the nitrogen remains as ammonium.

Denitrifying convert nitrate to nitrogen (N_2) or nitrous oxide (N_2O) gas. Denitrifiers are anaerobic, meaning they are active where oxygen is absent, such as in saturated soils or inside soil aggregates.

Actinomycetes are a large group of bacteria that grow as hyphae like fungi. They are responsible for the characteristically "earthy" smell of freshly turned, healthy soil. Actinomycetes decompose a wide array of hard-to-decompose compounds and are active at high pH levels. Fungi are more important in degrading these compounds at low pH. A number of antibiotics are produced by actinomycetes such as Streptomyces.

They are found where?

- It is more competitive when easy-to-metabolize substrates are present such as fresh, young plant residue and the compounds found near living roots.
- They are especially concentrated in the region next to and in the root.
- Some believe that plants produce certain types of root exudates to encourage the growth of protective bacteria.

One Celled Organisms that Encourage Plant Growth

Certain strains of the soil bacteria have anti-fungal activity that inhibits some plant pathogens (A disease-causing organism). Those strains can increase plant growth in several ways. They may produce a compound that inhibits the growth of pathogens or reduces invasion of the plant by a pathogen. They may also produce growth factors that directly increase plant growth.

The plant growth-enhancing organism occur naturally in soils, but may not be in high enough numbers to have a great effect. Soon, gardners may be able to coat seeds with this anti-fungal to ensure that the bacteria reduce pathogens around the seed and root of the crop.

Earthworms are the reason for a huge industry today. People are raising & selling earthworms all over the world, especially wholesale earthworms. The work of the earthworms in the soil is what earthworms eat. Producing castings that are equivalent to compost as an organic fertilizer and soil supplement.

Fungi Basically 2 types of fungi-mycorrhizal and normal. Fungi thrive in well-drained, neutral to acidic, aerated soils. Normal fungi help decompose the organic matter in litter and soil but play less of an overall role. Mycorrhizal fungi help develop healthy root systems by growing on plant roots. The fungus is actually a network of filaments that grow in and around the plant root cells, forming a mass that extends considerably beyond the plant's root system. This essentially extends the plant's reach to water and nutrients, allowing it to utilize more of the soil's resources.

Mode of Entry of Microbial Pathogens on Insect Pests and their Mechanism of Control

In Coffee

In the early 1870's the coffee leaf rust caused by the fungus Hemelia vastatrix, wiped out coffee plantations in Srilanka. In spite of the pioneering work of the English pathologist, H.Marshal Ward, no biological or chemical cure could be found to save the coffee plantations. Hence, the coffee production was abandoned in Srilanka and shifted to the Western hemisphere. In the early days, the British pioneers who opened up coffee plantations restricted themselves in growing the Arabica variety. Coffee cultivation reached its peak in the 1860's and it was at this point that the major outbreak of pests and diseases took place. Hemileia vastatrix—coffee leaf rust, root rot, berry blotch, collar rot, white stem borer –Xylotrechus quadripes, mealy bugs, green bugs, berry borer, shot hole borer, red borer, thrips etc. In search of remedies, number of toxic chemicals was tried out and even today the list of chemicals is ever growing.

Chlorinated hydrocarbons, mercury laced compounds, lead, Some of these chemicals were developed in the Western world and directly applied under Indian conditions without screening; resulting in resistance build up. Over the years the target organism has evolved new races or mutations which are tolerant to more powerful chemicals with inbuilt resistance. Planters never bothered to use pesticides diligently. We are suggesting a simple, yet effective approach of integrated pest management (IPM) in controlling and not eliminating pest and disease incidence.

Microbial Inoculants

Microorganisms can be effectively used as a safe and alternative method in controlling insect pests. Microorganisms that affect insects are termed ENTOMOPATHOGENS. Microbial inoculants or bio agents are Entomogenous microorganisms and or their products that cause pathologies and are usually fatal to their insect hosts. Almost 1500 insect pathogens have been isolated world wide but unfortunately less than one per cent has been exploited for commercial value. About 5 or 6 of them have been tried commercially. Their role includes disease initiation in target insects or to suppress harmful insect population directly or in combination with chemical insecticides or pesticides. Entomopathogens multiply in nature if the environment is conducive and cause infections on the target organism. Scientists have proved that microbial inoculants pose little human or environmental damage.

Prerequisities of Microbial Inoculants :

- Bio inoculants should be highly virulent to the target organism.
- Should be stable to the stress of nature.
- Virulence should be confined to the species of insects for which it is treated (Should be host specific).
- Should neither damage the crop, nor harm beneficial insects and micro flora and predators and animals.
- Should be easily mass culturable.
- Should be stable for a long period of time.

Advantages of Microbial Inoculants :

- They are environment friendly and leave behind no toxic residues.
- Most of them target the specific insect and in turn protect beneficial insects.

- Most inoculants are easily culturable in the lab, with minimum space.
- Inexpensive to produce large quantities of inoculum.
- Slowness in developing resistance to microbial pathogens.
- Can control insect in cavities where chemical insecticides cannot reach.
- In a way, we try to mimic nature by releasing them into an open environment.

Microbial inoculants can be used in two ways:

A. SHORT TERM CONTROL: For a particular season (annual) or by using highly virulent pathogens.

B. LONG TERM CONTROL: Generally used for perennial crops and less virulent pathogens are preferred.

Disadvantages of Microbial Inoculants :

- Necessity for careful and correct time of application.
- Host specificity of most pathogens, narrows down its use.
- Necessity of maintaining a pathogen in a viable condition, until the insect is contacted.
- Difficulty in producing some obligate and facultative pathogens on a large scale.
- Requirement of favourable environmental conditions for the pathogen to act, multiply and execute its mode of action.
- Tendency of dead insects remaining attached to the host.

Bacteria

There are approximately 100 species of bacteria pathogenic to insect pests but hardly one per cent among these is used as bio control agents. Bacteria belonging to the family Bacillaceae, Lactobacillaceae, Micrococacaceae, Pseudomonadaceae and Enterobacteriaceae are generally found pathogenic to insect hosts. For E.g. the most commonly used Bacteria in biological control of insect pests, Bacillus thuringiensis which is a spore former, was first isolated by a German by name Berliner in 1915 and was subsequently known to control several insect pests. Even today it is known as one of the most virulent bacterial pathogen which kills many larval forms of lepidopterous insects. It affects more than 175 larval species that include some of the most important economic pests. It can also be mixed with a number of commercial insecticidal formulations and is gaining ground in integrated pest management. There are many other species of bacteria which are used in biological control of insect pests such as Aerobacter aerogenes and Escherichia coli, Bacillus popillae, Bacillus sphaericus, Bacillus moritai, Bacillus lentimorbus are already used in commercial preparations. These bacteria form a toxic crystal protein which acts as a stomach poison inside the insect host resulting in the quick death of the insect. The above mentioned genus of bacteria attack mostly members of the insect family belonging to Hemiptera, Diptera, Coleoptera and Lepidoptera. Some species of bacteria in addition to forming toxic spores produce crystalline inclusions within the sporulating cell which aids in killing the target insect. Bacillus popillae and Bacillus lentimorbus when sprayed in the field invade the insect and gains entry into the blood system. This acts as a good medium for the growth and proliferation of the microorganism. The bacteria sporulates causing imminent death of the insect and once death occurs the bacterial spores are released into the atmosphere and soil and remain there for extended periods of time. These spores may in turn infect newly hatched beetle larvae and in turn infect it.

Fungi

Fungi belonging to the family Deuteromycetes and Entomopthora causes mycosis in insects. More than 800 fungal species comprising of 125 genera have been reported to infect insects. Probing a little deeper into this vast number of fungal species reveals that most harmful pests are associated with one fungi or the other. However, more work needs to be done in developing fungal biocontrol agents. They attack mostly members of the family Hemiptera, Lepidoptera and coleoptera. Approximately 100 genera of fungi are pathogenic to insects. Fungi penetrate the body directly. They invade insect's body walls and spread in the softer tissues of insect body. Fungi infected insects show a thick mycelial mat covering the entire body of the insect. In nature the insect may either be exposed to the thick mycelial mat found on plants, leaf or soil or may also be exposed to toxic fungal metabolites either by cuticular contact by ingestion or by injection into the blood stream known as haemocoel.Beauvaria bassiana the white muscardine fungus is a potential bio inoculant against 700 species of insect pests. The green muscardine fungus Metarhizium anisopliae has been reported to effectively control a large number of target organisms.

Viruses

Insect viruses fall into five major groups. They belong to the family Baculoviridae.

1. Nucleopolyhedrosis Viruses {NPV}
2. Granulosis Viruses {GV}
3. Cytoplasmic Polyhedrosis Viruses {CPV}
4. Non Inclusion Viruses {NIV}
5. Entomopox Viruses {EPV}.

More than 800 viruses have been isolated that target ARTHROPOD insect pests. Since insect viruses are absent in vertebrates and plants, they are safe not only to human beings, but also to no target organisms. Insect viruses are obligate parasites and hence need to be cultured in living hosts. The target pest is reared in the laboratory or factory and mass multiplying the virus by inoculating it. Nucleopolyhedrosis virus is the most commonly used for insect pest control. When viral biocide is sprayed in the field, the viral particles get attached to the foliage and the insect larvae feed and in turn ingest the polyhedral inclusion bodies (PIB). The pib in turn multiply inside the midgut and bring about larval mortality. These protein crystals are insoluble in water and are produced in geometrical proportions and after killing the insect, get released into the surrounding atmosphere. The crystals can reinfect live insects and retain their infectivity even after long storage outside the living tissue of plants.

Protozoa

Protozoa belonging to the family Nosematidae has shown great promise as a bio inoculant. Approximately 375 species of protozoa have been found to be pathogenic on various harmful insects. Nosema locustae has been tried out in the control of grasshoppers.

Nematodes

Nematodes parasitic on insects like rootworms, grubs, caterpillars have been recorded from early times.

Mode of Action

The two main factors in controlling the disease are by INVASIVENESS AND INITIATION of disease. The site of entry of bio inoculants may be broken or bruised integuments or by feeding habits of the insects or other openings. When attacking any insect the quantity and quality of inoculum is important. In essence the control of the disease depends on the capacity of the microbial inoculum to rapidly establish itself and multiply in large numbers, resulting in severe infection of the pathogen concerned. The time lag between the entrance of microorganisms into the body of the insect and the incubation period within varies from pathogen to pathogen and insect to insect. Proliferation and disease control are controlled by environmental factors to a large extent and internal host characteristics to a small extent. The mode of action may be due to the production of ENDO or EXO toxins.

Broadly bioagents which attack insect pests can be classified as follows.

1. *Obligate Bioagents:* These bio agents in nature are associated with a specific insect disease and are difficult to culture in artificial media. E.g. Bacillus popilliae causing milky diseases of white grubs and Bacillus larvae causing American foulbrood of bees.
2. ***Spore Forming Bio Agents:*** The protein crystals formed by these sporulating bio inoculants are highly toxic to the target agent. E.g. Bacillus cereus produces such toxins.
3. *Determinative Bioagents:* The bio agents get activated; once inside the host. They multiply in the haemocoele of insect and produce lethal septicemia.
4. *Facultative Bioagents:* The host tissue is damaged by the invasion of the bio agent but they are not obligate parasites.
5. *Predaceous Bioagents :* The bio agent either secretes a sticky substance or builds up.

Conclusion

We are certain that mankind has not provided adequate resources in terms of money, research and time in search for naturally occurring or safe alternate methods of pest control. It is simply because our mind is tuned towards shortcuts and profits. The use of bio inoculants or their by products is under exploited. The dangers of chemical poisons are there for all to see, right from water bodies to food chains, all of nature too is affected. Due to the accumulation of toxic metabolites in the air as well as soil system, it will result in creating immunity to dangerous pests. The genetic population of future generation will be the one, where less susceptible individuals are left to breed.

They leave behind untold miseries for the future children of this planet including wild life habitat, birds, fish, mammals etc. Their long term effect on the environment cannot be quantified by scientists sitting in the laboratories. We need a more intelligent way of using chemicals. On the other hand if we can only apply ourselves to the fact that evolution has all the answers to our problems, it will only be a short time before we discover natural agents which will aid in biological control of pests and diseases.

This approach will make the world a better place for all to live and share. Microbial Inoculants result in much greater protection for both flora and fauna. It is like a NATURAL system taking place in nature. We do not imply that modern technology should be abandoned. On the contrary we need to focus on Integrated Pest Management which will give us a better hope for tomorrow. This technology has been successfully tried out in many developed countries like the U.S.A., Canada, and Germany where it is used on several agricultural crops, trees and ornamental shrubs. It has been found to be economically viable and most importantly environment friendly. Microbial inoculants open up a world of opportunities. They create living spaces which bring about harmony in nature. We are blessed people to live on this bio diverse rich planet.

Chemical control of stored product insect pests has been the most efficient and effective means of protection of stored produce. However, with the increasing cost of inorganic chemicals and their known hazards to the environment an integrated means of control has been widely adopted. The search for botanical insecticides could supplement the expensive petroleum based chemicals. Botanical preparations have long been used for protection of stored produce by small scale farmers in oriental countries such as India where the neem tree has been used extensively.

The success of these small farmers serve as the impetus for exploring the utilization of indigenous resources for small scale product protection and for possible industrial scale product protection. Several species of plants are now being screened for source of active ingredients against stored product insect pests commonly used are made of polyethylene or polyvinyl chloride film having a thickness of 0.1 mm (.004 inch) weighing 100 grams per square meter.

After the completion of the fumigation it is necessary to collect the tablet residues or expended sachets. These materials can be disposed properly by swamping it in a soapy water until bubbling ceases (in open air) and then buried at least 30 cm below the soil surface. Under no circumstances should large quantities of expended residues be placed in a heap prior to being buried, because of the possible danger of ignition.

Formulation Available and Dosage Rate

Phosphine (PH3) is commonly available in the form of tablets, pellets and sachets containing aluminum phosphide.

- Each tablet weighs 3 g and after complete decomposition 1 g of phosphine is released.
- Each pellets weighs 0.6 g i.e., 1/5 of tablet weight) and on complete decomposition, 0.2 g of phosphine is released.
- Each sachet weights 34 g and on complete decomposition 11 g of phosphine is released.

The following equivalents, based on commercial formulations containing aluminum phosphide as the phosphine source are given to simplify dosage calculations.

1 g phosphine = 1 tablet = 5 pellets = 1/11 sachet

Under Philippine conditions, the following rates of phosphine application are recommended.

15-45 tablets*	2-5 tablets per	15-20 tablets
per 100 cu. Ft	metric ton	per 1000 cu ft
Or	or	or
3-5 sachets **	1 sachet per	2-3 sachet per
per 1000 cu ft	1-3 metric ton	1000 cu. ft.

* Recommended exposure time 3-5 day when tablet is used

** Recommended exposure time 3-6 days when sachet is used.

Identification of Active Principles

Neem Leaves and Fruits

The constituents of the neem leaf powder were obtained by Tirimanna (1985) using two-dimensional thin layer chromatography. As a result of a strong saponification procedure, it is only the very stable compounds, which were observed on the chromatoplate. The compounds violaxanthin, butoin, 3carotene, zeaxanthin, ant heroxanth in, cryptoxanth in and neoxanthen were identified.

Seven epoxy compounds have also been isolated from fruits of Melia azedarach (Kraws et al., 1981). These compounds were rather stable and maybe of significance to scientists in search of a stable compound having insecticidal properties. This is relevant in view of the fact the neem leaf loses insect repellent property with age.

One sterol and 3 stable ketones were found, together with six very stable phenolic compounds. The occurrence of phenolic compounds in the neem plant is already well documented. Phenolic constituents are also recorded as antihelminthic factors (Taniguchi, 1960). The identification of quercetin in the neem leaf accounts for its antibacterial and antifungal properties and hence the curative properties of leaves in cases of sores and scabies.

T. diversifolia (Wild Sunflower)

The active fraction D from this plant contains alpha-lactone with a hydroxyl group attached either to the lactone ring or to the alkyl substituent. The unsaturation may be present in the ring itself or alpha to the carbon as indicated by the strong IR absorption at 1670 cm' and by the failure of the sample to decolorize bromine in carbon tetrachloride.

Insecticidal Activity

Black Pepper (Piper Nigrum) and Red Pepper (Capsium Frutescens)

They found the crude and semi-purified extracts of black pepper topically applied on the insects toxic. The admixture treatment of grains showed the ground black pepper (GBP) more toxic than the semi-purified and crude extract as residual contact and stomach poisons against the corn weevil. It was also found resisually toxic for 2 months on the saw-toothed grain beetle, lesser grain borer and corn weevil.

The ground red pepper (GRP) and GBP were residually toxic to the bean weevils for 2 and 6 months, respectively. The effectiveness of red pepper on bag rice under warehouse condition (natural infestation) at average temperature of 25.70-27.90°C was further evaluated. The major stored product insect pests were corn weevil and red flour beetle. Three methods of applications were done: mixing powdered and whole fruit of RP with rice grain or in sachet with doses of 1200 ppm or 1800 ppm and spraying the bag with either 5, 10 or 20% of the extract. All methods of application and dose levels effectively protected the bagged rice from infestation for no longer than two months storage.

Neem (Azadirachta indica A. Juss)

The repellent and antifeedant effect of seed powder and oil of neem on five species of stored product pests was investigated by Akou-Edi (1985). Laboratory trials in Togo using red corn treated with neem oil at concentrations of 1, 2, 4 & 8 ml/kg or with seed powder at 20, 40, 80 g/kg infested with confused flour beetles and corn weevils showed significant difference between the treated and untreated samples. As a rule, the effect of neem oil and need powder increased with higher concentrations but difference between concentrations was not significant on both test insects. Atudy on the effect of neem seed kernel and leaf powder on the development of Callosobruchus maculatus (F.) on 3 bean seed species treated at the rate of 0.5, 0.1 and 2.0 parts per 100 parts of seeds showed reduced oviposition capacity, lengthened larval and pupal period, reduced percent adult emergence, longevity and growth index. Among mungbean, bush bean and cowpea bean, the latter was found to be the most preferred food for the development because it supported normal development and enhanced population growth. Studies on the effect on neem kernel extract (NSKE) on metamorphosis of Ephestia kuehniella (mill moth) and which established a concentration effect curve were undertaken by Maurer (1985). A methanolic extract of sundried neem seeds was applied at a concentration of 4 ppm on 4-5th instar larva. The stage of development that a larva had reached was determined by measuring the size of the head capsule. The average size of the head capsule at each larval stage was determined in a preliminary experiment on untreated larva under the same conditions (30°C, 82-86% RH). A 4 ppm concentration and less resulted in the insertion of an additional larval instar between the fourth and fifth larval stages. Hence, metamorphosis was prolonged after treatment with neem seed kernel extract. The mortality induced by concentrations of 4 ppm NSKE or higher, occurred first during the molting phase between different larval stages. No feeding deterency was observed.

Lageundi (Vitex Negundo)

The protectant effect of whole and powdered leaves of lagundi on stored corn against corn weevil was evaluated by Bhuijah (1988) for a period of 6 months. Two level of dosages were used (1 & 5%) for both whole and powdered leaves on 5 kg of corn. After 30 days of storage, the protectant effect of 5% lagundi leaf powder and 1% whole leaf was lost as shown by increased infestation of all lagundi treated samples. Only neem seed oil treated corn seeds remained protected and had low percent infestation throughout the study. Apparently the insecticidal constituent of the lagundi leaves were not stable, hence the repellent property were readily lost.

Vegetable Oils Controlling Storage Insect Pests The practice of adding a little vegetable oil to stored rice or legumes for protection against stored-insect pests is well known and well established in oriental countries like China, India and Indonesia. Recently the practice of protecting stored produce with oil has spread and has been adopted in Africa and South America. Recently Van Rheenen (1983) pointed out the applicability of this method of protecting storage grains to supplement safe chemical formulations. The mode of action, appropriate dosages and duration of efficacy of oils have been investigated by various workers on storage insect pests. Differences between crude and purified oil have been studied and crude oil has been found to be a better protectant while the triglyceride oleic acid combination was found the most effective (Schoonhoven and Hill (1981).) The amount of oil needed for the control of most storage pests vary from 2 cc/kg seed to 15 cc/kg seed (Cruz and Cardona, 1981) depending on the level of infestation.

Table : Percent infestation of treated corn grains at different storage periods at room temperature

Treatments	*DAYS*					
	30	60	90	120	150	180
Control	2.04a	10.60a	22.91a	61.97a	80.27a	88.51a
1% lagundi leaf powder	1.67a	11.24a	18.26a	58.51a	78.11a	90.22a
5% lagundi leaf powder	0.90b	6.79a	14.42a	65.73a	77.51a	90.15a
1% lagundi whole gried leaf	1.52ab	5.48ab	15.69a	62.20a	74.79a	38.58a
5% lagundi whole dried leaf	2.55a	10.80a	19.05a	57.12a	70.73a	88.81a
Actellic	0.38bc	0.33b	0.17b	5.45b	5.24b	14.19b
Neem seed oil	0.25c	0.06b	0.00b	1.30b	14.14b	31.75b

Means in a column followed by a common letter are not significantly different at the 1 % level by DMRT.

Table: Vegetable oils controlling storage pests

Oil	*Storage Pest*	*Author (S)*
Castor	C. chinensis	Sangappa, 1977
Coconut	C. maculatus	Varma and Pandey, 1978
Cotton seed	C. chinensis	Sangappa, 1977
	C. maculatus	Pandey, et al., 1981
	S. oryzae	de Oca et al., 1978
	S. granarius	Yun-Tai Qi and Burkholder, 1981
Groundnut	S. cerealella	de Oca et al., 1978
	C. maculatus	IITA, 1976
	S. granarius	Varma and Pandey, 1978
		Yun-Tai Qi and Burkholder, 1981
Maize	C. maculatus	Akelo-Tsegah, 1976
		Singh, et al., 1978
	A. obtectus	Magoya et al., 1982
		van Rheenenet al., in press
	C. chinensis	Cruz and Cardona, 1981
	S. granarius	Yun-Tai Qi and Burkholder, 1981
	S. oryzae	de Oca et al., 1978
	S. cerealella	de Oca et al., 1978
Mustard	C. chinensis	Sangappa, 1977
	C. maculatus	Varma and Pandey, 1978
Neem	C. chinensis	Pandey et al., 1976
		Sangappa, 1977
Palm	S. oryzae	de Oca et al., 1978
	S. cerealella	de Oca et al., 1978
Rice	C. maculatus	Pandey et al., 1981
Sunflower	C. chinensis	Sangappa, 1977
	C. maculatus	Pandey et al., 1981
Sesame	C. maculatus	Varma and Pandey, 1978
Soybean	C. chinensis	Cruz and Cardona, 1981
	S. granarius	Yun-Tai Qi and Burkbolder, 1981
	S. oryzae	de Oca et al., 1978
	S cerealella	de Oca et al., 1978
Sunflower	A obtectus	Magoya et al., 1982
		van Rheenen et al., in press
	C. chinensis	Sangppa, 1977
	C maculatus	Varma and Pandey, 1978
Paraffin	C. maculatus	Calderon, 1979

Table : Storage pests controlled by vegetable oils

Oil	*Storage Pest*	*Author (S)*
Cajanus cajan	Callosobruchus chinensis (L.)	Sangappa, 1977
Cicer arietinum	C. chinensis (L.)	Pandey et al., 1976
	C. maculatus (F.)	Calderon, 1979
Phaseolus vulgaris	A. obtectus (Say)	Magoya et al., 1982
		van Theenen et al., in press
	Zabrotes subfasciatus (Boh.)	Hill and van Schooven, 1981
		van Schooven, 1978
Sorghum vu/gare	Sitotroga cerealella (Ol.)	de Oca et al., 1978
	S. oryzae (L.)	de Oca et al., 1978
Triticum vu/gare	S. cerealella	de Oca et al., 1978
	S. granarius (1.)	Yun-Tai Qi and Burkholder, 1981
Vigna radiate	S. oryzae (L.)	de Oca et al., 1978
	C. maculatus (F.)	Pandey et al., 1981
		Varma and Pandey, 1978
Vigna unguiculata	C. chinensis (L.)	Cruz and Cardona, 1981
	C. maculatus (F.)	Akelo-Tsegah, 1976
		IITA, 1976
		Singh et al., 1978
Zea mays	S. cerealella (Ol.)	de Oca et al., 1978
	S. oryzae (L.)	de Oca et al 1978

Prospects of Insecticides from Plants

Acceptance of botanical insecticides for the control of storage insect pests by small scale farmers is influenced by the following parameters: availability, safety, quality and cost. Application of the control methods is simple and does not need sophisticated equipment add to the desirability of the method.

The search for more plant species with insecticidal properties is being pursued by scientists from all over the world. However, a systematic evaluation and identification of chemical and physical characteristics of active constituents should be accelerated. It is on this basis where knowledge on their efficacy in time, various levels of dosages and knowledge on toxicity on target pests will be valuable in field application.

Entomopathogens for the Control of Storage Pests

Several disease pathogens from insect pests of stored products have been studied. These diseases are caused by pathogenic bacteria, viruses, fungi, nematodes and protozoa and are frequently fatal to these pests especially during the larval stage. The young larvae are the most sensitive to the pathogens.

Insect pathogens can be isolated from infected insects and then cultured in the laboratory except for the viruses and protozoa which can only be cultured in living insects. Dust preparations or aqueous suspension of these pathogens can be applied to stored products in bulk in much the same way as conventional insecticides.

Another way of application of the pathogens is through food baiting with peromone or other attractants. The baited food may be placed close to the stored products. This method was effective for dissemination of pathogenic protozoa to kill beetles and also viral and bacterial pathogens of moths. Shapes et al., (1977) suggested, however, that to be very effective, the contaminated insects must leave the source of attractant and spread the pathogen within the pest population. This could be achieved by using a pheromone that fades out quickly and where there is little or no shelter so that the infected pesta are inclined to move out after feeding.

In this method, transmission of the pathogens may occur in one of these ways: 1) larval eating on cadavers on infected larvae or adults, 2) consumption of infected stored food, 3) contamination during mating, and 4) infection from the female during oviposition to its progeny.

General Considerations

In the considering the potential use of microbial insecticides, it may be generalized that the mode of infection or mode of action of entomogahhogens or their toxic by-products may be divided into two groups; according to their natural port of entry into their hosts.

The first group includes bacteria, protozoa and viruses. These pathogens must be ingested in order to cause infection and later on mortality. Viruses are quite specific in their sites of development and multiply only in certain tissues within the body of their host. Bacteria may multiply throughout the tissues and body fluids of the host causing septecemia. The crystalliferous bacteria (Bacillus thuringiensis) may kill their hosts purely on the basis of the activity of their associated toxins.

The second group includes the fungi and nematodes which enter their host through their integument. These microorganisms are more subject to regulations by the environment. If the environment favours them, they multiply tremendously and easily colonize their hosts. Physical factors like temperature and humidity affect their survival and/or ability to cause infection. These will also affect the progress of infection and the susceptibility or resustance of the host.

Bacteria

Bacillus thuringiensis Berliner, the most common bacteria used for the control of stored insect pests, was isolated from diseased larvae of the Mediterranean flour moth, Ephestia knehniella Zeller in 1911. It has been tested on a wide range of storage moths: Plodia interpunctella Hubner, E. elutella, E. cautella, S. cerealeela (Oliver) and Corcyra cephalonica (Stainton). It was effective against all the above species except S. cerealella. In this species the larva spends its life cycle within the single grain and hence probably does not have sufficient contact with the pathogen when applied on the surfaces (McGaughey, 1976).

In bulk wheat and corn infested by P. interpunctella and E. cautella, it has been demonstrated by McGaughey (1976, 1978) that good control (at least 92%), may be obtained if only the surface layers of the bulk are treated with dust or aqueous suspension of B. thuringiensis. The recommended depth of treatment was 100 mm at least and the pathogen had to be well-mixed to ensure an even treatment within this layer. The recommended dose rate for this particular preparation was 125 mg/kg. Admixture to surface layers of bulk grain during grain transfer and after the silo have been filled were found eaually effective. Viability of B. thuringiensis is slightly reduced after one year storage.

Viruses

A nuclear polyhedrosis and a granulosis viruses have been isolated from E. cautella. Both severely infect P. interpunctella; whereas granulosis virus from P. interpunctella does not cross-infect E cautella (Hunter, 1973). Another nuclear polyhedrosis virus has been isolated from C. cephalonica but it is not known whether it will cross-infect other moths. In all cases the young larvae are the most susceptible stage.

Granulosis virus controlled effectively moth on bulk wheat and corn when surface layer of bulk was treated with aqueous suspension or dust of the virus to a depth of at least 100 mm. The suspension or dust both contained 3.2 x 103 virus capsules/mg and the grain was treated at a rate of approximately 1.9 mg/kg.

The application of an aqueous suspension of the granulosis virus of P. interpunctella has also been found to be an effective protectant for stored in-shell almonds (Hunter, et al, 1973), and stored raisins (Hunter et al, 1979). High storage temperature, however, reduces the viability of the virus and hence lowers the efficiency of this method of control. At 27°C, the control, after 5 & 6 months, remained high at 93% and 77%, respectively. In contrast, at 32°C, control after four months was 87% but this dropped to 34% and then 16% after 5 & 6 months.

Protozoa

Several protozoa are known to be severe pathogens of the Coleopteran. The following protozoans were tested: Nosema whitei for Tribolium castaneum (Herbs") and T. confusum J. du Val, N. whitei and N. oryzaephili for Oryzaephilus surinamensis (Linn) and Mattesia trogoderma for Trogoderma spp. Another species, Nosema plodiae Kellen and Lindegreem, has been isolated from the moth Q interpunctella and its mode of transmission investigated. Its use in control has not been studied.

The admixture of protozoa spores in particular, N. whitei to stored grain has been discussed by Burges (1973). He suggests that it would be necessary to admix a pathogen cocktail to control the range of Coleoptera that might be encountered in a particular storage situation.

Fungi

There are many fungi (hyphomycetous species) which grow on insects. Some are saprophytic while others are parasitic. Those most widely encountered are members of the genera Beauveria, Metarhizium and Isaria, and to a lesser extent Aspergillus, Cephalosporium, Sorosporella, and Hirsutella.

The commonest species is Metarhizium anisopliae (Metch) (Sor., the cause of the green muscardine disease of river' insects). This fungus has been seen to penetrate the integuments of several insects. In a recent study by Quiniones (1986) on the control of stored pests of copra, particularly the copra beetle, diluted spores of M. anisopliae up to 4.8 x 10-6 when topically brushed on the surface of stored copra, was effective to control this beetle.

Insect Resistance : Its Impact on Microbial Control of Insect Pests

Insects are among the earliest and most successful group of animals that exist in a myriad of environment where the potential for infection by microorganisms and parasites is great. Insects have also demonstrated considerable ability to develop resistance to conventional insecticides; and more than 500 species have developed resistance to one or more chemicals. As part of a survival strategy, insects have evolved numerous and effective resistance and defence mechanism to most of the conventional chemical insecticides with possession of genes for high levels of oxidase, esterases, glutathione-s-transferases, "insensitive" acetylcholinesterase (AChE), and nerve insensitivity to pyrethroids. Similarly can an insect species susceptible to a pathogen become resistant to more microorganism ?. Host specificity observed with many insect pathogens demonstrates that insect species are naturally resistant to these microorganisms. Indeed, insects that are susceptible to a pathogen can show resistance to various entomopathogens and try to resist infection through morphological, behavioural, developmental (like maturation immunity), physiological, nutritional, biochemical and molecular genetic mechanism etc. In this review different aspects of insects immunity, viz. passive and active defence mechanism against foreign invaders in comparison with vertebrate immunity has been presented. How certain parasites and pathogens like fungus, bacteria,protozoa, nematodes and virus of insect pests evolved strategies for avoiding both the external barrier as well as the internal immune defence posed by the host insects has also been discussed. Understanding of current knowledge of insect haematology, and molecular basis of the insect biochemical and cellular defence has been stressed for the proper management of pests especially by using various biocontrol agents like parasites and pathogens, including Bt transgenic plants and also for the control of certain insect vectors which carries pathogens that cause human diseases by way of transforming insects themselves.

Insects vs. Vertebrate Immunity

Compared to vertebrates, insects do not possess the ability to produce antibodies (immunoglobulins) and do not use immunoglobulin as recognition molecules in the classical sense, against foreign antigen and hence antigenic memory appears to be lacking i.e. (non-memory type). Further, they do not produce alpha/beta interferons in response to viral infections. Nevertheless, they are capable of "immune" reactions which appears to be predominantly cellular in nature. Several haemolymph induced antibacterial proteins have been reported to be broad-spectrum in their action which are produced in insects in response to bacterial challenger and of shorter duration in nature. This would suggest that analogy to the phenomenon of immunity in vertebrate may be inappropriate, and hence immunity in insects is different from immunity in vertebrates. The defence mechanism in insect is broadly classified into two broad groups. The first one is non-specific immunity which consists of structural and passive barriers like cuticle, gut physio-chemical properties, and peritrophic membrane. The second one is specific immune system involving cellular and humoral immunity which includes the activation of phenoloxoidase cascade, phagocytosis, nodulation (haemocyte aggregation) and encapsulation especially with reference to, bateria,fungi, protozoa including nematode invaders.

Non-specific Defence Mechanisms

Morphological

An insect resists entomopathogenic nematode (EPN) belonging to Steinernema and Heterorhabditis infection through behavioural, physical, or physiological means, Unlike vertebrates which have extensive exposure of epithelial cells to the external environment, insects have extensive protection of their epithelial tissue. The chitinous cuticle of the insect covers virtually all external surface, even extending through the foregut, hindgut and tracheal tubes, constituting the first line of passive defence in insects. Unlike that of the fore and hindgut, the epithelium of the insect midgut does not have a cuticular lining. However, in many insects there is a membrane called the "peritrophic membrane" which apparently functions to protect cells of the midgut from injury from hard (or) sharp particles of food performing much the same function as mucus in the mammalian alimentary tract. Physical resistance occurs when the nematode cannot penetrate the integument or the cocoon of a host insect. *Romanomermis culicivorax* has difficulty in penetrating the integument of older mosquito larvae. Dauer juveniles of *Steinernema carpocapsae* cannot penetrate the silken cocoons of hymenopteran parasitoids. Spiracular openings are portals of entry for EPN, but sieve plates over the spiracles, especially with scarab larvae, may deny nematodes access through this entry point. Avoidence of nematodes due to the presence of thick peritrophic membrane can act as a morphological defences against EPN.

Behavioural

Behavioural resistance occurs when the insect actively avoids or repels the nematode. Extremely active mosquito species had a lower prevalence of infection by the mermithid *R.. culicivorax* than less active ones. Scarab larvae may avoid infection by wiping nematodes away from the mouth. Younger instars of black fly larvae are resistant to infection by *Steinernema carpocapsae* because the comparatively large nematode is excluded from the insect's mouth. Aggressive behaviour (grooming with legs, mouth parts, and rasker) of *Popillia japonica* larvae when nematodes are present on the cuticle thereby removing and/or killing nematodes on the cuticle has also been reported.

Physiological

Under defence mechanisms, the high gut pH, presence of protease etc; can be found to be detrimental to the infective juveniles of *H. bacteriophora*. Similarly low gut pH, absence of protease etc; in insect system though not play a key or important resistant mechanism to bacterial pathogen like *Bacillus thuringiensis* (Bt) and Baculoviruses comprising nucleopolyhedrovirus (NPV) and granuloviruses (GV); have a role to play for their low susceptibility to bacterial and viral pathogens. Grasshoppers/ locusts elevate their body temperatures higher than ambient via, habitat selection and/or orientation to solar radiations called 'basking' by way intercepting the solar radiation and raising internal thoracic temperature ranging from 380 to 420 c thereby showing 'behavioural fever' response.

Such a body temperature is predicted to inhibit fungal proliferation, thus giving the host immune system an edge in suppressing the fungus germination and growth which normally lakes place from 25 ° to 30 ° c, thereby reducing the infection. Thus, thermoregulation by grasshoppers, *Melanoplus sanguinipes* has been shown to reduce mycosis caused by *Beauveria bassiana* and *Metarhizium anisopliae* (Ouedraogo *et al.*,2003). Physiological resistance to infection involves the destruction of the nematode by digestive enzymes in the insect's alimentary tract and the melanization and encapsulation of the nematode within the hemocoel. Melanotic encapsulation of *Steinernema carpocapsae* has been reported in larvae of several mosquito species. Although the nematode is encapsulated, the majority of the larvae die of septicaemia caused by the bacterium *Xenorhabdus nematophilus*, which is mutualistically associated with this nematode.

Biochemical

Eicosanoids is a collective term for all biologically active, oxygenated metabolites of 20: 4n-6 and two other czo poly unsaturated fatty acids. Given the importance of eicosanoid signalling to the insect immune system, interference with eicosanoid metabolism would seem to be a sensible strategy, and thus a potentially important virulence factor for an entomopathogen. Mediation of bacterial clearance has been reported in insects by eicosanoids by way of nodulation on *Manduca sexta, Agrotis ipsilon, Pseudaletia unipuncta and Bombyx mori etc.*

Biotechnological (Molecular)

Very little is known concerning insect defence against virus infection although insect haemocytes can provide cell-mediated immunity to bacterial pathogens through phagocytes and encapsulation. Neither cell mediatod or humoral immunity has been demonstrated against virus infection in insects. "Apoptosis" – distinctive type of programmed cell death – a phenomenon evolved as a primitive viral defence in certain vertebrate animals and invertebrates lacking humoral immunity to function as antiviral defence mechanism is gaining importance in cellular defence against viral infections (Narayanan, 1997). However, insect baculoviruses like nuclear polyhedrosis virus, granulosis virus and other DNA viruses, of insects evolved methods apparently to bypass this defence phenomenon of Apoptosis by directly blocking this response with possession of *p35* gene. The above findings leads to the future possibility of blocking the apoptosis for increasing virulence of certain baculoviruses as well as for the determination of host range of certain baculovirus and development of robust cells for in vitro multiplication of insect viruses and development of genetically improved parasitoids.

Genetic

Genetic resistance to bacterial and viral (Briese 1986b;) pathogens has been recorded against pestiferous insects. High levels of resistance to the δ – endotoxin of *B. thuringiensis* subspecies *kurstaki* have been recorded for the Indian meal moth, *Plodia interpunctella*, a pest of stored-grain and cereal products. The resistant trait is incompletely autosomal, recessive, and several alleles or genes are believed to be involved and resistance in this insect is linked to an alteration in toxin-membrane binding of the midgut cells.

Resistance of the silkworm to viral diseases such as nuclear (NPV) and cytoplasmic polyhedrosis virus (CPV) and infections flacherie virus (IFV) is controlled by polygenes. The polygenes are supposed to be mainly concerned with defence mechanism of midgut such as antiviral activity of gut juice, characteristic of peritropic membrane, etc. On the other hand non –susceptibility to densonucleosis is controlled by recessive (*nsd* – 1, *nsd* – 2) or dominant (*nsd* –1) major genes. The major gene may cause a deficiency of an enzyme involved in viral multiplication or in the receptor synthesis within the midgut cell. Polygenic resistance can be introgressed into a silkworm variety by selection in a breeding programme of the silkworm variety. A hybrid of two strains usually shows high heterosis in polygenic resistance to viral disease. The breeding procedure for non-susceptible variety to DNV is much easier because the mechanism of non-susceptibility is controlled by a single major gene. The gene can be introduced into the breeding programme or transferred to an existing superior variety by back crossing

Specific Defence Mechanisms/Immunity

Cellular Immunity

In a cellular defence mechanism, unlike vertebrate which has red blood corpuscles (RBC) and white blood corpuscles (WBC) in a closed circulatory system, insects with open body cavity lack lymphocytes, the major source of vertebrate immunity to virus infection. But they have only free blood cells called haemocytes. Different types of blood cells have important roles in the protection of insects against invading microorganisms. Hence identification and classification of various types of insect blood cells based on the structure and function is important. Among the six major group of insect haemocytes in recognizing the "self" (Isografts) and non-self (Allografts), plasmocytes and granulocytes, are the major effector cells and they react to foreign invaders either by phagocytosing like microorganisms or nodulating and encapsulating the objects too large to be individually engulfed,viz. metazoan parasite by way of haemocytes attaching and forming many layers which often become melanotic, thereby causing the death of the parasitoid through starvation and or anoxia mechanism. Changes in total haemocytes (THC) during growth and development of healthy insects have been reported by a number of workers (Narayanan1976). Drastic reduction in number of haemocytes during various microbial infection has also been reported by several workers. Infection by *B. bassiana* results in a gradual suppression of the phagocytic competence of circulating haemocytes and alteration in both total and differential haemocyte counts (DHC) has been reported in the case of fungal, bacterial, viral (Narayanan, 1979), and parasitic infection.

Humoral Immunity

Humoral reactions require several hours for their full expression and involve induced synthesis of antibacterial proteins, including 'cecropins', 'attacins', 'diptericins' and 'defensins'. The detergent properties of their antibacterial proteins disrupt cell membranes of the invading bacteria. Insects also synthesizing lysozome which enzymatically attack bacteria by hydrolyzing their peptidoglycan cell walls. Haemocytic responses feature direct interactions between circulating haemocyte and bacteria, these typically take place immediately after infections.Humoral reaction involve synthesis and release of several antibacterial (immuno) proteins. The antibacterial nature of gut contents and partial characterization of haemolymph bacterial proteins has been reported.

There are several families of characterized antibacterial proteins like cecropins, attacins, diptericins and defensins including recent identification of "Hemolin" (Sun et al., 1990) which belongs to the immunoglobulin superfamily. These insect antibacterial proteins are the best characterized invertebrate antibacterial factors and they have counterparts in mammals. The role of these proteins in non-self recognition as well as acting both prokaryotic and eukaryotic cells including human infectious parasites has been well studied.

Bibliography

Babar, Md. : *Hydrogeomorphology : Fundamentals Applications and Techniques*, New India Pub, Delhi, 2005.

Bagdi, Madan Lal : *Physiology, Biochemistry and Biotechnology*, Manglam Pub, Delhi, 2007.

Banerjee, S : *Cell Biochemistry*, Dominant Pub, Delhi, 2009.

Banerjee, S. : *Cell and Resource Biology*, Dominant, Delhi, 2004.

Baruah, Akhil : *Advanced Morphology of Angiosperms*, Aavishkar, Delhi, 2008.

Bhagwan, Pooja : *A Handbook of Inorganic Chemistry : Nuclear, Atomic, Aqueous, Alkali and Alkaline*, ISPA, Delhi, 2005.

Bhatnagar, Vasudev : *Cell Science and Technology*, Campus Books, Delhi, 2009.

Bhutani, R.C.: *Fruit and Vegetable Preservation*, Biotech Books, Delhi, 2003.

Chadha K. L. and Pareek O. P.: *Advances in Horticulture: Fruit Crops,* New Delhi, Malhotra Publishing House, 1993.

Cronquist, A.: *The Evolution and Classification of Flowering Plants,* New York Botanical Garden, Bronx, New York., 1988

Dodia, DA ; IS Patel and GM Patel: *Botanical Pesticides for Pest Management*, Scientific, Delhi, 2008.

Dogra, S S : *Textbook of Environmental Chemistry and Pollution Control*, Swastik Pub, Delhi, 2007.

Garg, G.P. : *Microbial Biochemistry*, Discovery Pub, Delhi, 2010.

Geis, I.: *Chemistry, Matter, and the Universe*, Menlo Park, Ca., W.A. Benjamin, 1976.

Ghosh, G K : *Biopesticide and Integrated Pest Management*, APH, Delhi, 2009.

Goswami, C L : *Plant Physiology : Research Methods*, Scientific, Delhi, 2007.

Graham L. Patrick: *An Introduction to Medicinal Chemistry,* Oxford University Press, Delhi, 2009.

Gupta, S.K.: *Disease Problems in Vegetable Production*, Scientific, Delhi, 2006.

Harsh Bhaskar: *Basic Facts on Biochemistry*, Campus Books, Delhi, 2010.

Iqbal, S.A. and M. Satake: *An Introduction to Analytical Chemistry*, Discovery, Delhi, 1999.

Jagbir Sharma: *Advanced Environment Chemistry*, RBSA Pub, Delhi, 2006.

Jangir, M.L. : *Cell Biology : Fundamentals and Applications*, Agrobios, Delhi, 2009.

Kataria, T N : *Plant and Crop Physiology*, Pearl Books, Delhi, 2008.

Khalsa, Nisha : *Essentials of Biochemistry*, Aavishkar Pub, Delhi, 2008.

Kitchen, Herminie Broedel : *Soils and Crops : Diagnostic Techniques*, Satish Serial Publishing, Allahabad, 2004.

Kumar, N. : *Breeding of Horticultural Crops : Principles and Practices*, New India Pub, Delhi, 2006.

Marshall, F R : *Breeding Farm Animals*, Asiatic Pub, Delhi, 2006.

Mishra, Shubhrata R. : *Morphology of Fungi*, Discovery Pub, Delhi, 2005.

Mishra, U K : *Physiology of Animal Growth and Bioenergetics*, Satish Serial Pub House, Delhi, 2008.

Pareek O. P.: *Advances in Horticulture: Fruit Crops,* New Delhi, Malhotra Publishing House, 1993.

Pareek, Alka : *Environment and Nutritional Disorders*, Aavishkar Pub, Delhi, 2003.

Prakash, M. : *Cell Physiology*, Discovery, Delhi, 2010.

Prasad, Govind : *Trends and Techniques of Geomorphology*, Discovery, Delhi, 2007.

Purohit, S.S. : *Organic Farming : Biocontrol and Biopesticide Technology*, Agrobios, Delhi, 2008.

Raghunath Narvekar: *Handbook of Biochemistry*, Adhyayan, Delhi, 2008.

Ramann, E. : *The Evolution and Classification of Soils*, Asiatic Pub, Calcutta, 2006.

Reddy, P. Parvatha : *Fungal Diseases and Their Management in Horticultural Crops*, Scientific Pub, Delhi, 2010.

Saini, M.L. : *Plant Breeding and Crop Improvement*, CBS, Delhi, 1997.

Sharma, R.C. : *Diseases of Horticultural Crops: Fruits*, Indus, Delhi, 1999.

Singh, G : *Chemistry of Amino-Acids and Proteins*, Discovery, Delhi, 2007.

Singh, S K : *Biotechnology, Plant Propagation and Plant Breeding*, Campus Books, Delhi, 2008.

Srivastava, Seema : *Plant Physiology and Biochemistry*, Campus Books, Delhi, 2009

Steferud, Alfred : *Diseases of Fruits and Nuts*, Biotech Books, Delhi, 2005.

Suri, Nitin : *Molecular Biology and Biochemistry*, Oxford Book Company, Delhi, 2010.

Tomar, S.K. and Gunjan Goel : *Applied Dairy and Food Microbiology,* Agrotech, 2005.

Tripathy, S.N : *Food Biotechnology*, Dominant, 2004.

Tyagi, Prasum : *A Textbook of Animal Physiology*, Dominant, Delhi, 2010.

William Alec : *The Chemical Industry*, London: Longman Group Limited, 1971.

Yadav, Keshav Prasad : *Application of Morphometry in Geomorphology*, Radha Pub, Delhi, 2008.

Index

□□□